METHODS IN MOLECULAR BIOLOGY™

Series Editor
John M. Walker
School of Life Sciences
University of Hertfordshire
Hatfield, Hertfordshire, AL10 9AB, UK

For further volumes:
http://www.springer.com/series/7651

Two Hybrid Technologies

Methods and Protocols

Edited by

Bernhard Suter

Max-Delbrück-Centrum für Molekulare Medizin, Quintara Biosciences, Albany, CA, USA

Erich E. Wanker

Max-Delbrück-Centrum für Molekulare Medizin, Berlin-Buch, Germany

Humana Press

Editors
Bernhard Suter
Max-Delbrück-Centrum
für Molekulare Medizin
Quintara Biosciences
Albany, CA, USA
bernhard.suter@mdc-berlin.de

Erich E. Wanker
Max-Delbrück-Centrum
für Molekulare Medizin
Berlin-Buch, Germany
ewanker@mdc-berlin.de

ISSN 1064-3745 e-ISSN 1940-6029
ISBN 978-1-4939-5811-5 ISBN 978-1-61779-455-1 (eBook)
DOI 10.1007/978-1-61779-455-1
Springer New York Dordrecht Heidelberg London

Printed on acid-free paper

Humana Press is part of Springer Science+Business Media (www.springer.com)

Preface

Protein–protein interactions (PPIs) are strongly predictive of functional relationships among proteins in virtually all processes that take place in the living cell. Therefore, the comprehensive exploration of interactome networks is one of the major goals in systems biology. The development of "interactomics" as a field is largely driven by the development of innovative technologies and strategies for efficient screening, scoring, and validation of PPIs. The aim of this book is to provide a compendium of state-of-the art-protocols for the investigation of binary PPIs with the classical yeast two-hybrid (Y2H) approach, Y2H variants, and other in vivo methods for PPI mapping. Given the broad range of methodologies currently available, biochemical approaches like proteome-wide co-immunoprecipitation, and other in vitro and in vivo methodologies are not to be considered here. It needs to be emphasized, however, that alternative methods are very important for the complementation and validation of Y2H screens.

The book is structured into two sections. The first gives a survey of protocols that are currently employed for Y2H high-throughput screens by different expert labs in the field. Rather than detailing the principles of screening, which have been described previously, the focus is on different implementations of Y2H interactome mapping. First, two articles by Peter Uetz review the most important developments and applications of Y2H high-throughput screening. Then, Russ Finley, Ulrich Stelzl, Manfred Koegl, and coauthors describe their automated screening procedures in detail. A view on interactome research in pathogenic organisms is provided by Vincent Lotteau and Lionel Tafforeau (viral interactomes), and Douglas LaCount (interactomes of malaria parasites). Xiaofeng Xin and Thierry Mieg complement experimental protocols with their recently developed strategy of smart-pooling by shifted transversal design. Two more articles deal with bioinformatics for the analysis of Y2H data sets. Russ Finley and team discuss confidence scoring, whereas Gautam Chaurasia and Matthias Futschik describe the design of a database for high-throughput Y2H data (UniHI, Max Delbrueck Centrum, Berlin). John Reece-Hoyes and Albertha Walhout present a high-throughput yeast one-hybrid variant for the identification of proteins that bind-specific DNA segments. Finally, contributors from the lab of Young Chul Lee introduce their "one- plus two-hybrid system" for the efficient identification of PPIs altered by missense mutations.

The second part of the book considers innovative PPI detection methods that have the potential to emerge as alternative high-throughput methodologies. An important future role can be expected for systems that rely on the functional reconstitution (complementation) of reporter proteins by fused bait and prey proteins. A chapter on the split ubiquitin based system to screen for membrane protein interactions is provided by Igor Stagljar, whereas Mandana Rezwan and Daniel Auerbach of Dualsystems Biotech AG describe an approach to screen for interactors using the reconstitution of a split-TRP1 protein. For future human interactome studies, procedures that can reconstitute PPIs directly in mammalian cells could provide a better physiological context compared to yeast. A mammalian two-hybrid system based on the tetracycline-repressor system is presented by Kathryn Moncivais and Zhiwen

Zhang. A different principle in mammalian cells is used by Heinrich Leonhardt and team in their fluorescent two-hybrid approach, where bait and prey proteins are recruited to specific chromosomal locations. Perhaps the most advanced strategy for binary PPI mapping in mammalian cell culture is the mammalian protein–protein interaction trap (MAPPIT), developed by Jan Tavernier and his group. It is based on complementation of a cytokine receptor complex operating in mammalian cells. In the high-throughput ArrayMAPPIT application, prey proteins are arrayed in high-density microtiter plates to screen for interaction partners using reverse transfection into a bait-expressing cell pool. A variation of MAPPIT can be used to test substances that disrupt PPIs. Finally, Moritz Rossner provides a protocol for the use of uniquely expressed oligonucleotide tags (EXTs) that integrate complementation assays based on TEV protease and transcription factor activity profiling. Together, the protocols supply researchers with a comprehensive toolbox for the identification of biologically relevant protein interactions.

We are very grateful to all contributing authors for their great commitment to this project. We would like to express special gratitude to Dr. John M. Walker for his guidance and continuous support during the preparation of the manuscript.

Albany, CA, USA *Bernhard Suter*
Berlin, Germany *Erich E. Wanker*

Contents

Contributors

DANIEL AUERBACH • *Dualsystems Biotech Inc, Zurich, Switzerland*
LUKAS BAUMANN • *Dualsystems Biotech Inc, Zurich, Switzerland*
JAN MARTIN BERKE • *Tibotec Inc, Mechelen, Belgium*
CHARLES BOONE • *Terrence Donnelly Centre for Cellular and Biomolecular Research, University of Toronto, Toronto, ON, Canada*
ANNA BOTVINIK • *Research Group 'Gene Expression' Max-Planck-Institute of Experimental Medicine, Göttingen, Germany*
GAUTAM CHAURASIA • *Charité, Humboldt University, Berlin, Germany*
REGINALD CLAYTON • *Tibotec Inc, Mechelen, Belgium*
GÉRY DAMS • *Tibotec Inc, Mechelen, Belgium*
DIETER DEFEVER • *Department of Medical Protein Research, VIB and Department of Biochemistry, Ghent University, Ghent, Belgium*
RUSSELL L. FINLEY JR. • *Center for Molecular Medicine and Genetics, Wayne State University School of Medicine, Detroit, MI, USA*
MATTHIAS FUTSCHIK • *Centre for Molecular and Structural Biomedicine, University of Algarve, Faro, Portugal*
ARNDT GROSSMANN • *Max Planck Institute for Molecular Genetics (MPI-MG), Berlin, Germany*
ROMAN HÄUSER • *Karlsruhe Institute of Technology, Karlsruhe, Germany*
ANNA HEGELE • *Max Planck Institute for Molecular Genetics (MPI-MG), Berlin, Germany*
DIRK JOCHMANS • *Tibotec Inc, Mechelen, Belgium*
JI YOUNG KIM • *School of Biological Sciences and Technology, Chonnam National University, Gwangju, Republic of Korea*
MANFRED KOEGL • *Genomics and Proteomics Core Facility German Cancer Research Institute, Heidelberg, Germany*
DOUGLAS J. LACOUNT • *Department of Medicinal Chemistry and Molecular Pharmacology, Purdue University, West Lafayette, IN, USA*
YOUNG CHUL LEE • *School of Biological Sciences and Technology, Chonnam National University, Gwangju, Republic of Korea*
NICOLAS LENTZE • *Dualsystems Biotech Inc, Zurich, Switzerland*
HEINRICH LEONHARDT • *Center for Integrated Protein Science (CiPSM) and Department of Biology, Ludwig Maximilians University Munich, Planegg-Martinsried, Germany*
SAM LIEVENS • *Department of Medical Protein Research, VIB and Department of Biochemistry, Ghent University, Ghent, Belgium*
VINCENT LOTTEAU • *Université de Lyon, Lyon, France*
BERNARDO A. MANGIOLA • *Center for Molecular Medicine and Genetics, Wayne State University School of Medicine, Detroit, MI, USA*

KERSTIN MOHR • *Genomics and Proteomics Core Facility, German Cancer Research Institute, Heidelberg, Germany*
KATHRYN MONCIVAIS • *College of Pharmacy, University of Texas at Austin, Austin, TX, USA*
THILAKAM MURALI • *Center for Molecular Medicine and Genetics, Wayne State University School of Medicine, Detroit, MI, USA*
OK GU PARK • *School of Biological Sciences and Technology, Chonnam National University, Gwangju, Republic of Korea*
JODI R. PARRISH • *Center for Molecular Medicine and Genetics, Wayne State University School of Medicine, Detroit, MI, USA*
JULIA PETSCHNIGG • *Terrence Donnelly Centre for Cellular and Biomolecular Research (CCBR), University of Toronto, Toronto, ON, Canada*
CHANTAL RABOURDIN-COMBE • *Université de Lyon, Lyon, France*
SEESANDRA V. RAJAGOPALA • *J Craig Venter Institute (JCVI), Rockville, MD, USA*
JOHN S. REECE-HOYES • *University of Massachusetts Medical School, Worcester, MA, USA*
MANDANA REZWAN • *Dualsystems Biotech Inc, Zurich, Switzerland*
GEORGE G. ROBERTS III • *Center for Molecular Medicine and Genetics, Wayne State University School of Medicine, Detroit, MI, USA*
MORITZ J. ROSSNER • *Research Group 'Gene Expression' Max-Planck-Institute of Experimental Medicine, Göttingen, Germany*
ULRICH ROTHBAUER • *Center for Integrated Protein Science (CiPSM) and Department of Biology, Ludwig Maximilians University Munich, Planegg-Martinsried, Germany*
JAMIE SNIDER • *Terrence Donnelly Centre for Cellular and Biomolecular Research (CCBR), University of Toronto, Toronto, ON, Canada*
IGOR STAGLJAR • *Terrence Donnelly Centre for Cellular and Biomolecular Research (CCBR), University of Toronto, Toronto, ON, Canada*
THORSTEN STELLBERGER • *Karlsruhe Institute of Technology, Karlsruhe, Germany*
ULRICH STELZL • *Max Planck Institute for Molecular Genetics (MPI-MG), Berlin, Germany*
LIONEL TAFFOREAU • *Institute de biologie et de médecine moléculaires (IBMM), Université libre de Bruxelles (ULB), Gosselies, Belgium*
JAN TAVERNIER • *Department of Medical Protein Research, VIB and Department of Biochemistry, Ghent University, Ghent, Belgium*
NICOLAS THIERRY-MIEG • *Laboratoire Techniques de l'Ingénierie Médicale et de la Complexité - Informatique, Mathématiques et Applications de Grenoble (TIMC-IMAG), Faculte de Medecine, La Tronche, France*
PETER UETZ • *Center for the Study of Biological Complexity Virginia Commonwealth University, Richmond, VA, USA*
KOEN VAN ACKER • *Tibotec Inc, Mechelen, Belgium*
JOSÉ VAN DER HEYDEN • *Department of Medical Protein Research, VIB and Department of Biochemistry, Ghent University, Ghent, Belgium*
NELE VANDERROOST • *Department of Medical Protein Research, VIB and Department of Biochemistry, Ghent University, Ghent, Belgium*

Bertrand Van Schoubroeck • *Tibotec Inc, Mechelen, Belgium*
Albertha J.M. Walhout • *University of Massachusetts Medical School, Worcester, MA, USA*
Mareike Weimann • *Max Planck Institute for Molecular Genetics (MPI-MG), Berlin, Germany*
Victoria Wong • *Terrence Donnelly Centre for Cellular and Biomolecular Research (CCBR), University of Toronto, Toronto, ON, Canada*
Josephine M. Worseck • *Max Planck Institute for Molecular Genetics (MPI-MG), Berlin, Germany*
Xiaofeng Xin • *Terrence Donnelly Centre for Cellular and Biomolecular Research, University of Toronto, Toronto, ON, Canada*
Jingkai Yu • *National Key Laboratory of Biochemical Engineering, Chinese Academy of Sciences, Beijing, China*
Zhiwen Jonathan Zhang • *Bioengineering Program, School of Engineering, Santa Clara University, Santa Clara, USA*
Kourosh Zolghadr • *Center for Integrated Protein Science (CiPSM) and Department of Biology, Ludwig Maximilians University Munich, Planegg-Martinsried, Germany*

Chapter 1

Matrix-Based Yeast Two-Hybrid Screen Strategies and Comparison of Systems

Roman Häuser, Thorsten Stellberger, Seesandra V. Rajagopala, and Peter Uetz

Abstract

Today, matrix-based screens are used primarily for smaller and medium-size clone collections in combination with automation and cloning techniques that allow for reliable and fast interaction screening. Matrix-based yeast two-hybrid screens are an alternative to library-based screens. However, intermediary forms are possible too and we compare both strategies, including a detailed discussion of matrix-based screens. Recent improvement of matrix screens (also called array screens) uses various pooling strategies as well as novel vectors that increase their efficiency while decreasing false-negative rates and increasing reliability.

Key words: Protein–protein interactions, Pooling, Mating, PI-deconvolution, Smart pool array system, Shifted transversal design

Abbreviations

3-AT	3-Amino-1,2,4-triazole
AD	Activation domain
DBD	DNA-binding domain
GFP	Green fluorescent protein
GO	Gene ontology
ORF	Open reading frame
STD	Shifted transversal design
Y2H	Yeast two hybrid

1. Introduction: The Yeast Two-Hybrid Principle and Variations of It

Shortly after Stanley Fields and Ok-kyu Song invented the yeast two-hybrid (Y2H) system in 1989 (1), it was adapted for screens of random libraries. Like the original Y2H assay, matrix-based screens are usually carried out in living yeast cells although in theory any other cell could be used. This is a crucial advantage since it

Bernhard Suter and Erich E. Wanker (eds.), *Two Hybrid Technologies: Methods and Protocols*, Methods in Molecular Biology, vol. 812, DOI 10.1007/978-1-61779-455-1_1, © Springer Science+Business Media, LLC 2012

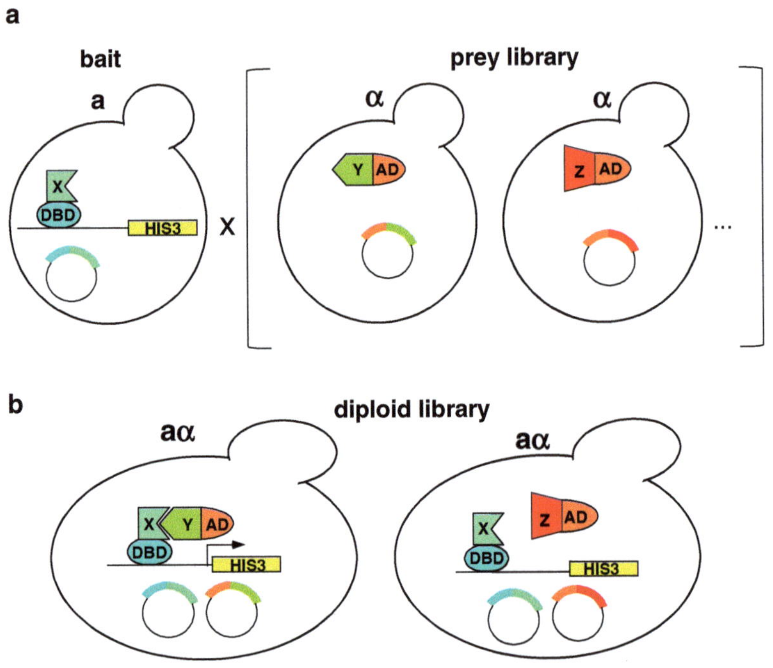

Fig. 1. The yeast two-hybrid principle. (**a**) Haploid yeast cells of mating type *a* are transfected with a bait plasmid and those of mating type α with prey plasmids. A single bait strain is mated with a prey library. (**b**) Resulting diploids (a/α) carry the genetic material of mated haploids. Interacting fusion proteins activate expression of the HIS3 reporter gene which assures survival on minimal medium that lacks histidine (diploid on the left); diploids with noninteracting fusions cannot grow (diploid on the right).

represents an "in vivo" situation. The proteins of interest are provided as plasmid-encoded recombinant fusion proteins (Fig. 1). The bait protein is often fused to a DNA-binding domain (DBD) of the yeast GAL4 transcription factor. The prey protein is tagged by the activation domain (AD) of GAL4. A physical contact of the bait and prey protein simulates the reconstitution of the GAL4 transcription factor. Once the bait protein is bound to its promoter sequence by its DBD, the interacting proteins recruit the basal yeast transcription machinery and thus activate the expression of a reporter gene. Note that other fusion proteins can be used too and have been established in other systems. For example, instead of the Gal4 components, the bacterial transcription factor LexA has been used. In general, any protein that can be split and reconstituted to form an active protein can be used (2).

For high-throughput screens, we routinely use the HIS3 auxotrophy marker. It encodes the essential enzyme imidazole-glycerol-phosphate dehydratase which catalyzes the sixth step of histidine biosynthesis. Hence, yeast growth on minimal medium that lacks histidine can be used to indicate an interacting protein pair.

Noninteracting pairs cannot support growth on minimal medium. This reporter system is very simple and easy to use because the presence of yeast colonies indicates an interaction. As for the fusion proteins, many other reporter genes are conceivable as long as they can be activated by the interacting fusion proteins. Before the binary tests are carried out, the bait and prey plasmids must be brought into the same yeast cell. This is conveniently done by mating. The bait and prey plasmids are separately transformed into haploid yeast cells of different mating types, *a* and α. Mating results in diploid yeast cells that carry the genetic material of both haploids, including the bait and prey plasmids. Although we focus in this chapter primarily on the GAL4 transcription factor and the usage of the HIS3 reporter gene, other DNA-binding proteins as well as reporter genes may be used.

Alternative reporter genes are LEU2 and URA3. They allow selection on readout medium that lacks leucine or uracil. Auxotrophy markers are not the only ones that can be used. The ADE2 reporter system changes colony color from red to white on adenine starvation medium when diploids express interacting proteins. Beta-galactosidase (*lacZ*) or green fluorescent protein (GFP) can be used as colorimetric or fluorescence reporters. Finally, transcription-independent two-hybrid systems have been developed. The split-ubiquitin system is based on the cleavage of the interacting fusion proteins by the proteasome (3). As long as the function of a protein can be used as reporter, the possibilities are as manifold as the nature of the proteins themselves.

2. Applications

It has become clear that the ability to conveniently perform unbiased library screens is the most powerful application of the Y2H system. With whole-genome arrays, such unbiased screens can be expanded to all proteins of an organism or any subset thereof. Arrays, like traditional two-hybrid screens, can also be adapted to answer many questions that involve protein–protein or protein–RNA interactions (Table 1).

Recent large-scale projects have been successful in systematically mapping whole or partial proteomes of various higher and lower organisms (Table 2). In addition to bacteria and eukaryotic genomes, several viral proteomes have been mapped as well, e.g., bacteriophage T7 (4) and herpesviruses (5, 6).

In combination with structural genomics, gene expression data, and metabolic profiling, the enormous amount of information in these networks helps us to model complex biological phenomena in molecular detail.

Table 1
Selected applications of two-hybrid assays (besides protein interaction screens)

Application	References
Identification of mutants that prevent or allow interactions	(22)
Screening for drugs that affect interactions	(23, 24)
Identification of RNA-binding proteins	(25)
Semiquantitative determination of binding affinities	(26)
Map interacting domains	(10, 11, 27)
Study protein folding	(28)
Map interactions within protein complexes	(29)

Table 2
Recent large-scale and comprehensive Y2H projects

Species	References
Saccharomyces cerevisiae	(7, 9, 30)
Drosophila melanogaster	(31)
Caenorhabditis elegans	(32)
Homo sapiens	(33, 34)
Helicobacter pylori	(27)
Campylobacter jejuni	(35)
Treponema pallidum	(36)

3. Matrix-Based Yeast Two-Hybrid Screens (One-on-One)

"Matrix" or "array based" means that preys are organized in a defined array format. For high-throughput purposes, preys can be arranged in 384 format on a single test plate. This was first demonstrated on a global scale by Uetz and colleagues (7). Each prey clone maps to an individual position. Preys may be organized as individual colonies, although we recommend duplicate or quadruplicate copies to ensure reproducibility (Fig. 2).

The whole array of haploid preys is usually mated against a single bait of the opposite mating type. Thus, each potential interaction pair is tested one-on-one (see Fig. 2). For high-throughput

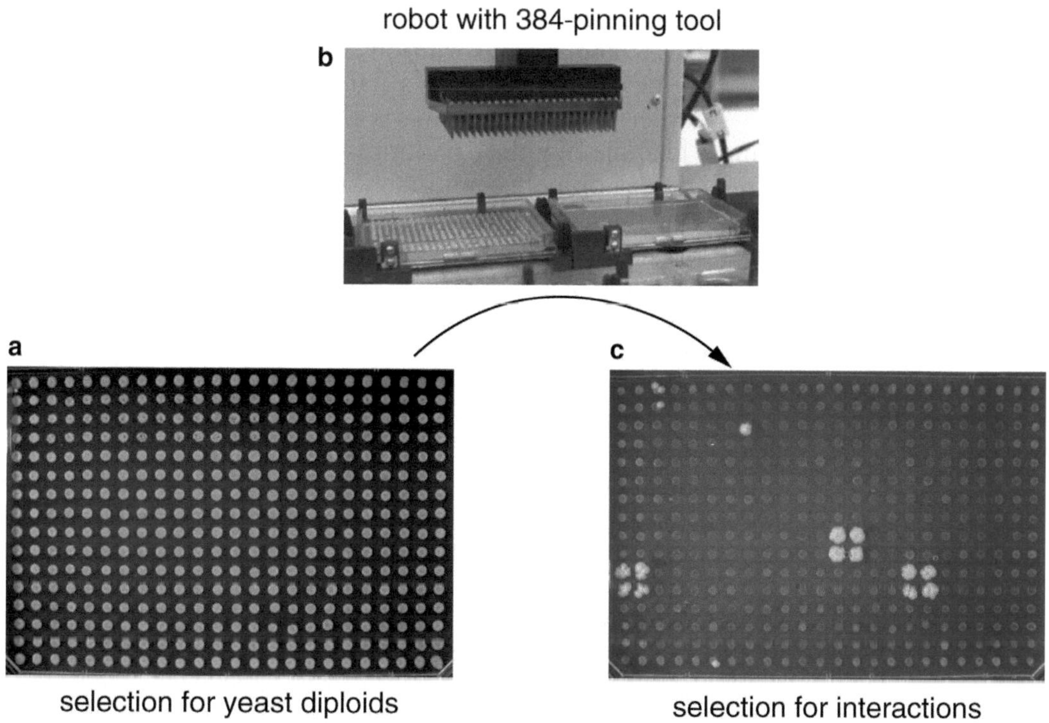

Fig. 2. A matrix-based screen. (**a**) Prey array mated against a single bait on diploid selective agar medium containing 96 individual preys. Single preys are replicated as quadruplicates to check interaction reproducibility. (**b**) 384-pinning tool of replication robot during pinning step of diploids onto readout medium. (**c**) Diploids on readout medium that lacks histidine. Diploids were grown on selective medium for 1 week at 30°C. Activation of the HIS3 reporter leads to growth on minimal medium indicating a pairwise interaction (quadruplicate spots). Noninteracting pairs do not support growth on minimal medium.

analysis, a replication robot should be used, typically with a 96- or 384-pin tool (Fig. 2). It automates the procedure by reproducibly stamping up to hundreds of array position in a single step, e.g., to transfer diploids onto readout plates.

3.1. Why Matrix-Based Screens?

Matrix-based screens are excellent to control experimental background signals. Background can be caused by self-activation of certain bait proteins. They lead to reporter gene expression and growth on readout medium without an interaction. In matrix-based screens, interactions can be identified even if background growth occurs. In a matrix screen of a single bait, the signal-to-noise ratio can be easily determined because all protein pairs are assayed under identical conditions. Furthermore, background of spontane ously appearing colonies caused by mutations or other random effects can be identified. The redundancy of two or more test positions helps to winnow random colonies.

The matrix-based strategy helps not only to control the background growth on readout medium, but also to check the previous screening steps. For instance, the mating efficiency can be controlled by just watching yeast growth on diploid selection medium and need not be determined by a separate experiment.

Another crucial advantage is that interacting preys can be simply identified by their positions. The matrix positions can be stored in a list or more comfortably in a database. Thus, identification of the interacting prey by sequencing is not required and time and costs can be minimized.

Finally, the matrix approach helps to distinguish strong from weak and spurious interactions since the size of growing yeast colonies is an indirect measure of binding affinity.

3.2. Matrix vs. Library Screens

Library screens are the classical way to screen for interaction partners. They are the fastest option. A single bait is mated with a library that contains all preys (see Fig. 3). Once mated, yeast can be plated directly onto readout medium plates and positives are selected. In contrast to the matrix-based strategy, this classical approach requires identification of the interacting prey by sequencing. However, this procedure may also produce more *false negatives* due to preys that are over- or underrepresented in the prey pool. Randomly generated prey plasmid libraries can be transformed directly into the haploid prey strain. Alternatively, prey libraries can be derived from a yeast prey matrix by pooling which ensures normalization (minimization of under- or overrepresentation). Since most library screens use randomized (cDNA) or even genomic libraries, *false positives* may result from fragments that do not fold properly or that expose protein sequences that are not exposed in vivo. On the other hand, certain *false negatives* are avoided that may arise in screens using full-length ORFs for the same reason. Clearly, both library and matrix screens do have advantages and disadvantages that should be considered when a project is planned.

3.3. Limitations of Matrix-Based Screens

Matrix-based screens do have certain disadvantages when compared to screens of random libraries.

Time considerations. Matrix-based screens can be time consuming, even when pooling strategies are used, given that individual clones

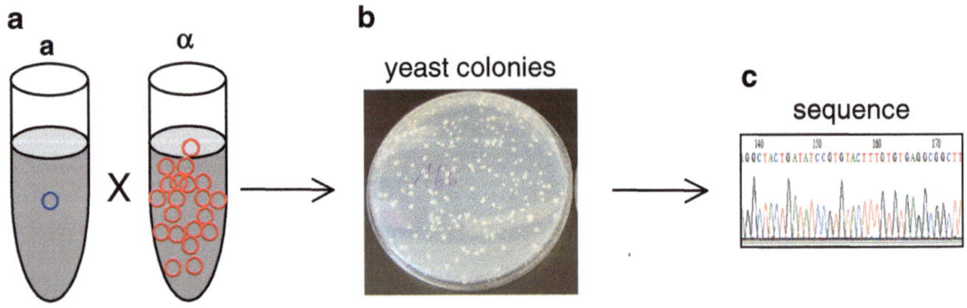

Fig. 3. Library screen. (**a**) Mating of a single bait strain (mating type a) with a prey clone library (mating type α). (**b**) Diploid selection on readout agar medium. (**c**) Identification of interacting prey by colony PCR and sequencing.

or relatively small numbers of clones are tested at a time. Also, the availability of robotics and/or sequencing should be considered.

Cost. The cost for robotic equipment can be prohibitive. In addition, a large number of screens require a similarly large number of plastic plates (e.g., Nunc Omnitrays). We typically use three plates per screen (i.e per 96 prey proteins): one for mating, one for testing the mating efficiency, and one for the actual Y2H selection. That is, a small bacterial genome with ~1,000 genes requires 1,000 [baits] × 10 plates [1,000 preys/~100 clones per plate] × 3 ≈ 30,000 plates. Omnitrays are on the order of 1–2 US$ per plate. In order to reduce cost, pooling is required in most cases (see below).

False negatives. Two-hybrid screens typically have a fairly high false-negative rate. This may have a number of reasons which also apply to the matrix-based approach. *First*, mating efficiency of some baits is lower than compared to others. Interactions of such proteins could be missed. *Second*, poorly understood random effects impose a sensitivity limit on screens so that certain interactions are only detected in a subset of assays (8, 9). This means that saturation may only be achieved if a screen is repeated three or more times. Only ~60% of the interactions may be detected within the first screen. *Third*, the fact that the Y2H system works with fusion proteins can also lead to missed interactions. The standard vectors work with N-terminally tagged fusions. If the interacting domain of a protein is near its N-terminus, the fusion of DNA-binding or activation domains may prevent an interaction. *Fourth*, screens with full-length ORF libraries can also result in false negatives. Several studies indicated that screens with protein fragments (as opposed to full-length proteins) yield more interactions, most likely because additional interaction surfaces are exposed (10, 11). Protein folding may play a role here too, as many proteins may undergo interactions while they are still folding. Similarly, protein processing may be required for interactions. For example, when defined mature proteins of hepatitis C virus were tested by Flajolet et al. (12), no interactions were found. When random fragments were used (possibly corresponding to exposed peptides of folding intermediates), a total of five interactions were found. There are a few other reasons why interactions may go undetected. However, they have little to do with the array format, e.g., proteins that are not properly localized to the nucleus, proteins that are unstable, or incorrectly folded proteins.

Because defined ORFs are often screened in a matrix format, matrix-based screens appear to have more false negatives than random libraries. Indeed, this problem may be alleviated by using random libraries, protein fragments, or alternative vector systems (see below).

False positives: As any other method, the Y2H system "detects" spurious interactions. Many reasons have been suggested, but few have been really shown experimentally. *First*, false positives can be caused by so-called "sticky" proteins that lead to unspecific interactions.

Heterologous overexpression in yeast may result in a certain fraction of unfolded proteins that expose hydrophobic patches which in turn may cause sticky behavior. Similarly, testing proteins in the absence of specific chaperones might result in incorrect folding. However, these hypotheses have never been rigorously tested. *Second*, the high sensitivity of the reporter system may detect weak interactions that occur in the living organism but might have no biological relevance.

Identifying false positives and false negatives. False-positive interactions can be identified in interaction datasets much more easily than false negatives. While we do not know what we are missing (unless we have known interactions as controls), false positives often share certain hallmarks (Table 3).

Contamination. Arrays are prone to cross-contaminations as plates have to be kept open when pinned. Sterile conditions of the pinning tools and plates are thus needed. The array has to be watched attentively.

To exclude false positives, simple filter mechanism can be applied, e.g., the bait and prey count (number of interaction partners of a single bait or prey) or logistic regression (13) that uses validated training sets, respectively. Strength of interactions can indicate their biological relevance and spurious interactions can be identified by the yeast colony size. Subsequent retest experiments and the involvement of alternative approaches, like pull downs or alternative reporter genes, can help to exclude potential false-positive interactions. Background growth control makes the matrix-based approach an excellent way to prevent or identify false positives, especially since randomly appearing colonies and growth caused by self-activation can be easily excluded.

Table 3
Criteria to identify false-positive interactions in Y2H screens

Stickiness	A bait interacts with many prey proteins and vice versa
Specificity	Interactions are highly unspecific, i.e., a protein interacts with highly unrelated proteins (e.g., proteins of different GO annotation, localization, etc.)
Reproducibility	An interaction cannot be reproduced by repeating the same Y2H assay or by other assays (see also Subheading 5 below)
Signal strength	Weak reporter gene activation may be spurious, especially when other background is present

4. Pooling Strategies

The capacity of matrix-based screens is limited by the size of the clone set to be tested. For instance, a small proteome that encodes for 1,000 proteins requires at least $1,000^2$ (one million) individual pairwise tests in a comprehensive screen. For large genomes, such as the human, $23,000^2$ (over half a billion) one-on-one tests would be necessary to test all possible combinations! Genome-wide screens face four main issues: *cost*, *efficiency* (the number of assays, speed), *specificity* (detecting false positives), and *sensitivity* (avoiding false negatives).

Solutions to make large-scale matrix screens more efficiently require *pooling* (14–17) which may dramatically reduce the number of individual Y2H tests as well as the need for sequencing while keeping the advantages of matrix-based screens. "Smart" pooling and arrangements of prey as well as bait clones can help to speed up the screening procedure drastically, resulting in interaction detection with (almost) the same sensitivity and specificity as one-on-one Y2H screens.

4.1. Mini-Pool Screens

In matrix-based pooling screens, several preys share a position. In the simplest case, a prey array that consists, for example, of 960 individual preys can be collapsed into a single 96-well plate with 10 clones in each position (Fig. 4). This minimizes the required mating operations with a single bait by 1/10. The disadvantage of this strategy is that interacting preys cannot be identified immediately as it is possible for the matrix-based screens. They must be identified by yeast colony PCR and sequencing or retesting of individual bait–prey pairs. Retests (as opposed to sequencing) have the advantage that potential interaction partners are retested positively if a pool contains more than one interacting prey. When sequenced, two or more PCR products may lead to unreadable sequencing results. Another point is that certain preys might be over- or under-represented once pooled as in the library screen strategy. In the pooling strategy, it is hard to attain equal prey cell numbers and thus underrepresentation of preys can lead to additional false negatives.

4.2. Two-Phase Mating

Zhong et al. (17) went one step further and showed that single pools can contain more than 96 different preys and that interacting baits and preys can be identified without a retest experiment or sequencing. The authors estimated that screening the yeast genome (ca. 6,000 proteins) by using their *two-phase mating* protocol requires only 1/24 of time and effort since only a fraction of mating operations and replication steps are necessary compared to one-on-one matrix-based screens. With increasing genome sizes,

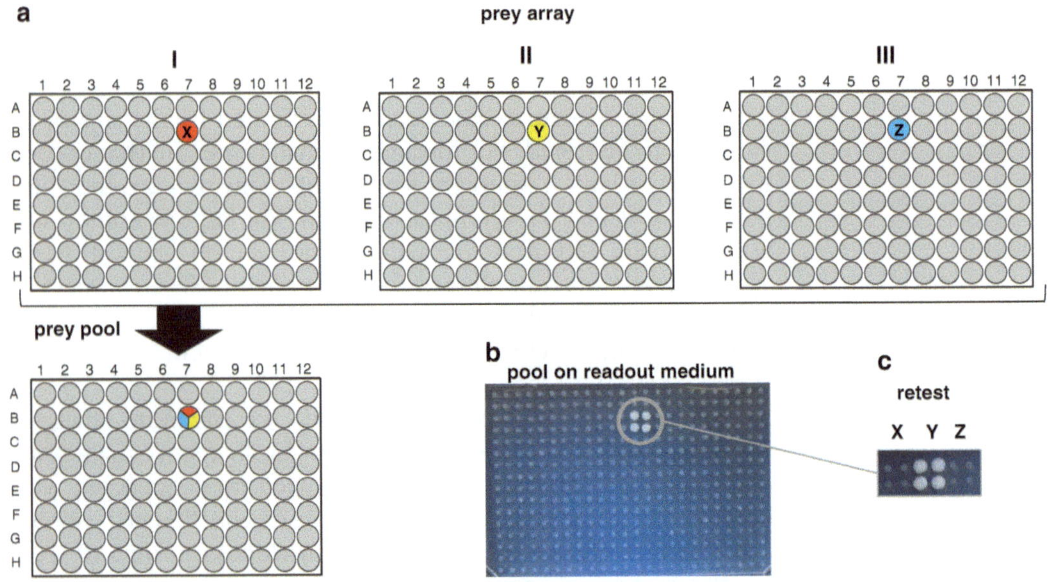

Fig. 4. Principle of mini-pool screen. (**a**) A full-matrix prey array that consists of three plates (I, II, III), each with 96 individual preys. The three plates are merged into a single prey pool plate. Pooling results in mini-pools that consist of three different preys (e.g., X, Y, and Z). (**b**) Mini-pool on readout medium (as quadruplicates). After mating with a bait clone and selection, [pool B7 …] Pool B7 exhibits an interaction. Preys X, Y, and Z are potential interaction partners. (**c**) Determination of interaction partner by a one-on-one retest assay. Prey Y is identified as the interaction partner, whereas X and Y do not interact.

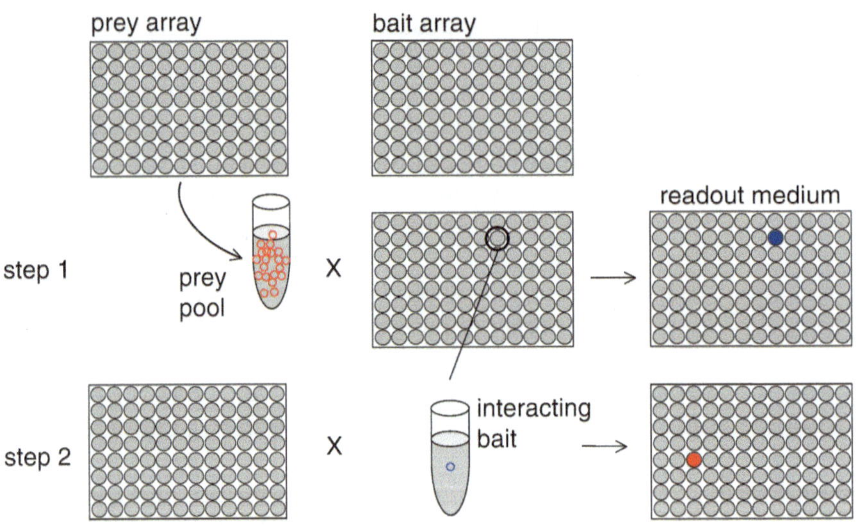

Fig. 5. Two-phase screening according to Zhong et al. (17). Step 1: A prey pool is mated against a bait array. A positive bait shows up on readout medium (in *blue*). Step 2: The positive bait from step 1 is mated against the prey array. Thus, the interacting prey can be identified (in *red*).

this strategy becomes even more efficient. For example, to detect interactions among the ~14,000 predicted *Drosophila* proteins, the two-phase strategy would require only 1/40 of the mating operations. The principle is based on two steps (Fig. 5). First, a prey array of, e.g., 96 different preys is pooled as a single 96-prey pool. Then, the pool is mated against an array that consists of 96 individual baits. On readout medium, interacting baits can be found by their positions. However, at this time point, the interacting prey is still unknown. In the second step, only positive baits are mated against the nonpooled prey array. Thus, the corresponding interacting prey can be identified.

4.3. PI-Deconvolution

Jin and colleagues (15) developed a strategy called *pooling with imaginary tags followed by deconvolution (PI-deconvolution)* which is applicable not only to Y2H screens, but also useful for other kinds of biological array-based screens like drug or protein microarray screens. They criticized that pooling strategies like the two-phase mating method are more prone to produce false negatives and false positives since interactions can pass the primary screen. The PI-deconvolution gives each bait an imaginary tag and allows screening of $2n$ baits in $2n$ pools and minimizes potential false positives and negatives because of the experimental redundancy: screens are carried out on a prey matrix in a single screen (not to be mistaken with quadruplicate or duplicate experiments) (see Fig. 6 for details). Nevertheless, the PI-deconvolution cannot resolve all interactions at once, e.g., in cases, where two or even more baits are possible interaction partners or a false positive or a false negative shows up. In such cases, retest experiments are necessary. But the PI-deconvolution identifies such experimental errors.

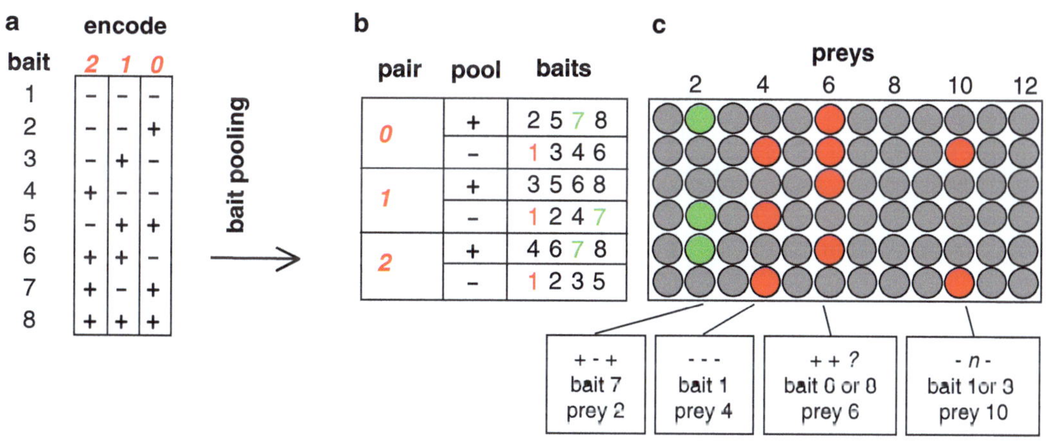

Fig. 6. PI-deconvolution scheme according to Jin et al. (15). In this example, a sample of eight baits is used. (**a**) Each of the eight tested baits is given an individual 3-bit coding tag (because $8=2^3$) by using "+" or "–" symbols (*n* bits can encode for 2^n baits and thus the size of the bait pool can be increased). (**b**) According to the mapping in (**a**), the baits are pooled in six samples ($2 \times n$) consisting of three different pool pairs, named 2, 1, and 0. Each pool pair includes a "+" and a "–" pool with the corresponding bait code. (**c**) Each bait pool is screened against a prey matrix, here consisting of 12 preys repeated in all 8 rows, i.e., each column contains the same prey and thus represents the interaction profile of that prey. Positive positions are labeled in *red*. The pattern can be tracked by the string code and interacting baits can be identified at once, e.g., bait 7 binds to prey 2, and bait 1 to prey 4. Ambiguous interaction profiles can occur, including false positives, or a prey could interact with more than one bait in the pool. For example, prey 6 might interact with bait 6 *or* 8 or 6 *and* 8 ("?"). Similarly, the absence of a signal for prey 10 indicated by "*n*" makes the identification of the interacting bait not possible because of a false-negative test position. In any case, such cases indicate immediately irregularities which may be still partially deconvoluted or may need further retesting.

The pooling strategies proposed by Zhong et al. (17) and Jin et al. (15) involve screening against bait arrays or bait pools. Due to self-activation behavior of single baits, this approach is not a trivial task and thus might be prone to produce additional false positives. Self-activating baits can be identified by an activation pretest and we recommend to exclude such baits from bait pools or screening with the two-phase mating. Furthermore, in our experience, pooling of nonactivating baits can lead to self-activation in the pool. This has to be tested for each individual bait pool in advance for the pooling method used.

4.4. Smart Pool Array

Jin and colleagues enhanced the PI-deconvolution strategy by a smart pool array (SPA) system in which, instead of individual preys, well-designed prey pools are screened in an array format that allows built-in replication and prey–bait deconvolution (14). It increases Y2H screening efficiency by an order of magnitude. Screening individual baits against prey pools avoids the above-mentioned self-activation issue of bait pools and makes the screens less error prone.

4.5. Shifted Transversal Design

The *shifted transversal design* (STD) as demonstrated by Xin et al. (16) is one more enhancement of smart pooling strategies. It achieves similar levels of sensitivity and specificity as one-on-one array-based screens, but can lower the costs and workloads three-fold. In STD, a large redundancy can be chosen but the extra redundancy is actually utilized, therefore providing high noise correction capabilities. However, this power comes at a price: despite its clean mathematical construction, the design is complex and difficult to visualize.

A simple example illustrates the STD design (see Fig. 7a). Initially, 18 preys are split into two groups of nine preys (group A and B). Each of these groups is pooled independently according its corresponding STD subdesign to obtain two sets of micropools (set A and B). Each micropool includes three different preys, and each prey is represented in three different micropools. So each prey has its own signature and is represented with an experimental redundancy of three. Two positive micropools are adequate to identify the interacting prey and one extra redundant experiment is left. Finally, each pair of same-numbered micropools from set A and B is superimposed to obtain one batch of STD pools (i.e., the micropools are pooled one more time). These still possess a redundancy of three test positions, but they contain now six preys in total instead of three. Each prey still has its unique signature, although the extra redundancy is now zero because all three pools are required to identify the interacting prey. By increasing the number of preys in the micropools and the number of STD pools, the extra redundancy can be increased again as demonstrated by the authors up to ten or even higher (see Fig. 7b). Thus, a very high noise correction can be achieved and false positives and false negatives can be minimized.

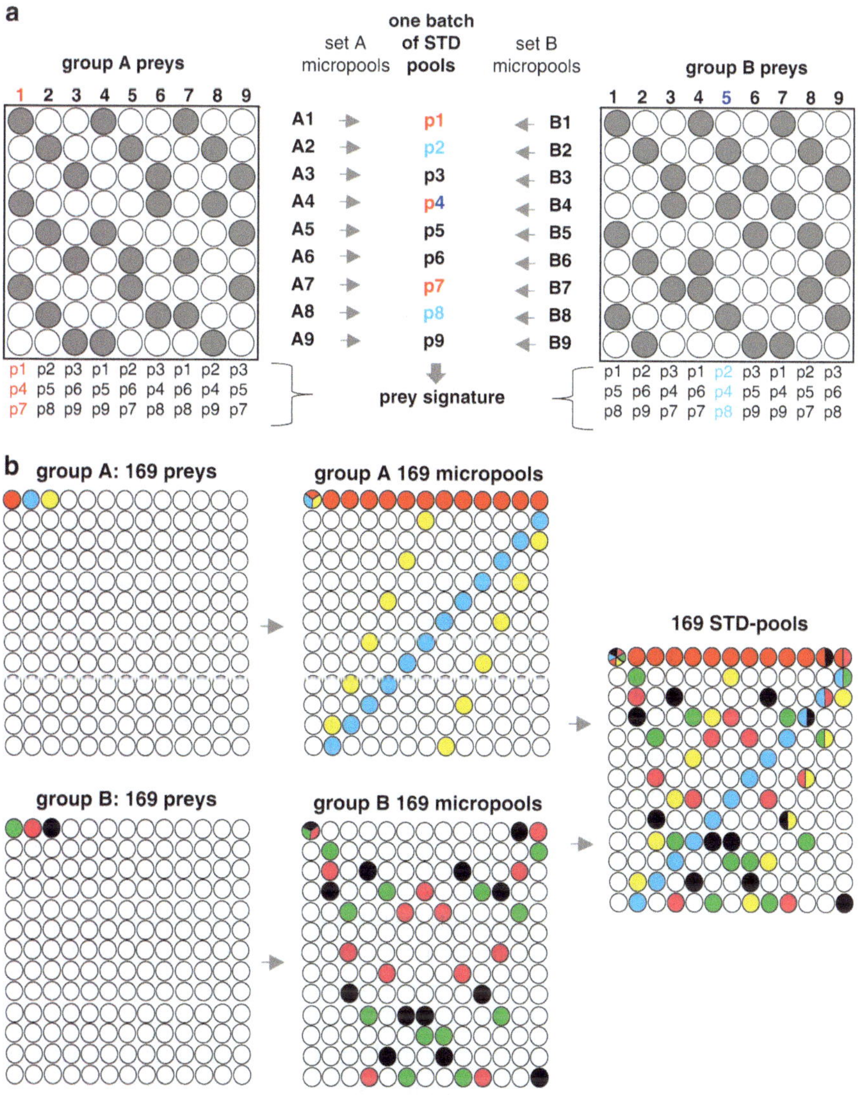

Fig. 7. Shifted transversal design. (**a**) A simple example of STD pools: 18 preys are split into two prey groups A and B with nine preys each. Those are pooled into 2 × 9 micropools. Each micropool contains three different preys (A1–A9 and B1–B9; included preys are indicated by gray circles) with a unique prey composition. Micropools with the same numbers from group A and B are superposed into nine STD pools. An interacting prey from each group can be identified by its specific pattern (e.g., prey 1 from group A (labeled in *red bold*) and prey 5 from group B (labeled in *blue bold*)). (**b**) Extensive STD pools as demonstrated by Xin et al. (2003) (16) for a proteome-wide *C. elegans* array: here, group A and B preys contain each 169 different preys. From these, the micropools are generated (set A and B). Each prey is distributed to 13 different positions resulting in a unique pattern profile with an experimental redundancy of 13. Two preys co-occur at most in one micropool. Thus, the prey can be identified by any 2 of the 13 test positions. The micropools have an extra redundancy of 11. Moreover, preys from group A and B are arranged very differently. Two preys from the two different groups co-occur in at most two common STD pools. Thus, each prey can be identified by any 3 of the 13 test positions by still maintaining a very high extra redundancy of 10 experiments. After Xin et al. (16).

5. Comparison of Y2H Systems

As soon as Y2H screens were used for library screens, it became clear that even the same bait protein can produce completely different results in independent screens. While random library screens are difficult to compare because each library is different, matrix screens allow for stringent control of screening parameters.

5.1. Random Library Screens

As an example for a random library screen, Fromont-Racine et al. (18) screened two proteins, Lsm2 and Lsm8, as both Gal4 and LexA fusions. While Gal4-Lsm2 found 33 interactions, LexA-Lsm2 found only 13 (Table 4). A comparison of these screens shows that it remains difficult to assign clear advantages to one system or another.

5.2. Comparing Vector Systems

More recently, we have followed up this issue by comparing several bait and prey vector pairs (21). In order to compare vector pairs, exactly the same ORFs were cloned into different vectors and then tested in pairwise array screens. In the first example, a genome-wide array containing all ORFs from *Treponema pallidum* was screened with 49 motility-related baits cloned into two different bait vectors, namely, pLP-GBKT7 and pAS1-LP. These two vectors yielded 77 and 165 interactions, respectively, including 21 overlapping interactions (Fig. 8). Since the bait proteins and the prey library were exactly identical, the differences must have been caused by the bait vectors. pAS1-LP expresses the Gal4 fusion from a full-length ADH promoter while pLP-GBKT7 has a truncated promoter that may have lower transcriptional activity. The only other significant difference is a shorter linker region between Gal4 DBD and the bait ORF in pAS1-LP (46 vs. 57 amino acids). However, it remains to be seen whether such differences can account for the differences in screening results.

In addition to the whole-genome arrays, we have also tested 90 motility-related proteins from *Escherichia coli* in all pairwise

Table 4
Reproducibility in random library Y2H screens (based on (18))

Bait	Preys	Shared preys	Percentage of preys shared
Gal4-Lsm2	33	2	6
LexA-Lsm2	13		15
Gal4-Lsm8	19	8	42
LexA-Lsm8	36		22

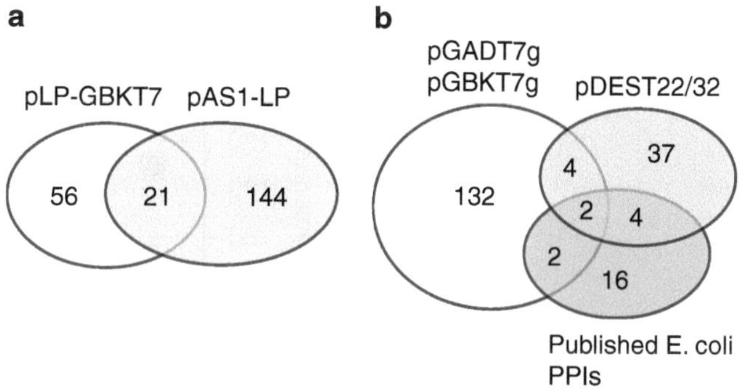

Fig. 8. Overlapping interactions between different data sets. (**a**) Overlap between the total numbers of interactions from 49 screens using motility proteins as baits (in bait vectors pLP-GBKT7 and pAS1-LP) against the whole-genome *T. pallidum* prey array (with ~1,000 individual preys). (**b**) Overlap between *E. coli* motility array screens using bait/prey vector pairs pGBKT7g/pGADT7g and pDEST32/pDEST22. Note that exactly the same set of proteins pairs (i.e., *E. coli* flagellum proteins) was tested. Twenty-four published interactions among *E. coli* flagellar proteins (from MPIDB (20)) are included as gold-standard data set ("Published *E. coli* PPIs"). Note that despite the significant difference in total interactions, the overlap with the gold-standard set is very similar. From Rajagopala et al. (21).

Table 5
Y2H vectors

		Gal4 fusion		Selection			
Vector	**Promoter**	**DBD**	**AD**	**Yeast**	**Bacterial**	**Ori**	**Source**
pDEST22	fl-ADH1	–	X	Trp1	Amp	CEN	Invitrogen
pDEST32[a]	fl-ADH1	X	–	Leu2	Gentam	CEN	Invitrogen
pGBKT7g	t-ADH1	X	–	Trp1	Kan	2μ	(37)
pGADT7g	fl-ADH1	–	X	Leu2	Amp	2μ	(37)
pAS1-LP[a]	fl-ADH1	X	–	Trp1	Amp	2μ	(36)
pLP-GADT7	fl-ADH1	–	X	Leu2	Amp	2μ	Clontech
pLP-GBKT7	t-ADH1	X	–	Trp1	Kan	2μ	Clontech

Baits contain DNA-binding domains (DBDs) and preys contain activation domains (ADs). From ref. 21
[a]Also encodes CYH2; fl-, t-ADH1 = full-length and truncated ADH1 promoters. The bacterial origin in all cases is from pUC (also called ColE1). The pDEST, pGBKT7g, and pGADT7g vectors are Gateway compatible (as indicated by the "g") while "LP" indicates loxP sites for recombinational insertion of bait and prey ORFs

combinations using two different vector pairs, namely, pGBKT7g/pGADT7g and pDEST32/pDEST22 (Table 5). Again, while the protein pairs were identical, the vectors were different. pGBKT7g/pGADT7g yielded 140 interactions, but pDEST32/pDEST22 yielded only 47 interactions (Fig. 8). It is still not clear which are

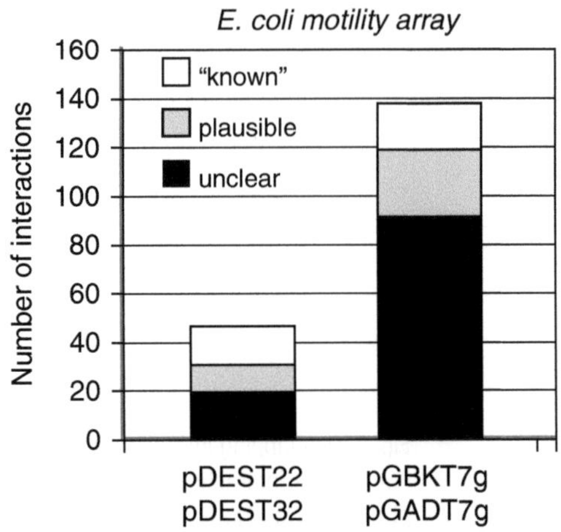

Fig. 9. Validation of two-hybrid interactions. Interactions were validated by interologs (i.e., homologous interactions) and expert evaluation. Interactions from motility array screens were classified into one of three classes: "known," plausible, and unclear (unknown). Most interactions (34% + 23% = 57%) detected with pDEST22/pDEST32 were either known or plausible while only 34% (14% + 20%) of the interactions detected with pGBKT7/ pGADT7 were assigned to these classes. From Rajagopala et al. (21).

the most important factors determining these differences, but certain patterns emerge. *First*, the Gal4 AD is slightly truncated in pDEST22 as opposed to pGADT7g. We do not know what the consequence is of this truncation. *Second*, the linkers between Gal4 (AD or DBD) and the fused ORF are significantly different between different constructs, ranging from 14 amino acids in pDEST22 to 56 amino acids in pLP-GADT7, with a similar range in the bait vectors. Given the many fewer interactions detected using the pDES22/pDEST32 pair, their shorter linker may reduce the flexibility of the fusion protein and thus result in fewer interactions. However, this hypothesis needs to be tested by increasingly longer linker sequences and additional Y2H assays. *Third*, fusion proteins encoded by both pDEST as well as the pGBKT7g/ pGADT7g vectors generate C-terminal tail sequences of 13–29 amino acids appended to bait and prey proteins, which may affect their interactions.

Interestingly, the pDEST22/ pDEST32 vectors appear to produce a higher fraction of interactions that are conserved and that are biologically relevant when compared with the pGBKT7/ pGADT7-related vectors, but the latter appear to be more sensitive and thus detect more interactions overall (Fig. 9).

5.3. N- and C-Terminal Fusions of Bait and Prey Proteins

More recently, we have taken another approach to vary Y2H vectors, based on the fact that Y2H assays by definition use two hybrid proteins. Most fusion proteins use N-terminal fusions, but it is clear that this would block any interactions involving regions

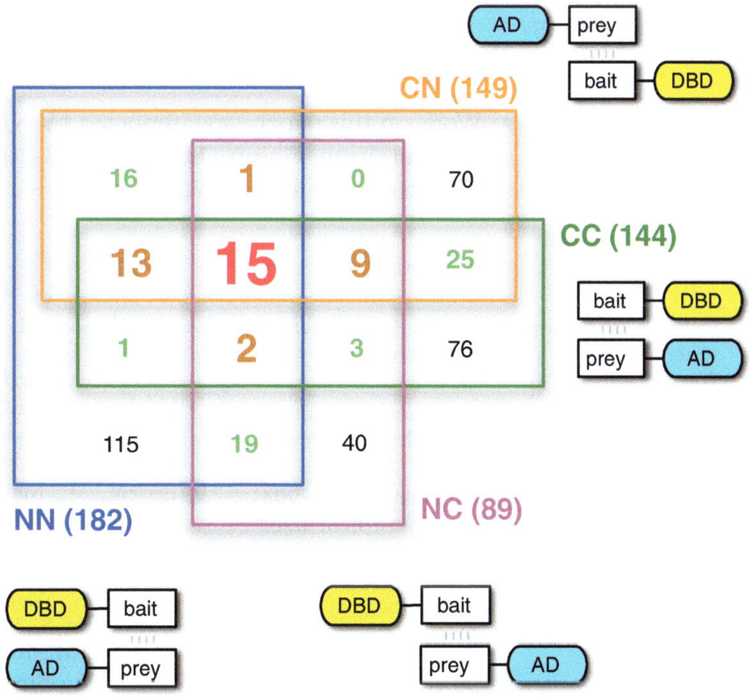

Fig. 10. Interactions found with combinations of N- and C-terminal fusions. For example, 182 interactions were found with N-terminal bait and prey fusions (*blue rectangle*) of which 115 were only found in this combination. Fifteen interactions (largest type) were found in all four combinations and are thus considered the most reliable. The box provides summaries of how many interactions were found in one, two, three, or four combinations. From Stellberger et al. (19).

around the N-terminus of these proteins. Thus, we tested C-terminal fusions of the DNA-binding and activation domains and also combined N- with C-terminal fusions. Stellberger et al. (19) tested all pairwise interactions among the ~70 ORFs of Varicella Zoster Virus using both N- and C-terminal vectors as well as combinations thereof (Fig. 10). About ~20,000 individual Y2H tests resulted in 182 NN, 89 NC, 149 CN, and 144 CC interactions. Overlap between screens ranged from 17% (NC-CN) to 43% (CN-CC). Performing four screens (i.e., permutations) instead of one resulted in more than twice as many interactions and thus much fewer false negatives. In addition, interactions that are found in multiple combinations confirm each other and thus provide a quality score. This study was the first systematic analysis of such N- and C-terminal Y2H vectors and suggested that future large-scale Y2H studies should routinely use multiple vectors, given the significantly increased number of interactions detected.

Acknowledgments

Work on this paper was supported by NIH grant RO1GM79710, the European Union (HEALTH-F3-2009-223101), and by the Landesstiftung Baden-Württemberg.

References

1. Fields, S., and Song, O. (1989) A novel genetic system to detect protein-protein interactions. *Nature 340*, 245–6.
2. Drees, B. L. (1999) Progress and variations in two-hybrid and three-hybrid technologies. *Curr Opin Chem Biol 3*, 64–70.
3. Johnsson, N., and Varshavsky, A. (1994) Ubiquitin-assisted dissection of protein transport across membranes. *Embo 13*, 2686–98.
4. Bartel, P. L., Roecklein, J. A., SenGupta, D., and Fields, S. (1996) A protein linkage map of *Escherichia coli* bacteriophage T7. *Nat Genet 12*, 72–7.
5. Fossum, E., Friedel, C. C., Rajagopala, S. V., Titz, B., Baiker, A., Schmidt, T., Kraus, T., Stellberger, T., Rutenberg, C., Suthram, S., Bandyopadhyay, S., Rose, D., von Brunn, A., Uhlmann, M., Zeretzke, C., Dong, Y. A., Boulet, H., Koegl, M., Bailer, S. M., Koszinowski, U., Ideker, T., Uetz, P., Zimmer, R., and Haas, J. (2009) Evolutionarily conserved herpesviral protein interaction networks. *PLoS Pathog 5*, e1000570.
6. Uetz, P., Dong, Y. A., Zeretzke, C., Atzler, C., Baiker, A., Berger, B., Rajagopala, S. V., Roupelieva, M., Rose, D., Fossum, E., and Haas, J. (2006) Herpesviral protein networks and their interaction with the human proteome. *Science 311*, 239–42.
7. Uetz, P., Giot, L., Cagney, G., Mansfield, T. A., Judson, R. S., Knight, J. R., Lockshon, D., Narayan, V., Srinivasan, M., Pochart, P., Qureshi-Emili, A., Li, Y., Godwin, B., Conover, D., Kalbfleisch, T., Vijayadamodar, G., Yang, M., Johnston, M., Fields, S., and Rothberg, J. M. (2000) A comprehensive analysis of protein-protein interactions in *Saccharomyces cerevisiae*. *Nature 403*, 623–627.
8. Koegl, M., and Uetz, P. (2007) Improving yeast two-hybrid screening systems. *Brief Funct Genomic Proteomic 6*, 302–12.
9. Yu, H., Braun, P., Yildirim, M. A., Lemmens, I., Venkatesan, K., Sahalie, J., Hirozane-Kishikawa, T., Gebreab, F., Li, N., Simonis, N., Hao, T., Rual, J. F., Dricot, A., Vazquez, A., Murray, R. R., Simon, C., Tardivo, L., Tam, S., Svrzikapa, N., Fan, C., de Smet, A. S., Motyl, A., Hudson, M. E., Park, J., Xin, X., Cusick, M. E., Moore, T., Boone, C., Snyder, M., Roth, F. P., Barabasi, A. L., Tavernier, J., Hill, D. E., and Vidal, M. (2008) High-quality binary protein interaction map of the yeast interactome network. *Science 322*, 104–10.
10. Boxem, M., Maliga, Z., Klitgord, N., Li, N., Lemmens, I., Mana, M., de Lichtervelde, L., Mul, J. D., van de Peut, D., Devos, M., Simonis, N., Yildirim, M. A., Cokol, M., Kao, H. L., de Smet, A. S., Wang, H. D., Schlaitz, A. L., Hao, T., Milstein, S., Fan, C. Y., Tipsword, M., Drew, K., Galli, M., Rhrissorrakrai, K., Drechsel, D., Koller, D., Roth, F. P., Iakoucheva, L. M., Dunker, A. K., Bonneau, R., Gunsalus, K. C., Hill, D. E., Piano, F., Tavernier, J., van den Heuvel, S., Hyman, A. A., and Vidal, M. (2008) A protein domain-based interactome network for C-elegans early embryogenesis. *Cell 134*, 534–545.
11. Vollert, C. S., and Uetz, P. (2004) The phox homology (PX) domain protein interaction network in yeast. *Mol Cell Proteomics 3*, 1053–64.
12. Flajolet, M., Rotondo, G., Daviet, L., Bergametti, F., Inchauspe, G., Tiollais, P., Transy, C., and Legrain, P. (2000) A genomic approach of the hepatitis C virus generates a protein interaction map. *Gene 242*, 369–79.
13. Bader, J. S., Chaudhuri, A., Rothberg, J. M., and Chant, J. (2004) Gaining confidence in high-throughput protein interaction networks. *Nat Biotechnol 22*, 78–85.
14. Jin, F. L., Avramova, L., Huang, J., and Hazbun, T. (2007) A yeast two-hybrid smart-pool-array system for protein-interaction mapping. *Nature Methods 4*, 405–407.
15. Jin, F. L., Hazbun, T., Michaud, G. A., Salcius, M., Predki, P. F., Fields, S., and Huang, J. (2006) A pooling-deconvolution strategy for biological network elucidation. *Nature Methods 3*, 183–189.
16. Xin, X., Rual, J. F., Hirozane-Kishikawa, T., Hill, D. E., Vidal, M., Boone, C., and Thierry-Mieg, N. (2009) Shifted Transversal Design smart-pooling for high coverage interactome mapping. *Genome Res 19*, 1262–9.

17. Zhong, J., Zhang, H., Stanyon, C. A., Tromp, G., and Finley, R. L., Jr. (2003) A strategy for constructing large protein interaction maps using the yeast two-hybrid system: regulated expression arrays and two-phase mating. *Genome Res 13*, 2691–9.
18. Fromont-Racine, M., Mayes, A. E., Brunet-Simon, A., Rain, J. C., Colley, A., Dix, I., Decourty, L., Joly, N., Ricard, F., Beggs, J. D., and Legrain, P. (2000) Genome-wide protein interaction screens reveal functional networks involving Sm-like proteins. *Yeast 17*, 95–110.
19. Stellberger, T., Hauser, R., Baiker, A., Pothineni, V. R., Haas, J., and Uetz, P. (2010) Improving the yeast two-hybrid system with permutated fusions proteins: the Varicella Zoster Virus interactome. *Proteome Sci 8*, 8.
20. Goll, J., Rajagopala, S. V., Shiau, S. C., Wu, H., Lamb, B. T., and Uetz, P. (2008) MPIDB: the microbial protein interaction database. *Bioinformatics 24*, 1743–4.
21. Rajagopala, S. V., Hughes, K. T., and Uetz, P. (2009) Benchmarking yeast two-hybrid systems using the interactions of bacterial motility proteins. *Proteomics 9*, 5296–5302.
22. Schwartz, H., Alvares, C. P., White, M. B., and Fields, S. (1998) Mutation detection by a two-hybrid assay. *Hum Mol Genet 7*, 1029–1032.
23. Vidal, M., and Endoh, H. (1999) Prospects for drug screening using the reverse two-hybrid system. *Trends Biotechnol 17*, 374–81.
24. Vidal, M., and Legrain, P. (1999) Yeast forward and reverse 'n'-hybrid systems. *Nucleic Acids Res 27*, 919–29.
25. SenGupta, D. J., Zhang, B., Kraemer, B., Pochart, P., Fields, S., and Wickens, M. (1996) A three-hybrid system to detect RNA–protein interactions in vivo. *Proc Natl Acad Sci USA 94*, 8496–8501.
26. Estojak, J., Brent, R., and Golemis, E. A. (1995) Correlation of two-hybrid affinity data with in vitro measurements. *Mol Cell Biol 15*, 5820–9.
27. Rain, J. C., Selig, L., De Reuse, H., Battaglia, V., Reverdy, C., Simon, S., Lenzen, G., Petel, F., Wojcik, J., Schachter, V., Chemama, Y., Labigne, A., and Legrain, P. (2001) The protein-protein interaction map of Helicobacter pylori. *Nature 409*, 211–215.
28. Raquet, X., Eckert, J. H., Muller, S., and Johnsson, N. (2001) Detection of altered protein conformations in living cells. *J Mol Biol 305*, 927–38.
29. Cagney, G., Uetz, P., and Fields, S. (2001) Two-hybrid analysis of the *Saccharomyces cerevisiae* 26 S proteasome. *Physiol Genomics 7*, 27–34.
30. Ito, T., Chiba, T., Ozawa, R., Yoshida, M., Hattori, M., and Sakaki, Y. (2001) A comprehensive two-hybrid analysis to explore the yeast protein interactome. *Proc Natl Acad Sci USA 98*, 4569–74.
31. Giot, L., Bader, J. S., Brouwer, C., Chaudhuri, A., Kuang, B., Li, Y., Hao, Y. L., Ooi, C. E., Godwin, B., Vitols, E., Vijayadamodar, G., Pochart, P., Machineni, H., Welsh, M., Kong, Y., Zerhusen, B., Malcolm, R., Varrone, Z., Collis, A., Minto, M., Burgess, S., McDaniel, L., Stimpson, E., Spriggs, F., Williams, J., Neurath, K., Ioime, N., Agee, M., Voss, E., Furtak, K., Renzulli, R., Aanensen, N., Carrolla, S., Bickelhaupt, E., Lazovatsky, Y., DaSilva, A., Zhong, J., Stanyon, C. A., Finley, R. L., Jr., White, K. P., Braverman, M., Jarvie, T., Gold, S., Leach, M., Knight, J., Shimkets, R. A., McKenna, M. P., Chant, J., and Rothberg, J. M. (2003) A protein interaction map of *Drosophila melanogaster*. *Science 302*, 1727–36.
32. Li, S., Armstrong, C. M., Bertin, N., Ge, H., Milstein, S., Boxem, M., Vidalain, P. O., Han, J. D., Chesneau, A., Hao, T., Goldberg, D. S., Li, N., Martinez, M., Rual, J. F., Lamesch, P., Xu, L., Tewari, M., Wong, S. L., Zhang, L. V., Berriz, G. F., Jacotot, L., Vaglio, P., Reboul, J., Hirozane-Kishikawa, T., Li, Q., Gabel, H. W., Elewa, A., Baumgartner, B., Rose, D. J., Yu, H., Bosak, S., Sequerra, R., Fraser, A., Mango, S. E., Saxton, W. M., Strome, S., Van Den Heuvel, S., Piano, F., Vandenhaute, J., Sardet, C., Gerstein, M., Doucette-Stamm, L., Gunsalus, K. C., Harper, J. W., Cusick, M. E., Roth, F. P., Hill, D. E., and Vidal, M. (2004) A map of the interactome network of the metazoan *C. elegans*. *Science 303*, 540–3.
33. Rual, J. F., Venkatesan, K., Hao, T., Hirozane-Kishikawa, T., Dricot, A., Li, N., Berriz, G. F., Gibbons, F. D., Dreze, M., Ayivi-Guedehoussou, N., Klitgord, N., Simon, C., Boxem, M., Milstein, S., Rosenberg, J., Goldberg, D. S., Zhang, L. V., Wong, S. L., Franklin, G., Li, S., Albala, J. S., Lim, J., Fraughton, C., Llamosas, E., Cevik, S., Bex, C., Lamesch, P., Sikorski, R. S., Vandenhaute, J., Zoghbi, H. Y., Smolyar, A., Bosak, S., Sequerra, R., Doucette-Stamm, L., Cusick, M. E., Hill, D. E., Roth, F. P., and Vidal, M. (2005) Towards a proteome-scale map of the human protein protein interaction network. *Nature* 437, 1173–1178.
34. Stelzl, U., Worm, U., Lalowski, M., Haenig, C., Brembeck, F. H., Goehler, H., Stroedicke, M., Zenkner, M., Schoenherr, A., Koeppen, S., Timm, J., Mintzlaff, S., Abraham, C., Bock, N., Kietzmann, S., Goedde, A., Toksoz, E., Droege, A., Krobitsch, S., Korn, B., Birchmeier, W.,

Lehrach, H., and Wanker, E. E. (2005) A human protein-protein interaction network: A resource for annotating the proteome. *Cell 122*, 957–968.

35. Parrish, J. R., Yu, J., Liu, G., Hines, J. A., Chan, J. E., Mangiola, B. A., Zhang, H., Pacifico, S., Fotouhi, F., Dirita, V. J., Ideker, T., Andrews, P., and Finley, R. L., Jr. (2007) A proteome-wide protein interaction map for *Campylobacter jejuni*. *Genome Biol 8*, R130.

36. Titz, B., Rajagopala, S. V., Goll, J., Hauser, R., McKevitt, M. T., Palzkill, T., and Uetz, P. (2008) The binary protein interactome of Treponema pallidum – the syphilis spirochete. *PLoS ONE 3*, e2292.

37. Rajagopala, S. V., Titz, B., and Uetz, P. (2007) Array-based yeast two-hybrid screening for protein-protein interactions. In: *Yeast Gene Analysis, Second Edition*, 2nd Edn., Stark, M., Stansfelid. I., (Eds), 2007, Elsevier Amsterdem, *36*, 139–163.

Chapter 2

Array-Based Yeast Two-Hybrid Screens: A Practical Guide

**Roman Häuser, Thorsten Stellberger,
Seesandra V. Rajagopala, and Peter Uetz**

Abstract

Yeast two-hybrid screens are carried out as random library screens or matrix-based screens. The latter have the advantage of being better controlled and thus typically give clearer results. In this chapter, we provide detailed protocols for matrix-based Y2H screens and give some helpful instructions how to plan a large-scale interaction screen. We also discuss strategies to identify or avoid false negatives and false positives.

Key words: ORFeome, Mating, Pooling, Protein–protein interactions, Yeast two hybrid, Array, Vectors, Yeast strains

Abbreviations

3-AT	3-Amino-1,2,4-triazole
AD	Activation domain
DBD	DNA-binding domain
ORF	Open reading frame
Y2H	Yeast two hybrid

1. Introduction

The construction of an entire proteome array of an organism that can be screened in vivo under uniform conditions is a challenge. When proteins are screened at a genome scale, automated robotic procedures are necessary. The protocols described here were established for yeast proteins, but they can be applied to any other genomes or subsets thereof; for example, viral and bacterial genomes have been screened for interactions in our lab. Different high-throughput cloning methods used to generate two-hybrid

Bernhard Suter and Erich E. Wanker (eds.), *Two Hybrid Technologies: Methods and Protocols*, Methods in Molecular Biology, vol. 812, DOI 10.1007/978-1-61779-455-1_2, © Springer Science+Business Media, LLC 2012

clones, i.e., proteins with AD fusions (preys) and DBD fusions (baits), are presented below. The steps of the process involve the construction of the array and screening of the array by either manual or robotic manipulation, including the selection of positives and scoring of results.

High-throughput screening projects deal with a large number of proteins; therefore, hands-on time and amount of resources become an important issue. Options to reduce the screening effort are discussed. A prerequisite for array-based genome-wide screens is the existence of a cloned *ORFeome* (typically defined as full-length ORF sets) or at least a number of protein-coding clones; we briefly mention strategies how to create such ORFeomes.

1.1. General Screen Requirements

1.1.1. Strategic Planning

Before starting an array-based screen, the size and character of the array must be designed and the ultimate aims of the experiment need to be considered. Factors that may be varied include the format of the array (e.g., full-length protein or single domain, choice of epitope tags, etc.). Similarly, the arrayed proteins may be related (e.g., a family or pathway of related proteins, orthologs of a protein from different species, the entire protein complement of a model organism). In our experience, certain protein families work better than others (e.g., splicing proteins, bacterial flagellum proteins, and proteins involved in DNA replication) while others do not appear to work at all (e.g., many metabolic enzymes and membrane proteins). We recommend to carry out a small-scale pilot study, incorporating positive and negative controls, before committing to a full-scale project.

Although high-throughput screening projects can be performed manually, automation is strongly recommended. Highly repetitive tasks are not only boring and straining, but also error prone when done manually. If you do not have local access to robotics, you may have to collaborate with a laboratory that does.

1.2. Generation of a Protein Array Suitable for High-Throughput Screening

Once the set of proteins to be included in the array is defined, the coding genes need to be PCR amplified and cloned into Y2H bait and prey vectors. In order to facilitate the cloning of a large number to ORFs, site-specific recombination-based systems are commonly used (e.g., Gateway or Univector cloning (1, 2)) (Fig. 1). Some of these systems require expensive enzymes and vectors, although both may be produced in the lab.

1.2.1. Cloning by Homologous Recombination in Yeast

An alternative to site-directed systems is the cloning by homologous recombination directly in yeast (3). A two-step PCR protocol is used to make DNA with sufficient homology to vector DNA at the terminal ends to allow homologous recombination in the yeast cell (Fig. 1). In the "first-round" PCR reaction, the ORF is amplified with primers that contain ~20 nucleotide tails which are homologs to sequences in the two-hybrid vectors. In the

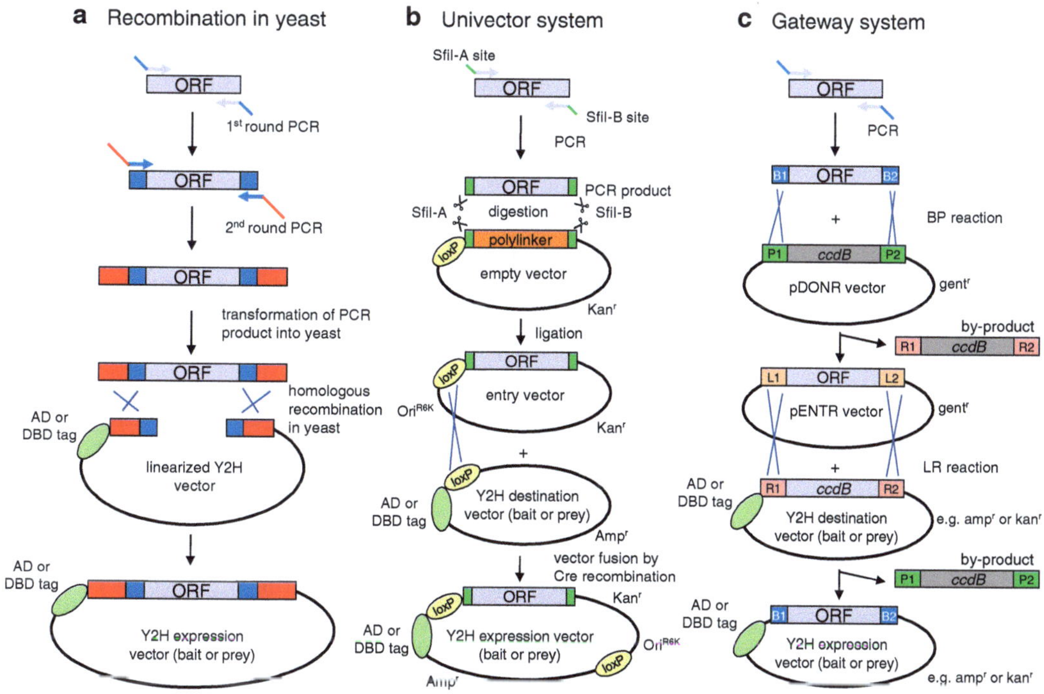

Fig. 1. ORFeome cloning systems. (**a**) Homologous recombination in yeast: ORFs are amplified (first PCR) with gene-specific primers that generate a product with common 5′ and 3′ 20-nucleotide tails. A second PCR generates a product with common 5′ and 3′ 70-nucleotide tails. The common 70-nucleotide ends allow cloning into linearized two-hybrid expression vectors by cotransfection into yeast. The endogenous yeast homologous recombination machinery performs the recombination reaction and results in a circular plasmid. (**b**) Univector plasmid-fusion system: ORFs are amplified with gene-specific primers that generate a product with common 5′ and 3′ rare-cutting restriction sites. The PCR product is cloned into a pUNI entry vector by DNA ligation. Cre–*loxP*-mediated site-specific recombination fuses the pUNI entry clone and yeast two-hybrid expression plasmids (bait/prey) at the *loxP* site. As a result, the gene of interest is placed under the control of the yeast two-hybrid expression vector promoter. (**c**) Gateway cloning: The ORFs are amplified with gene-specific primers that generate a product with common 5′ and 3′ recombination sites (*attB1* and *attB2*). The entry clones are made by recombining the ORFs of interest with the flanking *attB* sites into the *attP* sites of a suitable Gateway entry vector (e.g., pDONR201 or pDONR207) mediated by the Gateway BP Clonase II Enzyme Mix (Invitrogen). Subsequently, the fragment in the entry clone can be transferred to any yeast two-hybrid destination vector that contains the *attR* sites by mixing both plasmids and using the Gateway LR Clonase II Enzyme Mix.

second-round PCR, ~50-nucleotide tails (homologous to the destination vector-cloning site) are attached to the first-round PCR product (Fig. 1). The PCR product is then transfected into the yeast cells together with the linearized vector and the recombination event between them takes place inside the yeast cell. The advantage of this strategy is its much reduced cost. The disadvantage is that plasmids have to be recovered from yeast which can be time consuming and inefficient.

1.2.2. Univector Plasmid-Fusion System

Similar to the Gateway system, the Univector Plasmid-Fusion System (UPS) requires an entry vector containing the ORF. The UPS uses *Cre–loxP*-based site-specific recombination to catalyze plasmid fusion between the entry "univector" and destination vectors containing, e.g., specific promoters, fusion proteins, and selection markers. Cre is a site-specific recombinase, which catalyzes the recombination between two 34 bp *loxP* sequences (Fig. 1).The pUNI plasmid is the entry vector of this system, the vector into which the gene of interest is inserted. The pHOST plasmid is the recipient vector containing the appropriate transcriptional regulatory sequences that eventually control the expression of the gene of interest in the designated host cells. A recombinant expression construct is made through Cre–*loxP*-mediated site-specific recombination that fuses pUNI and pHOST into a dimeric fusion plasmid. A crucial feature of the pUNI plasmid is its conditional origin of replication derived from the plasmid R6Kγ that allows its propagation only in bacterial hosts expressing the *pir* gene (encoding the essential replication protein p). Thus, only dimeric pUNI–pHOST vectors are selected and propagated (1) (Fig. 1).

1.2.3. Gateway® Cloning

Gateway® (Invitrogen) cloning provides another fast and efficient way of cloning ORFs (2). It is based on the site-specific recombination properties of bacteriophage lambda (4); recombination is mediated between the so-called attachment sites (*att*) of DNA molecules: between *attB* and *attP* sites or between *attL* and *attR* sites. The first step to Gateway® cloning is inserting your gene of interest into a specific entry vector. This entry clone is a plasmid containing your gene of interest flanked by *attL* recombination sites. These *attL* sites can be recombined with *attR* sites on a destination vector resulting in a plasmid for functional protein expression in a specific host. One way of obtaining the initial entry clones is by recombining a PCR product of the ORF flanked by *attB* sites with the *attP* sites of a pDONR vector.

Site-specific recombination systems like the Gateway® or UPS system have got some crucial advantages in comparison to classical ligation cloning: the recombination reaction is highly efficient and fast to perform. The entry vector library can not only be transferred into yeast two-hybrid destination vectors, but also in any other compatible vector system that carries the recombination sites. For instance, the Gateway® technology provides plenty of commercially available destination vectors that can be used for further downstream experiments like protein purification or in vivo expression analysis. Furthermore, bait and prey plasmids can be created simultaneously within the same recombination reaction as long as they contain different bacterial selection markers.

1.3. ORFeome Cloning

The starting point of an array-based Y2H screen is the construction of an ORFeome array. An ORFeome represents all ORFs of a genome or a subset thereof – in our case: the selected gene set individually cloned into entry vectors of a recombination-based cloning system. More and more ORFeomes are available and can be directly used for generating the Y2H bait and prey constructs. Alternatively, they can be cloned into the entry vector by multiple strategies, such as classical ligation or recombination. Both entry vector construction and the subsequent destination vector cloning can be done for multiple ORFs in parallel. The whole procedure can be parallelized using 96-well plates so that whole ORFeomes can be processed in parallel.

1.4. Prey Array

The Y2H array is made from an ordered set of AD-containing strains (preys), rather than DBD-containing strains (baits), because the former do not generally result in self-activation of transcription. The prey constructs are assembled by transfer of the ORFs from entry vectors into specific prey vectors by recombination. Several prey vectors for the UPS and the Gateway® system are available. In our lab, we preferentially use the Gateway®-adapted pGADT7g (a derivative of pGADT7 from Clontech) and pDEST22 (Invitrogen) vectors (Fig. 2). An alternative is the direct cloning of prey constructs by homologous recombination in yeast (see above). These prey constructs are transformed into haploid yeast cells (e.g., the Y187 strain, *see protocol 3.1*). Finally, individual yeast colonies, each carrying one specific prey construct, are arrayed on agar plates in a 96 format. By a second pinning step, the preys are copied as quadruplicates or duplicates to yield the final prey array that can be used for the screening procedure.

1.5. Baits

Baits are also constructed by recombination-based transfer of the ORFs into specific bait vectors or, alternatively, directly by homologous recombination in yeast. Bait vectors used in our lab are pGBKT7 (Clontech) modified for Gateway® cloning (pGB-KT7g) and pDEST32 (Invitrogen) (Fig. 2). The bait constructs are also transformed into haploid yeast cells; we use the AH109 strain (*protocol 3.1*). After self-activation testing, the baits can be tested for interaction screening against the Y2H prey array.

Bait and prey plasmids must be transformed into haploid yeast strains of opposite mating type to combine bait and prey plasmids by mating. To our knowledge, it does not make a difference whether baits or preys are transformed into either *a* or alpha cells, respectively.

1.6. Self-Activation Test of Baits (One-Hybrid Assay)

Prior to the two-hybrid analyses, the bait yeast strains should be examined for self-activation. Self-activation is defined as a detectable bait-dependent reporter gene activation in the absence of any prey interaction partner. Weak to intermediate-strength self-activator baits can be used in two hybrid array screens because the corresponding bait–prey interactions confer stronger signals than the self-activation background. If the HIS3 reporter gene is used, the self-activation background can be suppressed by adding 3-AT, a competitive inhibitor of HIS3.

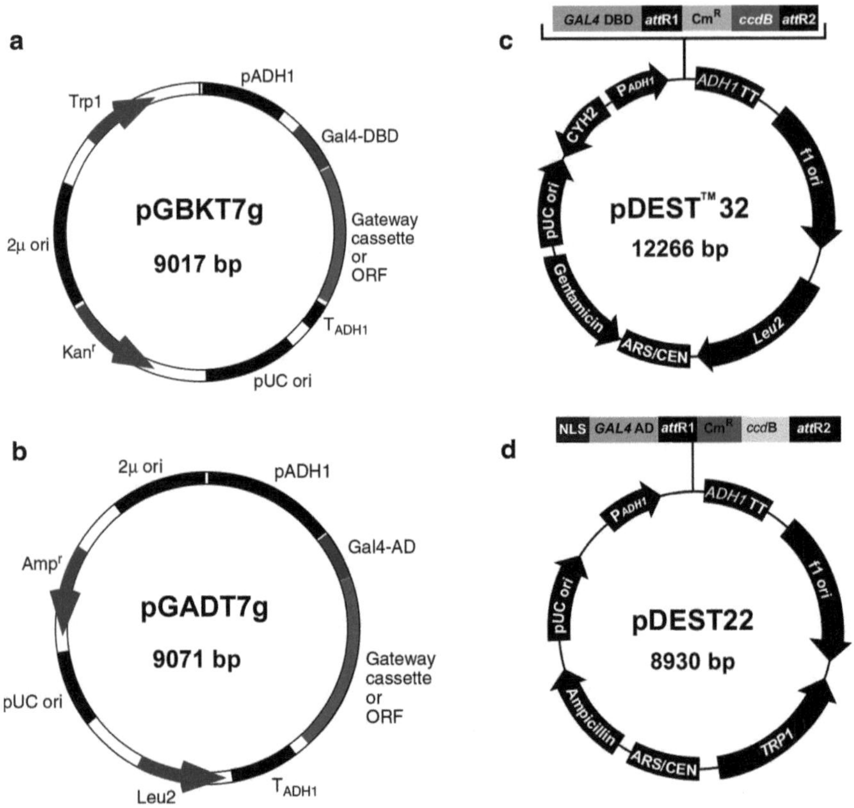

Fig. 2. Commonly used yeast two-hybrid expression vectors. The vectors shown here are all Gateway® system-compatible expression vectors and carry the Gateway cassette which is flanked by the recombination sites *attR1* and *attR2*. The cassette contains *ccdB* whose product inhibits *E. coli* gyrase. Thus, after recombination, positive *E. coli* clones can be directly selected on the respective antibiotic selection medium, whereas negative recombination products are automatically deselected. (**a**, **b**) Bait vector pGBKT7g and prey vector pGADT7g. (**c**, **d**) Bait vector pDEST32 and prey vector pDEST22 (Invitrogen). Here, the Gateway® cassette is shown in closer detail.

Self-activation of all the baits should be examined simultaneously on plates containing different concentrations of 3-AT (see *protocol 3.2*). For instance, a titration series with 3-AT concentrations of 0, 1, 2, 4, 8, 16, …, 128 mM can be used. The lowest concentration (minimal inhibitory concentration) of 3-AT that suppresses growth in this test is used for the interaction screen because it avoids background growth, whereas true interactions are still detectable.

1.7. Screening Procedure

The Y2H prey array can be screened for protein interactions by a mating procedure that can be carried out manually or using robotics (see *protocol 3.3*). Since the screening procedure used here is based on yeast mating, the bait and prey strains can be mated by manual mixing or by a robotic device that essentially replica plates preys on an array of baits. For large numbers of strains, automation is obviously desirable. Typically, these mating steps are carried out either in a 96 or 384 format so that colonies can be picked up from equivalent 96- or 384-well microtiter plates and then copied onto solid agar (see step 1 of *protocol 3.3* for details).

1.8. Time Considerations in One-on-One Yeast Two-Hybrid Screens

In the simplest case, a set of baits is tested individually against a set of preys. For ten baits and ten preys, this results in $10 \times 10 = 100$ individual tests (e.g., when all components of a protein complex are tested against each other). For a viral genome of 100 genes, already 10,000 tests are required. Thus, the number of tests grows exponentially with the number of baits and preys.

As a consequence, automation is required for larger projects. For example, in our laboratory, a single Biomek 2000 robot was sufficient for testing about 50 baits against a bacterial genome of 1,000 ORFs per week or all 100 proteins of a viral genome against each other. Note that each interaction also should be tested at least twice as duplicates, just to make sure that the result is reproducible. This doubles the number of tests to be done. In fact, for smaller projects, we recommend to do each test four times, e.g., by spotting quadruplicates of each prey.

In larger projects, all tests can be done once, but then each positive protein pair needs to be retested later, ideally in a coordinated effort to verify all positives. This time, quadruplicates can be used.

In theory, the colony density of the array can be increased as well, e.g., from 384 to 768 or even to 1,536 colonies per plate. However, this approach requires a higher precision of the robot, smaller colony sizes, and thus can reduce the number of detected interactions, e.g., due to a smaller number of transferred cells. While we have used 768-spot arrays on microtiter-sized plates, 1,536 spots turned out to be too error prone with our equipment. In our experience, the higher the number of test positions is, the more noisy becomes the signal since the single colonies start to compete for nutrients and thus clearly slow down growth.

1.9. Pooling Strategies

In an independent chapter, we described a couple of pooling strategies that can be used to speed up high-throughput screening. For very large genome-wide screens, pooling is recommended. The most critical point of pooling is that equal cell numbers of different clones in the pool cannot be adjusted perfectly well (causing over- and underrepresentation) and thus pooling is prone to produce false negatives. Single replication steps must be watched more carefully to yield high mating efficiency and preparation of pool plates takes additional time compared to one-on-one matrix screens. But once the experimental setup is well-established, pooling strategies can yield the same sensitivity as one-on-one screens by lowering cost and time (see *protocol 3.4*).

1.10. Retests

A major consideration when using the Y2H system is the number of false positives. The major source of false positives are nonreproducible signals which arise through little-understood mechanisms. In array screens and probably in random library screen, more than 90% of all interactions can be nonreproducible background (5). Thus, simple retesting by repeated mating can identify most false positives. We routinely use quadruplicate retesting. It is done by mating the interaction pair to be tested and by comparing the activation strength of this pair with the activation strength of a control, usually the bait mated with the strain that contains the empty prey vector. Retests can also be used to identify interacting preys of positively tested pools (see *protocol 3.5*).

1.11. Evaluation of Raw Screen Data

Filtering of raw results significantly improves the data quality of the protein interaction set. For filtering, at least three parameters should be considered. *First*, protein interactions that cannot be reproduced in the retest experiment should be discarded. *Second*, for each prey, the number of different interacting baits is calculated. Preys interacting with a large number of baits are assumed to be nonspecific and thus may have no biological relevance. However, the concrete cutoff number depends also on the nature of baits that are screened: if a large family of related proteins is screened, it is not surprising that many of them find the same prey. As a rough guideline, the number of baits interacting with a certain prey should not be larger than 5% of the bait number. The *third* parameter is the background activation activity of the tested bait. The activation strength of interaction pairs must be significantly higher than with all other (background) pairs. In principle, at least with the HIS3 reporter, no activation (i.e., no colony growth) should be observed in noninteracting pairs.

In addition to these parameters, more sophisticated statistical evaluations of the raw results have been suggested. For instance, filtering the raw interaction dataset by logistic regression (which uses positive and negative training sets of interactions) can help to qualify the most reliable data (6, 7).

2. Materials

2.1. Yeast Media

2.1.1. YEPD

1. *YEPD liquid medium*: 10 g yeast extract, 20 g peptone, 20 g glucose. Make up to 1 L with sterile water and autoclave.
2. *YEPD solid medium*: 10 g yeast extract, 20 g peptone, 20 g glucose, 16 g agar. Make up to 1 L with sterile water and autoclave. After autoclaving, cool media to ~60°C and add 4 ml of 1% adenine solution (1% in 0.1 M NaOH). Pour 40 ml into each sterile Omnitray plate (Nunc) under sterile hood and let them solidify (see Note 1).

2.1.2. Selective Media

Dropout mix (-His, -Leu, -Trp): 1 g methionine, 1 g arginine, 2.5 g phenylalanine, 3 g lysine, 3 g tyrosine, 4 g isoleucine, 5 g glutamic acid, 5 g aspartic acid, 7.5 g valine, 10 g threonine, 20 g serine, 1 g adenine, 1 g uracil. Mix all components and store under dry conditions at room temperature.

1. *Medium concentrate (5×):* 8.5 g yeast nitrogen base, 25 g ammonium sulfate, 100 g glucose, 7 g dropout mix. Make up to 1 L with water and sterile filter (e.g., Millipore sterile filter). Store at 4°C (see Notes 2 and 3).
2. *Amino acid stock solutions* (see Note 4):
   Histidine (His): Dissolve 4 g of histidine in 1 L water and sterile filter.
   Leucine (Leu): Dissolve 7.2 g of leucine in 1 L water and sterile filter.
   Tryptophan (Trp): Dissolve 4.8 g of tryptophan in 1 L water and sterile filter.
3. *3-Amino-triazole (3-AT) stock solution*: 0.5 M. Sterile filter (see Note 4).

For one liter of minimal medium, autoclave 16 g of agar in 800 ml of water, cool the medium to ~60°C, then add 200 ml 5× medium concentrate, and mix. Pour ca. 40 ml into each sterile Omnitray plate under sterile hood and let them solidify (see Note 1). Depending on the required selective plates, you have to add the missing amino acids or 3-AT. Liquid minimal media can be prepared without adding agar. Corresponding amino acids are added from the amino acid stock solutions as follows (see Note 5):

Selection of baits (-Trp plates): 8.3 ml leucine and 8.3 ml histidine.

Selection of preys (-Leu plates): 8.3 ml tryptophan and 8.3 ml histidine.

Selection of diploids (-Leu-Trp plates): 8.3 ml histidine.

Readout medium (-Leu-Trp-His plates): Add 3-AT from 0.5 M stock solution as needed for screening self-activating baits.

2.2. Yeast Transformation

1. *Carrier DNA* (salmon sperm DNA): Dissolve 7.75 mg/ml salmon sperm DNA (e.g., Sigma D1626) in water and store at −20°C following a 15 min 121°C autoclave cycle.
2. *96 PEG solution* (100 ml): Mix 45.6 g PEG, 6.1 ml of 2 M LiOAc (lithium acetate), 1.14 ml of 1 M Tris-HCl, pH 7.5, and 232 μl 0.5 M EDTA; make up to 100 ml with sterile water and autoclave. Store at room temperature.
3. *CT110*: Mix 20.73 ml 96 PEG, 0.58 ml boiled salmon sperm DNA (boil frozen salmon sperm DNA at 95°C for 5 min), and 2.62 ml DMSO. Add DMSO last and mix quickly after adding by shaking vigorously and vortexing for 30 s (see Note 6).

2.3. Screen Procedure, Retests, and Bait Self-Activation Test

1. 96-well microtiter plates, round bottom.
2. 1-well plates (Nunc OmniTray plates, Nunc).
3. Bleach solution (20%): Dilute a 12% sodium hypochlorite solution 1:5 with water (see Note 7).
4. 95% ethanol solution, industrial.
5. Autoclaved water.
6. Replication tool or robot (e.g., Biomek, Beckman Coulter), 96- and 384-pinning tool.
7. 1% (w/v) adenine solution (1% in 0.1 M NaOH), sterile filter.
8. YEPD and selective media as liquids and agar plates as described (see Subheading 2.1).

2.4. Vectors (Examples)

1. Bait plasmid(s): pGBKT7 (Clontech), pOBD2 (8), pDEST32 (Invitrogen), pGBKT7g (9), pGBKCg (10), pAS1, pAS2-1 (Clontech).
2. Prey plasmid(s): pGADT7 (Clontech), pDEST22 (Invitrogen), pGADT7g (9), pGADCg (10).

Any other vectors can be used as long as they are compatible with each other and the yeast strains.

2.5. Yeast Strains (Examples)

1. AH109: Genotype (MAT a, trp 1-901, leu2-3,112, ura3-52, his3-200, Δgal4, Δgal80, LYS2: $GAL1_{UAS}$-$GAL1_{TATA}$-HIS3, $GAL2_{UAS}$-$GAL2_{TATA}$-ADE2, URA3: $MEL1_{UAS}$-$MEL1_{TATA}$-lacZ) (11, 12).
2. Y187: Genotype (MAT α, ura3-52, his3-200, ade2-101, trp1-901, leu2-3, 112, Δgal4, met, Δgal80, URA3: $GAL1_{UAS}$-$GAL1_{TATA}$-lacZ) (13).

3. Methods

The following protocols are described for the *HIS3* reporter and the pGBKT7g and pGADT7g vector system. The protocols are applicable for the pDEST32/pDEST22 system and others as well. However, the different yeast and *E. coli* selection markers have to be considered during the selection steps and the selection media have to be exchanged.

3.1. Yeast Transformation for Bait and Prey Construction

This protocol is suitable for 100 yeast transformations, and may be scaled up or down as needed. Selection of the transformed yeast cells requires leucine- or tryptophan-free media ("-Leu" or "-Trp," depending on the selective marker on the plasmid). Moreover, at least one of the haploid strains must contain a two-hybrid reporter gene under GAL4 control.

1. Inoculate 50 ml YEPD liquid medium with ~200 μl liquid stock of yeast strains (e.g., AH109, Y187, or any other appropriate yeast strain; we use Y187 strains for preys and AH109 for baits) in a 250-ml flask and grow overnight with shaking at 30°C (minimum 15 h, max. 24 h).
2. Spin down cells in 50-ml conical tube (3,500 × *g*, 5 min at room temperature), discard supernatant, dissolve the pellet by adding 2 ml LiOAc (0.1 M), and transfer resuspended yeast to two 1.5-ml microfuge tubes. Spin and pellet yeast and resuspend in a total volume of 1.8 ml LiOAc (0.1 M).
3. Prepare CT110 solution.
4. Add all the competent yeast cells prepared above and mix vigorously by hand or by vortexing for 1 min. Immediately pipette 245 μl into each of 96 wells of a 96-well plate.
5. Add 50–100 ng of plasmid or 5 μl of PCR products (in case of cotransformation and homologous recombination in yeast) and positive control (empty vector) and negative control (only CT110). Seal the 96-well plate with plastic or aluminum tape and vortex for 4 min.
6. Incubate at 42°C for 30 min.
7. Spin the 96-well plate for 10 min at 2,000 × *g*; discard the supernatant and aspirate with eight-channel wand or by tapping on cotton napkin for a couple of times. Add 150 μl of sterile water to all 96 wells, resuspend, and plate cells on selective agar plates (e.g., standard Petri dishes) with -Leu for pGADT7g or -Trp for pGBKT7g.
8. Incubate the plates at 30°C for 3 days. After 2 days, the colonies start to appear; pick colonies after 3 days.
9. Rearray baits and preys in 96-well plates. Grow them up again for 1–2 days in -Leu or -Trp liquid minimal medium at 30°C.
10. The bait and prey plate can now be used to make a couple of copies on selective agar medium, to back up the arrays as glycerol (25%) stocks for −80°C long-term storage, and to use the baits directly for the self activation test (see below) (see Note 8).

3.2. Bait Self-Activation Test

The aim of this test is to measure the background reporter activity (here: *HIS3*) of bait proteins in the absence of an interacting prey protein. This measurement is used for choosing the selection conditions used during the interaction screen and can be achieved by mating individual bait strain with a single prey strain that carries the empty prey plasmid. Ninety-six individual bait activation tests can be carried out on one plate simultaneously.

1. Load a 96-well plate (round button) with ~200 μl YEPD liquid medium.
2. Inoculate plate with baits by replicating the 96-format bait plate from solid medium into the destination plate by using a sterile 96-pinning tool (see step 1 of Subheading 3.3 for sterilization details).
3. Inoculate the yeast strain Y187 which carries the empty prey vector in 30–50 ml YEPD liquid medium.
4. Grow yeast for ~18 h at 30°C (it is not necessary to shake the 96-well plate, whereas shaking of the prey strain in a flask is recommended).
5. Pellet yeast in the 96-well plate by centrifugation for 10 min at 2,000 × *g*; discard the supernatant and aspirate with eight-channel wand or by tapping on cotton napkin for a couple of times.
6. Use 96-replication tool to pin baits from 96-well source plate onto a YEPD single-well agar plate as quadruplicates.
7. Pour the yeast strain with the empty prey vector into a single-well plate.
8. Use 384-replication tool to pin yeast onto the YEPD single-well agar plate that harbors the baits already.
9. Mating occurs at 30°C for 1–2 days.
10. Replicate from mating plate on -Leu-Trp agar single-well plates to select diploids.
11. Incubate for 2–3 days at 30°C.
12. Pin diploids on -Leu-Trp-His agar medium in single-well plates with different concentrations of 3-AT (e.g., 0, 1, 2, 4, 8, …, 128 mM).
13. Select yeast for 7 days at 30°C.
14. Determine minimal inhibitory concentration of 3-AT which is needed for each bait to suppress self-activation growth for use in the interaction screen.

3.3. Yeast Two-Hybrid Screen

1. Preparations
   (a) *Sterilization steps*: Sterilize the pinning tool by dipping the pins into 20% bleach for 20 s, sterile water for 1 s, 95% ethanol for 20 s, and sterile water again for 1 s. Repeat this sterilization after each transfer (see Note 9).

   (b) *Prepare prey array for screening*: Use the sterile replicator to transfer the yeast prey array (e.g., 384 format) from selective plates to single-well plates containing solid YEPD medium and grow the array overnight in a 30°C incubator (maximum 24 h). Ideally, the template prey array should be kept on selective plates (see Note 10).

(c) *Prepare bait liquid culture* (DBD fusion-expressing yeast strain): Inoculate 20–30 ml of liquid YEPD medium in a 50-ml conical flask with a bait strain from plates with selective medium and grow overnight in 30°C shaker for 18–22 h (see Note 10).

2. Mating Procedure

(a) Add a corresponding volume adenine from a 1% adenine stock solution to a final concentration of 0.004% into the bait liquid culture (see Note 11).

(b) Pour the overnight liquid bait culture into a sterile Omnitray plate. Dip the sterilized pins of the pin replicator (thick pins should be used to pin baits) into the bait liquid culture and place directly onto a fresh single-well (Omnitray) plate containing solid YEPD media. Repeat with the required number of plates (see Note 12).

(c) Pick up the prey array yeast colonies with sterilized pins (thin pins (≤1-mm diameter] should be used) and transfer them directly onto the baits pinned onto the YEPD plate so that each of the 384 bait spots per plate receives different prey yeast cells (i.e., a different AD fusion protein) (see Note 13).

(d) Incubate for 1–2 days at 30°C to allow mating (see Note 14).

3. Selection of Diploids

For the selection of diploids, transfer the colonies from YEPD mating plates to single-well plates containing -Leu-Trp medium using the sterilized pinning tool (thin pins should be used in this step). Grow for 2–3 days at 30°C until the colonies are >1 mm in diameter (see Note 15).

4. Interaction Selection

Transfer the colonies from -Leu-Trp plates to a single-well plate containing solid -His-Leu-Trp agar using the sterilized pinning tool. If the baits are self-activating, they have to be transferred to -His-Leu-Trp + the specific concentration of 3-AT which was determined in the self-activation assay (*see* Subheading 3.2). Incubate at 30°C for 6–10 days.

Score the interactions by looking for growing colonies that are significantly above background by size and that are present as duplicate (or quadruplicate) colonies (see Note 16). Scoring can be done manually or using automated image analysis procedures. When using image analysis, care must be taken not to score contaminated colonies as positives.

3.4. Mini-Pooling Screens

This protocol describes pooling of a prey array that contains 960 different clones resulting in one pool plate with 10-prey mini pools. The protocol can be adjusted to bait pools as well or to create larger pools than 10. For this, the corresponding volumes have to be adjusted.

1. Load ten 96-well plates with ~200 μl liquid YEPD medium per well.
2. Inoculate these plates with preys from the prey array grown on -Leu selective agar medium with a sterilized 96-pinning tool.
3. Grow yeast for 18–22 h at 30°C.
4. Resuspend yeast in all wells by pipetting with a 12-channel pipette (see Note 17).
5. Transfer from each source plate 20 μl yeast suspension in the destination pooling plate. Pool, therefore, all preys from source positions A1, A2, …, H12.
6. Once the transfer is done, mix the wells of the pooling plate by pipetting (see Note 17).
7. Spin the pooling plate for 10 min at 2,000×*g*.
8. Discard the supernatant and aspirate with eight-channel wand or by tapping on cotton napkin a couple of times.
9. Pin yeast with a sterile 96-pinning tool from the prey pool plate onto YEPD solid medium as, e.g., quadruplicates.

Mating of a single bait and selection steps follow as described in the previous protocol (Subheading 3.3). The prey pooling plate can be used directly to be mated against ten different baits by using thick pins (≥1 mm). Then, the pools are depleted. The pooling plate can be stored as −80°C glycerol culture and grown up again as master plate for further screens. Although the preparation time is quiet intensive (because of the pipetting steps), we recommend always to prepare fresh master plates. So, equal cell numbers of single clones in the pooling plates can be assured.

Preys can be pooled in different ways, e.g., pooling columns, rows, or similar positions from different source plates. In the simplest case of pooling screens, the interacting prey of a positively tested pool can be identified by retesting by one-on-one retest assays (see the next protocol).

The situation becomes more complicated if preys must be rearrayed for "smart" pooling as described in a separate chapter. For large libraries and to ensure equal cell numbers, preys can be picked by a robot which has a picking and rearraying function (e.g., Q-bot, Genetix).

3.5. Retests

Testing for reproducibility of interactions greatly increases the reliability of the interaction data. This protocol is used for specifically retesting interaction pairs detected in a one-on-one or pooling screen.

1. Rearray bait and prey strains or positively tested prey pool of each interaction pair to be tested in 96-well microtiter plates. Use an individual 96-well plate for the baits, as well as for the preys. For each retested interaction, fill one well of the bait plate and one corresponding well of the prey plate with ~200 μl YEPD.

2. For each retested interaction, inoculate the bait strain into a well of the 96-well bait plate and the prey strain at the corresponding position of the 96-well prey plate.

   For example, bait "X" is transferred at positions B1, B2, and B3 of the bait plate. The preys to be tested are arrayed into B1 (prey "Y"), B2 (prey "Z"), and the prey strain that carries the empty prey vector into B3 of the prey plate. The B3 test position is the control that helps to verify the background/self-activation.
3. Incubate the plates O/N at 30°C.
4. Spin the bait and prey plates for 10 min at 2,000 × *g*.
5. Discard the supernatant and aspirate with eight-channel wand or by tapping on cotton napkin a couple of times.
6. Pin baits with a sterile 96-pinning tool on -Trp and preys on -Leu selective agar medium as quadruplicates.
7. Allow baits and preys to grow at 30°C for 2–3 days.
8. Mating: First, transfer baits with a sterile 384-pinning tool on YEPD mating plates. Second, transfer preys onto baits.

The rest of the procedure can be done according to the screening protocol (Subheading 3.3). For interaction retesting, diploids are pinned on -Leu-Trp-His selective media plates with different concentration of 3-AT (see Note 18). The control test position has to be compared to bait self-activation background signals. Reproducible interactions should show up on different concentrations of 3-AT, whereas the activation control test position indicates clearly no colony growth.

4. Notes

1. Prepared agar plates should be stored for 1–2 days with closed lid under a sterile hood before used. Fresh solidified media is often wet and cannot be used directly.
2. Medium concentrate can be stored at 4°C up to 6 months.
3. Some components of the medium concentrate (e.g., amino acids) are not well soluble in water. The solution has to be stirred before the filtration step for up to 5 h until all components are dissolved. Thereby, heating is not recommended because of the heat sensitivity of amino acids.
4. Stock solutions can be stored up to 6 months at 4°C. Alternatively, the stock solutions can be frozen as aliquots at −20°C for long-time storage.
5. Selection media may differ due to the used Y2H expression vector system and have to be adapted. For instance, in the

pDEST32/pDEST22 system, the selection markers for baits and preys are interchanged (baits are selected on -Leu and preys on -Trp) while selection of pGBKT7g baits is done on -Trp. pGADT7g preys must be selected on -Leu medium.

6. CT110 has to be prepared fresh before yeast transfection and should not be stored.
7. Sodium hypochlorite solution is not very stable and has to be prepared every day. Alternatively, other disinfection solutions with a bleaching effect can be used. We do not recommend to use higher final concentrations than 2.4% since the steel pins of the replication tool might stain.
8. Yeast on agar medium can be stored for ~2 months at 4°C. The plates should be sealed with parafilm to avoid drying out. Baits and preys should be stored on the corresponding selective media since loss of plasmids can occur on nonselective medium.
9. Sterilization steps have to be established to the robotic system and sterilization solutions that are used. For example, in advance, it should be tested what the minimal time for the sterilization is since this speeds up the whole screen. However, it must be ensured that no cross contamination occurs.
10. The needed baits and prey arrays can also be used for the mating procedure when grown on/in selective medium. To our knowledge, this does not influence the mating efficiency much, but we recommend to use here YEPD medium since yeast grows faster and higher cell numbers can be achieved.
11. Adenine achieves a higher mating efficiency. Many yeast strains (e.g., AH109, Y187) are deficient in synthesizing adenine since they can carry an additional adenine selection marker.
12. After transfer from the liquid culture, allow the plates to dry for ~30 min. The positions should be dry when the preys are copied onto the bait spots. Also the plate should be checked if enough bait cells were transferred. Reasonable amounts were transferred when each spot occurs cloudy. This is critical for a good mating efficiency.
13. Thick pins can be used as well. We use thin pins since more replication steps can be done from a single source plate. If only a replication tool with thick pins is available, more prey array plates have to be prepared since only a couple of transfer steps can be done regarding source plate depletion.
14. Mating takes place in <15 h, but a longer period is recommended because some bait strains show poor mating efficiency.
15. This step is an essential control step because only diploid cells containing the *Leu2* and *Trp1* markers on the prey and bait vectors, respectively, grow on this medium. This step also helps recovery of the colonies and increases the efficiency of the interaction selection step.

16. We normally score interactions after 7 days. But the plates should be examined every day. Most two-hybrid positive colonies appear within 3–5 days, but occasionally positive interactions can be observed later. Very small colonies are usually designated as background; however, there is no absolute measure to distinguish between the background and real positives. When there are many (i.e., >30) large colonies per array of 6,000 positions, we consider these baits as "random" activators. In this case, the screen should be repeated to ensure that these positives are reproducible (unless the screen is done already in duplicate or quadruplicate).
17. Alternatively, the cells can be resuspended by vortexing. To save pipet tips the cells can also be resuspended by vortexing. However, if vortexing is used we recommend to seal the plate with adhesive tape since rigorous vortexing may cause cross-contamination.
18. Pinning the retest onto readout medium with various concentrations, 3-AT can be used to semiquantify interactions. This helps, e.g., to distinguish between "strong" and "weak" signals and might also help to separate spurious ones.

Acknowledgments

Work on this paper was supported by NIH grant RO1GM79710, the Landesstiftung Baden-Württemberg (Germany), and the Seventh Research Framework Programme of the European Union (grant HEALTH-F3-2009-223101).

References

1. Liu, Q., Li, M. Z., Leibham, D., Cortez, D., and Elledge, S. J. (1998) The univector plasmid-fusion system, a method for rapid construction of recombinant DNA without restriction enzymes. *Curr Biol 8*, 1300–9.
2. Walhout, A. J., Temple, G. F., Brasch, M. A., Hartley, J. L., Lorson, M. A., van den Heuvel, S., and Vidal, M. (2000) GATEWAY recombinational cloning: application to the cloning of large numbers of open reading frames or ORFeomes. *Methods Enzymol 328*, 575–92.
3. Cagney, G., Uetz, P., and Fields, S. (2000) High-throughput screening for protein-protein interactions using two-hybrid assay, in *Applications of Chimeric Genes and Hybrid Proteins, Pt C.* pp 3–14.
4. Landy, A. (1989) Dynamic, structural, and regulatory aspects of lambda site-specific recombination. *Annu Rev Biochem 58*, 913–49.
5. Uetz, P. (2002) Two-hybrid arrays. *Curr Opin Chem Biol 6*, 57–62.
6. Bader, J. S., Chaudhuri, A., Rothberg, J. M., and Chant, J. (2004) Gaining confidence in high-throughput protein interaction networks. *Nat Biotechnol 22*, 78–85.
7. von Mering, C., Jensen, L. J., Kuhn, M., Chaffron, S., Doerks, T., Kruger, B., Snel, B., and Bork, P. (2007) STRING 7-recent developments in the integration and prediction of protein interactions. *Nucleic Acids Res 35*, D358–62.
8. Uetz, P., Giot, L., Cagney, G., Mansfield, T. A., Judson, R. S., Knight, J. R., Lockshon, D., Narayan, V., Srinivasan, M., Pochart, P., Qureshi-Emili, A., Li, Y., Godwin, B., Conover, D., Kalbfleisch, T., Vijayadamodar, G., Yang, M., Johnston, M., Fields, S., and Rothberg, J. M. (2000) A comprehensive analysis of protein-protein

interactions in *Saccharomyces cerevisiae. Nature* 403, 623–627.

9. Uetz, P., Dong, Y. A., Zeretzke, C., Atzler, C., Baiker, A., Berger, B., Rajagopala, S. V., Roupelieva, M., Rose, D., Fossum, E., and Haas, J. (2006) Herpesviral protein networks and their interaction with the human proteome. *Science* ***311***, 239–42.
10. Stellberger, T., Hauser, R., Baiker, A., Pothineni, V. R., Haas, J., and Uetz, P. (2010) Improving the yeast two-hybrid system with permutated fusions proteins: the Varicella Zoster Virus interactome. *Proteome Sci* ***8***, 8.
11. James, P. (2001) Yeast Two-Hybrid Vectors and Strains., in *Two-Hybrid Systems. Methods and Protocols.* (MacDonald, P. N., Ed.), Humana Press, Totowa, New Jersey.
12. James, P., Halladay, J., and Craig, E. A. (1996) Genomic libraries and a host strain designed for highly efficient two-hybrid selection in yeast. *Genetics* ***144***, 1425–36.
13. Harper, J. W., Adami, G. R., Wei, N., Keyomarsi, K., and Elledge, S. J. (1993) The p21 Cdk-interacting protein Cip1 is a potent inhibitor of G1 cyclin-dependent kinases. *Cell* 75, 805–16.

Chapter 3

High-Throughput Yeast Two-Hybrid Screening

George G. Roberts III, Jodi R. Parrish, Bernardo A. Mangiola, and Russell L. Finley Jr.

Abstract

Charting the interactions among proteins is essential for understanding biological processes. While a number of complementary technologies for detecting protein interactions are available, the yeast two-hybrid system is one of the few that have been successfully scaled up. Two-hybrid screens have been used to construct extensive protein interaction maps for humans and several model organisms, and these maps have proven invaluable for studies on a variety of biological systems. These maps, however, have not come close to covering all proteins or interactions detectable by yeast two-hybrid. This is due in part to the difficulty of using library screening methods to sample all possible binary combinations of proteins. Ideally, every binary pair of proteins would be tested individually to ensure that every detectable interaction is identified. For organisms with large proteomes, however, this is not economically feasible and instead efficient pooling schemes must be implemented. The high-throughput two-hybrid screening methods presented here are designed to efficiently maximize coverage for selected sets of proteins or entire proteomes.

We present two high-throughput screening protocols. Both methods are designed to identify interactors for any number of bait proteins expressed as DNA-binding domain (BD) fusions. The choice of which protocol to use depends largely on the nature of the available library of proteins fused to an activation domain (AD). The first protocol is appropriate for screening a library of AD clones, such as a cDNA library, a domain library, or a large pool of AD clones. By contrast, the second protocol is appropriate for screening a large array of individual sequence-verified AD clones. This protocol screens small pools of AD clones from the array in a two-phase scheme. Although the methods presented were developed using the LexA version of the yeast two-hybrid system, we include notes as appropriate to accommodate users of other versions.

Key words: High throughput, Yeast two-hybrid, Interactome mapping, Protein–protein interaction, Protein networks

1. Introduction

Both of the high-throughput yeast two-hybrid (1) methods described here use interaction mating to screen for interactions between sets of yeast clones expressing BD fusions and yeast clones

Bernhard Suter and Erich E. Wanker (eds.), *Two Hybrid Technologies: Methods and Protocols*, Methods in Molecular Biology, vol. 812, DOI 10.1007/978-1-61779-455-1_3,

expressing AD fusions (3–5). In interaction mating, the vectors for expressing AD and BD fusions are put into different haploid yeast strains of opposite mating type. To test for an interaction, the BD and AD strains are mated with each other by mixing them together and growing on solid rich medium. Cells from each strain will fuse to form diploids, which are then tested for activation of the two-hybrid reporters. An ideal way to carry out large-scale screening is to first construct separate arrays of AD and BD yeast strains by cloning individual open reading frames (ORFs) into the two-hybrid vectors and transforming haploid yeast with the resulting plasmids. The AD- and BD-expressing yeast clones can then be mated individually or in pools to screen for interactions. Such arrays have been constructed to represent most of the genes for yeast (6–8), *Caenorhabditis elegans* (9), *Drosophila* (10, 11), several viruses (12–14), and bacteria (2, 15). Yeast two-hybrid arrays representing human proteins are also being developed (16, 17). The methods used to construct these arrays are not discussed in detail here, but have included recombination in yeast and in vitro recombination systems. Similar methods could be used to construct arrays for other organisms or arrays that cover more protein variants or fragments.

The first screening method that we describe (Protocol 1, Subheading 3.1) can be used to screen a library or large pool of AD clones to find interactors for a set of BD clones. The BD yeast clones are individually constructed and, for the protocol described here, arrayed on one or more 96-well plates. The AD library can be constructed by inserting cDNA into the AD vector and transforming yeast to create a single pool of all the cDNA clones. An AD library can also be constructed by pooling all members of a proteome-wide array of individual AD clones. Each screen involves mating 96 BD strains with a single AD library on one 96-well plate (Fig. 1a). The resulting diploids are then plated onto media to detect reporter activity. Colonies in which the reporters are active are then picked and the AD fusion is identified by sequencing. This approach is very rapid and particularly useful when no proteome-wide AD array is available. A drawback of the high-throughput library screen, however, is the limited sensitivity. Many interactions are missed because only a fraction of the AD clones in the library successfully mate with each BD yeast clone. One way to circumvent this problem is to perform multiple screens for each BD bait until saturation is achieved.

In contrast to library screening, the most sensitive approach to detect interactions is to mate individual BD strains with individual AD strains. However, for organisms with significant numbers of proteins, testing every binary pair would be prohibitively inefficient. A compromise is to divide an array of AD clones into small pools that maintain a high sampling efficiency. Pooling schemes of varying complexity have been devised to minimize the number of tests required for full coverage (18–20). The second screening

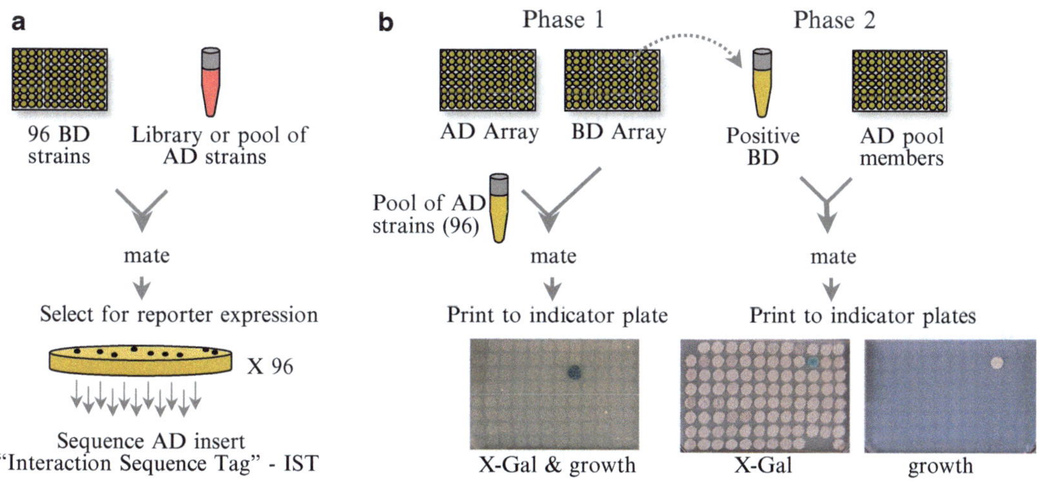

Fig. 1. Two high-throughput yeast two-hybrid screening methods. (**a**) High-throughput library screen. BD strains arranged in a 96-well format are mated with a library of AD strains pooled together. Diploids from each well are plated on media to assay for reporter gene expression and individual AD interactors are identified by sequencing. (**b**) High-throughput 2-phase screen. In phase 1, an AD array representing all of the ORFs for an organism is divided into pools of 96AD strains. BD strains in a 96-well format are then screened against each pool of ADs. In phase 2, individual BDs are screened against each member of the AD pool with which they were positive to identify the AD interactors.

method that we present (Protocol 2, Subheading 3.2) is a two-phase approach utilizing small pools of 96AD clones (Fig. 1b). This method was designed to maximize the number of interactions detected between two large sets of clones (20). A major advantage of this approach is its applicability to genomes of any size; the number of procedures performed scales with increasing genome size without sacrificing sensitivity. The downside is that it requires more time than library screens.

The yeast two-hybrid system comes in different versions. The original version uses the well-studied Gal4 transcription factor to provide both the DNA-binding domain and the transcription activation domain (21). The protocols presented here are based on a version of the two-hybrid system that uses bacterial LexA as the DNA-binding domain and an *Escherichia coli*-encoded ORF called B42 for the AD (22). A novel feature of this system is that the AD fused protein is expressed from the inducible yeast *GAL1* promoter, which is inactive in glucose and activate in galactose media. This facilitates the testing of proteins that inhibit the growth of yeast and also allows for easy validation of the dependence of an interaction on AD expression. The LexA/B42 system uses two reporter genes with upstream LexA binding sites (operators): a genomic *LEU2* gene, expression of which enables growth in the absence of leucine, and the *lacZ* gene, which is generally carried on a multicopy plasmid. While the protocols described below focus on use of the LexA/B42 system, they could readily be adapted for

Gal4-based systems as noted. Adapting to different versions of the two-hybrid system requires cognizance of the media that will be required to maintain selection of the different BD and AD plasmids, which have different markers in the different systems, and the media used to detect expression of the different reporters. It should also be noted that galactose is required for AD expression in the LexA/B42 system.

2. Materials

Materials listed are common to both protocols discussed in this chapter unless otherwise noted.

2.1. Equipment

1. Liquid-handling robots. While handheld multichannel pipettes would be sufficient for these protocols, robots can be used to increase accuracy and efficiency. We use a BIOMEK robot (Beckman Coulter, Fullerton, CA) with a 96-channel head and a SPAN-8 head, though similar robots may be used. Key requirements are the ability to pipette 96 samples at a time in the format of a 96-well microtiter plate, with volumes from 5 to 125 μl. The SPAN-8 head is capable of pipetting a sample from any position in one 96-well plate to any position in another 96-well plate and is used for rearraying liquid cultures from one set of 96-well plates to another.
2. 30°C incubator for growth of agar plates.
3. 30°C shaking incubator for growth of liquid cultures.
4. Centrifuge with a microtiter-plate rotor.
5. A handheld electronic 8-channel or 12-channel repeat pipettor capable of pipetting 50–1,200 μl is useful.
6. An electronic pipettor for use with 10 and 25 ml glass or serological pipettes.
7. E-gel electrophoresis apparatus or equivalent for running DNA samples on 96-well agarose gels.
8. Thermal Cycler with 96-well capacity.

2.2. Yeast Media

2.2.1. Media Overview and Designations

Synthetic defined (SD) media is made with dropout powder containing essential nutrients, but lacking various combinations of the following: uracil (u), histidine (h), tryptophan (w), or leucine (l). The media will also contain a carbon source consisting of either glucose, also known as dextrose (D), for routine growth, or galactose (G) to activate expression of the AD fusion, raffinose (R) to enhance growth, and maltose (M) if the LexA fusion is expressed from the *MAL* promoter (20, 23); if LexA is expressed from the *ADH1* promoter, maltose can be left out of all media. Liquid media also contains 15% glycerol (g) so that cultures can be stored at −70°C. Strains bearing plasmids must be maintained on selective media appropriate for the marker on each plasmid. For the LexA/B42 system, the BD strains should be grown on SD-uh/D to maintain selection for both the *HIS3*-marked BD plasmid and the *URA3*-marked *lacZ* reporter plasmid. The AD strains should be grown on SD-w/D to maintain selection for the *TRP1*-marked AD plasmid. Diploids resulting from mating the strains will be grown on SD-uhw with an appropriate carbon source. To test for *LEU2* reporter activity, diploids will be grown on media lacking leucine. To test for *lacZ* reporter activity, the media will contain X-Gal (X).

2.2.2. Media Constituents

1. X-Gal 40 mg/ml: Dissolve 1.6 g of 5-bromo,4-chloro,3-indolyl β-D galactopyranoside in 40 ml N, *N*′-dimethylformamide in a 50 ml Falcon tube. Cover tube with foil to protect from light and store at −20°C.
2. 10× BU salts: Add 70 g monobasic sodium phosphate heptahydrate and 34.5 g dibasic sodium phosphate to 500 ml dH_2O. Bring pH to 7 with NaOH, raise volume to 1 l, and autoclave.
3. 20% Sterile glucose: Add 200 g glucose per liter dH_2O final volume and autoclave.
4. 20% Sterile galactose: Add 200 g galactose per liter dH_2O final volume and autoclave. Low heat is required to completely dissolve the galactose. Galactose is required specifically for the LexA/B42 system.
5. 20% Sterile maltose: Add 200 g maltose per liter dH_2O final volume and autoclave. Maltose is required specifically for the LexA/B42 system, but only if LexA expression is driven by the *MAL* promoter (23). Some vectors for expressing LexA use the constitutive *ADH1* promoter, which does not require maltose.
6. 20% Sterile raffinose: Add 200 g raffinose per liter dH_2O final volume and filter sterilize. Do not autoclave. Raffinose is required specifically for the LexA/B42 system to enhance growth on galactose or maltose media. It is not necessary for glucose media.

2.2.3. Solid Media (see Note 1)

1. YPD: Add 10 g yeast extract, 20 g peptone, 25 g agar, one NaOH pellet (see Note 2), and a stir bar to 900 ml dH_2O. Mix, autoclave, and then add 100 ml 20% sterile glucose. Allow to cool to ~50°C before pouring into plates. Leave plates on bench for 2–3 days to remove excess moisture before use. It is recommended that all plates are dried in this fashion before use (see Note 3).

2. SD-uhwl/GRM: Add 6.7 g (yeast nitrogen base without amino acids YNB w/o AA), 2 g dropout powder lacking the appropriate amino acids, 25 g Bacto Agar, a stir bar, and one NaOH pellet to 865 ml dH_2O. Mix, autoclave, and then add 100 ml 20% sterile galactose, 25 ml 20% sterile maltose, and 10 ml 20% sterile raffinose. Cool to ~50°C before pouring into plates.
3. SD-uhwl/D: Same as item 2 above (SD-uhwl/GRM) except use 900 ml dH_2O, and after autoclaving add 100 ml 20% sterile glucose in place of galactose, maltose and raffinose.
4. SD-uhwl/GRM/X: Add 6.7 g YNB w/o AA, 2 g dropout powder lacking the appropriate nutrients, 25 g Bacto Agar, a stir bar, and one NaOH pellet to 765 ml dH_2O. Mix, autoclave, and then add 100 ml 20% sterile galactose, 25 ml 20% sterile maltose, and 10 ml 20% sterile raffinose. Cool to ~55°C, then add 100 ml 10× BU salts (see Note 4) and 4 ml of 40 mg/ml X-Gal (see Note 5). Stir for at least three additional minutes before pouring into dishes (see Note 3). For Protocol 1, pour into Petri dishes. For Protocol 2, pour into Nunc OmniTrays (Nalge Nunc International).
5. SD-uhw/GRM/X: Same as item 4 above (SD-uhwl/GRM/X) except use the appropriate dropout powder. Both Protocols 1 and 2 require this medium in Nunc OmniTrays.
6. SD-uhw/D/X: Add 6.7 g YNB w/o AA, 2 g dropout powder lacking the appropriate nutrients, 25 g Bacto Agar, a stir bar, and one NaOH pellet to 800 ml dH_2O. Mix, autoclave, and then add 100 ml 20% sterile glucose. Cool to ~55°C, then add 100 ml 10× BU salts (see Note 4) and 4 ml of 40 mg/ml X-Gal (see Note 5). Stir for at least three additional minutes before pouring into Nunc OmniTrays (see Note 3). Protocol 1 only.
7. SD-uhw/D: Add 6.7 g YNB w/o AA, 2 g dropout powder lacking the appropriate nutrients, 25 g Bacto Agar, a stir bar, and one NaOH pellet to 900 ml dH_2O. Mix, autoclave, and then add 100 ml 20% sterile glucose. Cool to ~50°C before pouring into Nunc OmniTrays. Protocol 1 only.

2.2.4. Liquid Media

Liquid media are made just like solid media, except that the agar and the NaOH pellet are omitted. 15% Glycerol is added to liquid media so that cultures can be stored frozen and to aid in 96-well spotting (transfer of liquid to solid medium) onto agar plates.

1. SD-w/D/g: Add 6.7 g YNB w/o AA and 2 g dropout powder lacking the appropriate nutrients to 750 ml dH_2O. Add 150 ml glycerol and autoclave. Add 100 ml 20% sterile glucose after autoclaving. Used for AD strain growth and storage.
2. SD-uh/D/g: Same as item 1 above (SD-w/D/g) except use appropriate dropout powder. Used for BD strain growth and storage.

3. SD-uhw/GRM/g: Add 6.7 g YNB w/o AA and 2 g dropout powder lacking the appropriate nutrients to 715 ml dH_2O. Add 150 ml glycerol and autoclave. Add 100 ml 20% sterile galactose, 25 ml 20% sterile maltose, and 10 ml 20% sterile raffinose.
4. SD-uhwl/GRM/g: Same as item 3 above (SD-uhw/GRM/g) except use appropriate dropout powder.
5. YPD: Add 10 g yeast extract and 20 g peptone to 900 ml H_2O. Add 100 ml 20% sterile glucose following autoclaving. Protocol 1 only.

2.3. Labware/ Miscellaneous

2.3.1. Labware

1. Deepwell plates. 96-Well, 2.2 ml volume per well (do not exceed 1.2 ml to lessen risk of cross-contamination from splashing).
2. 96-Well plates (BD Falcon) and lids. The plates must have round-bottom wells and be made of polypropylene to accommodate growth and recovery of yeast.
3. Grooved reservoirs (e.g., cat. no. 1064-05-8, Thermo Scientific Matrix, Hudson, NH). Grooves facilitate consistent recovery of small volumes of reagent.
4. 300 ml Robotic reservoir. These can be reautoclaved indefinitely despite being labeled disposable. High-volume reservoirs for maximum efficiency in filling plates with media.
5. Pipette basin (e.g., 13-681-501, Thermo Fisher Scientific, Pittsburgh, PA). Sterile, single use, V-shaped basins for use with an 8- or 12-channel pipettor.
6. Nunc OmniTray (Nalge Nunc International) or equivalent. These are rectangular Petri dishes with the same dimensions as a standard 96-well microtiter plate.
7. 48-Well dishes. Square, 22 cm × 22 cm dishes with a 48-well divider, such as Qtrays (Genetix, Hampshire, UK) or equivalent. These dishes are robot-compatible (see Subheading 3.1.4b). Protocol 1 only.
8. PCR Plate, 96 well, full skirted, 0.2 ml thin wall. Protocol 1 only.
9. Filter Plate, 96-well (e.g., MAHVS4510, Millipore, Billerica, MA). Protocol 1 only.
10. Petri dishes. Protocol 1 only.
11. Square 22 cm × 22 cm dish, such as the BioAssay Dish (Corning). Protocol 1 only.
12. Glass beads for culture spreading on Petri plates.
13. Hand roller, sometimes called a brayer, for use with aluminum plate and PCR plate seals.

2.3.2. Replica Plating

1. Sterile rectangular velvets (160 mm × 117 mm) (see Fig. 2). Between uses, velvets should be washed thoroughly and dried in a dryer before autoclave sterilization.
2. A rectangular block replica plating apparatus (115 mm × 72 mm) is used to transfer strains in a 96-well format between Nunc OmniTray indicator plates. The replica plating apparatus can be fashioned from a wood block slightly smaller than OmniTray dimensions (Fig. 2). Use skirt cut from a 96-well plate to hold velvet onto the block.

2.4. Consumables

1. BIOMEK Span-8 robot tips.
2. BIOMEK multichannel robot tips.
3. 14 ml Falcon tubes.
4. Nalge screw-cap cryovials (Thermo Fisher Scientific, Rochester, NY).
5. Aluminum plate seals. Deepwell and 96-well plates are sealed with aluminum adhesive seals.
6. Aluminum foil for autoclaving.
7. PVC lab wrap for agar plates to be incubated.
8. Avery 5267 labels and Cryo labels (see Note 6).
9. 96-Well and 48-well agarose gels (e.g., E-gel Ready gels, Invitrogen, Carlsbad, CA). Protocol 1 only.
10. PCR reagents. Taq DNA polymerase, 2.5 mM each dNTPs, 50 mM $MgCl_2$, 10× Taq buffer, 10 μM each primers (see Subheading 3.1.8). Protocol 1 only.
11. *Alu*I enzyme. Protocol 1 only.
12. PCR Plate. 96-well skirted PCR plates. Protocol 1 only.
13. PCR plate seals.

Fig. 2. Velvet transfer block assembly. The wooden block is 115 mm × 72 mm. Sterile velvets are held to the block by a skirt cut from a 96-well library plate or similarly sized polypropylene plate. A tighter fit can be accomplished by gluing scrap plastic to the inside perimeter of the skirt.

2.5. Strains and Plasmids

The protocols presented here use the LexA/B42 two-hybrid system, first described by Gyuris et al. (22), with updated strains and plasmids (5, 24). The BD bait plasmid (pEG202 or derivative) has the 2 μm origin, *HIS3* as a marker, and expresses LexA fusions from the constitutive *ADH1* promoter. Alternatively, a derivative of pEG202 that expresses LexA from *MAL62p*, pHZ5, can be used (23). BD plasmids are put into the BD strain, RFY206 (MATa, *trp1::hisG his3Δ200 leu2-3 lys2Δ201 ura3-52*) along with a *lacZ* reporter plasmid such as pSH18-34 (2 μm, *URA3*). The BD strains are maintained in the absence of histidine and uracil. The AD plasmid (pJG4-5 or derivative) has the 2 μm origin, *TRP1* as a marker, and expresses B42 AD fusions from the *GAL1* promoter. The AD plasmids are put into a reporter strain such as EGY48 (MATα, *his3, trp1, ura3, 3LexAop-LEU2::leu2*) or RFY231 (Matα, *his3, trp1::hisG, ura3, 3LexAop-LEU2::leu2*). These strains have a *LEU2* reporter driven by upstream LexA binding sites. The strains transformed with the AD plasmid must be maintained in the absence of tryptophan.

3. Methods

3.1. Protocol 1: High-Throughput Library Screening

Although less comprehensive than the two-phase mating approach (Protocol 2), high-throughput library screening (Fig. 1a) is a simple and effective way to discover protein interactions and allows the user to avoid the large time investment of constructing an AD array. This protocol requires that the user has already constructed a set of bait strains expressing BD fusions to the proteins of interest and is designed to screen 96 such bait strains at a time. The user must also have an AD library constructed in an appropriate plasmid vector (e.g., pJG4-5 or derivative) and transformed into an appropriate yeast strain (e.g., EGY48 or RFY231). This library could be a cDNA library or a large pool of AD strains from an AD array. A disadvantage of this approach is that some members of the AD library may fail to mate with a bait strain due to the relatively small number of cells that fit on one well of a 96-well filter plate. One way to address this concern is to perform multiple screens for each bait strain in an effort to approach saturated coverage of the AD library.

3.1.1. BD Culture Growth

1. Start with fresh cultures of BD strains in a 96-well plate or thaw a 96-well plate from a BD array on ice.
2. Fill two deepwell plates with 900 μl SD-uh/D from a 300 ml reservoir using an electronic multichannel pipette or a Biomek robot.

3. Replicate the BD array by pipetting 5 μl of well-mixed culture from the thawed 96-well plate into each of the two deepwell plates. Seal plates with an aluminum seal and hand roller.
4. Incubate at 30°C, 250 rpm to an A_{600} of ~2. This takes 4 days for RFY206 BD strains.

3.1.2. Filter Plate Mating

1. Thaw a single-use aliquot of AD library on ice.
2. Centrifuge both BD deepwell plates at 800 × *g* then carefully remove media manually with a multichannel pipette.
3. Dilute AD library to reach a viable cell count of ~7×10^6 per ml in a final volume of at least 12 ml YPD (see Note 7).
4. Pour diluted AD library into a pipette basin.
5. Resuspend the BD pellets in one deepwell plate with 100 μl of diluted AD culture and transfer to corresponding well of filter plate; use a manual 12-well multichannel pipette.
6. Place filter plate on vacuum station and vacuum media out. Stop suction as soon as medium has been removed – do not overdry. Be sure to clean vacuum station with 70% ethanol before use and keep sterile packaging from filter plate for storage of underdrain for diploid recovery in Subheading 3.1.3. Note the orientation of underdrain and the plate it came from.
7. Remove the underdrain from the filter plate being careful not to rip the delicate filters and place on a YPD agar plate made in a rectangular assay dish (Nunc OmniTray). Place lid on agar plates and set in incubator with filter plate on top. Weigh down lid with a few issues of Science or other journal of similar thickness (see Note 8).
8. Repeat steps 3–7 substituting the AD library strain containing the empty vector in place of the AD library. This control will be used to calculate background for each BD (see Note 9).

3.1.3. Diploid Recovery

1. Remove the filter plate from the agar and replace the underdrain in the same orientation. Resuspend cells in each filter well with 200 μl SD-uhw/GRM/g using a multichannel pipettor; transfer resuspended cells to deepwell plate. Repeat twice for a total of 600 μl/well. Care must be taken not to puncture the fragile filters.
2. Centrifuge deepwell plate for 2 min at 800 × *g*.
3. Remove supernatant without disturbing pellet.
4. Add 800 μl SD-uhw/GRM/g and mix to resuspend pellet.
5. Incubate 3–5 h at 30°C while shaking (galactose induction).
6. Centrifuge 2 min at 800 × *g*, then remove as much media as possible and replace with 800 μl SD-uhwl/GRM/g. Repeat once for a total of two washings to remove as much leucine as possible.

7. Remove 120 μl into a 96-well plate for Subheading 3.1.4a (or, alternately Subheading 3.1.4b). Freeze the original deepwell plate at −80°C in case it is needed for wells that do not produce enough colonies in the standard plating procedure, Subheading 3.1.4a.
8. Plate cells manually (see Subheading 3.1.4a), or with SPAN-8 Robot into 48-well dishes (see Subheading 3.1.4b).

3.1.4. Plating Diploids

a. Manual Plating
   1. Prepare a 10^{-3} dilution of the full mating panel in a second deepwell plate.
   2. Plate 100 μl of the 10^{-3} dilution of diploids on SD-uhw/GRM plates (100 mm Petri plates) to determine the number of diploid colony forming units (CFU) per aliquot of mated yeast.
   3. Plate 100 μl of undiluted diploids from the library mating and the vector control mating on separate SD-uhwl/GRM/X 100 mm Petri plates to select interactors and determine BD background, respectively.
   4. Incubate SD-uhw/GRM plates 2 days at 30°C (SD-uhwl/GRM/X plates requires 5-day incubation).
   5. Count colonies on both plates for each mating.
   6. Proceed to Subheading 3.1.5.

b. High-Throughput Plating Using SPAN-8 Robot and 48-Well Dishes
   1. Prepare 10^{-3} and 10^{-4} dilutions of the diploids in a deepwell plate.
   2. Label 48-well dishes and prepare for inoculation by dropping a single sterile glass bead into each well before mounting plate on robot. Two 48-well dishes are required for each BD deepwell plate.
   3. Determine the number of diploids plated in step 4 by plating 40 μl of a 10^{-3} dilution of the diploids on separate SD-uhw/GRM 48 well dishes using SPAN-8 robot (requires importing custom Labware into the BIOMEK software and custom brackets to adapt the dishes to the automated Labware positioners (ALP)). Replace dish covers and shake until glass beads completely disperse culture. Remove beads by inverting plate. To ensure that the number of resulting colonies are within a countable range, repeat this step with a 10^{-4} dilution.
   4. Plate 100 μl of undiluted diploids from the library mating on SD-uhwl/GRM/X plates to select interactors.
   5. Determine background by plating 50 μl of the empty vector mating on 48-well dishes as in step 3.

6. Incubate SD-uhw/GRM plates 2 days at 30°C (SD-uhwl/GRM/X plates require 5-day incubation).
7. Count colonies on both sets of plates for each mating applying dilution factors as appropriate to account for different volumes and dilutions plated.
8. Proceed to Subheading 3.1.5.

3.1.5. Pick Leu⁺ Diploids

1. Exclude plates from BDs with high transactivation potentials (see Note 9).
2. From the remaining SD-uhwl/GRM/X library mating plates, pick as many colonies as possible (making sure to select colonies of different sizes and shades of blue if there is a variety) into deepwell plates containing 150 μl SD-uhw/D (turns off *GAL1* promoter). These are referred to as "pick plates" hereafter.
3. Incubate 2 days at 30°C.
4. Make backup copies of all pick plates by transferring 5 μl culture to 96-well plates containing 100 μl SD-uhw/D/g followed by incubation at 30°C for 2 days. Store copies at −80°C. These plates can be safely discarded after rearraying clones with a unique restriction fragment class (RFC) (see Subheading 3.1.8, step 4).

3.1.6. Determine AD Dependence by Assaying for Galactose Inducibility

1. Wash pick plate diploids twice with 800 μl/well sterile water by centrifugation (2 min at 800 × *g*) followed by aspiration. Replace with 200 μl/well sterile water and mix.
2. Inoculate 5 μl/well onto the following solid media indicator plates using a 96-channel robot head: SD-uhw/D/X, SD-uhw/GRM/X, SD-uhwl/D, and SD-uhwl/GRM.
3. Store the pick plates at 4°C.
4. Incubate indicator plates for 5 days at 30°C.
5. For each 96-well pick plate, take a single picture of all four indicator plates.
6. Record growth on the −leu plates and the level of blue on the X-Gal plates. We use the following scheme: *LEU2*: 0 (no growth) through 3 (full growth). *lacZ*: 0 (white) through 5 (dark blue).
7. Identify Gal-dependent diploids (spots with higher reporter score on galactose *versus* glucose medium).
8. Rearray Gal-dependent diploids by transferring entire well volumes from the pick plates into new 96-well plate using SPAN-8 robot or equivalent.
9. Make copies of Gal dependent positives in uhw/D/g and freeze at −80°C.
10. Plate 5 μl/well of Gal-dependent diploids onto w/D OmniTray for colony PCR.

3.1.7. Colony PCR

Colony PCR uses a small amount of a yeast colony rather than purified DNA as template. Colony PCR of ADs is used to determine the identity of the AD fusions by sequencing the colony PCR product. BD colony PCR can be used to verify the presence and identity of the bait plasmid if there is any doubt about tracking. Use primers derived from the appropriate vector sequence flanking the site of insertion.

1. Grow yeast clones that you plan to use as PCR template on plates, not in liquid culture. It is preferable to use fresh (less than 1 week old) colonies. After assembling reaction mix in a PCR plate, transfer culture from the w/D OmniTray to the PCR plate with a multichannel pipettor. Use the edge of the tips to accomplish transfer making sure to take a very small amount of yeast (about as little as you can see), as too much yeast inhibits the PCR.
2. Colony PCR mix:

Template	0 μl
ddH_2O	29 μl
10× Buffer	4 μl
50 mM $MgCl_2$	1.6 μl
2.5 mM dNTP	4 μl
Primer 1 (10 μM)	0.5 μl
Primer 2 (10 μM)	0.5 μl
Taq polymerase	0.4 μl
Total	40 μl

3. Cycle 25 times in a Thermal Cycler using and annealing temperature appropriate for the primers.

3.1.8. Identify Unique RFCs

This step can reduce the number of clones that need to be sequenced by identifying the clones that have the same AD fusion. It is possible that many or all of the picks for a particular BD bait will have the same AD fusion. These will have the same distinctive pattern of restriction fragments, i.e., they will belong to the same restriction fragment class or RFC.

1. Perform restriction digest (~3 h at 37°C) using an aliquot of the AD colony PCR product from Subheading 3.1.7:

PCR reaction	5 μl
AluI (10 μm/μl)	0.4 μl
10× Buffer	2 μl
dH_2O	12.6 μl
Total	20 μl

2. Run the full digest reaction on 48-lane 4% agarose gels and photograph.
3. Assign a number to each unique restriction pattern (RFC) and record this number in a database or spreadsheet. Record how many of each RFC was obtained for each BD bait.
4. Consolidate the PCR products to be sequenced by rearraying an appropriate amount (usually about 5 μl) of one exemplar of each RFC to a plate appropriate for sequencing.
5. Sequence the PCR products.

3.1.9. Confirm Interactions by Retesting

Retesting interactions discovered in a library screen is highly advisable due to the large potential for false positives resulting from mutations amongst the large numbers of cells plated (see Note 9). Retesting interactions by binary assays is particularly efficient if AD arrays are available, as described in Subheading 3.2.4. Alternatively, the AD library plasmid from positive clones can be isolated and reintroduced into the AD strain, which can then be mated with the original BD bait strain to confirm the interaction.

3.2. Protocol 2: High-Throughput Two-Phase Screen

The two-phase pooling screen (Fig. 1b) utilizes a complete AD array representing the proteome of interest and arrayed in 96-well plates. For the bait, only the BDs of interest need to be cloned. However, the protocol is designed for screening plates of 96 BD strains at a time. During phase 1, BD strains arrayed on a 96-well plate are mated with each pool of ADs. This identifies the pools that contain interactors for a given BD. In phase 2, individual BD strains are mated with each member of the AD pools that were positive in phase 1. This is sufficient to determine the identity of the AD interactor(s), since their positions in the AD array are known beforehand. Finally, putative interactors identified in phase 2 are retested in a binary fashion to confirm the interactions.

In the protocol described here each AD pool will be constructed from 96AD yeast strains taken from one plate of the AD array. Larger pools can be used, though the larger the pool size, the more chance of missing interactions (20). Smaller pools, on the contrary, increase the number of plates that need to be mated in phase 1. The one to one correspondence of AD array plates and pools employed in the mating strategy outlined here, while sacrificing some efficiency, minimizes time spent constructing pools, allowing screening to begin almost immediately following AD array generation. Furthermore, the results can be interpreted intuitively, without formulas to deconvolute results from theoretically more efficient pooling schemes.

3.2.1. Construction of AD Pools

Once all of the AD strains are available in a 96-well format with their identities verified, preparations for phase 1 can begin. In phase 1, pools of 96AD strains are generated, and then mated with the BD arrays. AD pools must be constructed and frozen ahead of time in media containing 15% glycerol. All liquid transfers are performed robotically.

1. Thaw AD array 96-well plates on ice.
2. Pipette 5 μl of well-mixed culture from each AD array plate into a corresponding deepwell plate containing 1.2 ml of SD-w/D/g in each well. Incubate with shaking for 3 days at 30°C to an $A_{600} \sim 3$.
3. Centrifuge deepwell plates at 800 × *g* for 2 min. Remove 1.1 ml of supernatant from each well. Resuspend the cell pellets in the remaining 100 μl of culture medium.
4. Transfer all of the resuspended culture from each of the 96 wells of the deepwell plate into a single-well robotic reservoir. Mix well. This will result in ~9.6 ml of pooled culture with an $A_{600} \sim 36$. Dilute with SD-w/D/g to an $A_{600} = 30$.
5. Aliquot 200 μl into cryovials. Store at −70°C.
6. Similarly, make cryovials containing the AD strain with the empty AD vector (without an insert). These will be used for determining the background reporter activation by the BD clones in the absence of an interactor. We find it beneficial to determine the background by triplicate assays on different days.

3.2.2. Phase 1

In phase 1, we identify AD pools that interact with a particular BD. AD pools must be prepared in advance (see Subheading 3.2.1). Fresh cultures of BD and AD strains are needed, since direct use of frozen plates in place of fresh cultures can substantially reduce the number of interactors that can be identified in a screen.

1. Prepare fresh BD cultures: Thaw the BD array 96-well plate on ice. Transfer 10 μl of well-mixed culture into a deepwell plate containing 1 ml SD-uh/D/g in each well (see Note 10). This volume is sufficient to screen close to 200 AD pools. Grow at 30°C with shaking at 200 rpm to an $A_{600} \sim 2$. For RFY206 strains this requires 4 days.
2. Prepare AD pools: Thaw the necessary number of AD pool aliquots and transfer 167 μl into a 14 ml Falcon tube containing 10 ml SD-w/D/g (A_{600} of resulting culture will be 0.5). Incubate tubes diagonally in a shaker at 30°C at 200 rpm until A_{600} of 2.5–3.5 is reached. For RFY231 strains, this will take 3 days. At this time, also start a 10-ml culture of the control strain, which contains the empty AD vector without insert.
3. Concentrate BD cells prior to mating by centrifugation and aspiration to achieve roughly similar numbers of the BD and AD strains in each spot.
4. Set up mating. Transfer 5 μl of BD culture to prelabeled YPD agar plates. Include a plate for each pool and one for the empty

vector control. Allow spots to dry at room temperature. At this time, also spot the BD culture onto a SD-uh/D plate for use in phase 2.

5. Pour entire 10 ml volume of AD pool or control into a grooved reservoir.
6. Transfer 5 μl of AD pool or control culture on top of BD spots on the agar plates. Allow plates to dry, invert, and wrap with PVC wrap to prevent desiccation. Incubate inverted for 2 days at 30°C to allow growth of diploids. Also, perform this step using the AD vector only control culture.
7. Transfer yeast diploids from YPD plates to a sterile velvet secured onto a rectangular block. Then transfer diploid cells from the velvet to the indicator plate SD-uhwl/GRM/X by pressing the plate onto the velvet block. Wrap plates with PVC wrap and incubate for 5 days at 30°C. For the control plates (BD strains mated with the AD strain carrying vector without insert) also transfer the diploids to SD-uhwl/GRM and SD-uhw/GRM/X for determination of background activity for phase 2.
8. Photograph using a photography stand for reproducibility (see Note 11).
9. Score each mating position for its reporter activity. We use the following scheme for phase 1: *lacZ* ranges from 0 (white) to 5 (dark blue). *LEU2* ranges from 0 (no growth) to 5 (full growth). Higher scores correlate with increased interaction strength (25).
10. Enter scores into a spreadsheet or database and subtract the background activity determined for each BD. This background activity was determined by mating the BD plates with the strain containing the empty AD vector with no insert. Extract records above background to determine the BD-AD pool pairs that need to be tested in phase 2, Subheading 3.2.2 (see Note 12). Note that we recommend measuring background by averaging the results from triplicate independent matings.

3.2.3. Phase 2

In the second phase, each BD that was positive with an AD pool is mated against all AD fusions in that pool by mating the BD strain with the original arrayed 96AD strains. It is useful to order the phase 2 matings such that all of the matings for a particular AD array plate are done at the same time to minimize AD array thaws. BD strains will be grown by inoculating from the spotted SD-uh/D agar plates generated in phase 1 (Subheading 3.2.2, step 4). Note that these plates will need to be regenerated if they are older than 3 months.

1. Make a BD culture for each mating to be performed (see Note 13). Inoculate from the SD-uh/D agar plates into a 15 ml Falcon culture tube containing 10 ml SD-uh/D. Incubate in a

shaker at 30°C until the culture reaches saturation (A_{600} ~ 2), which takes 4 days for slow growing strains such as RFY206.

2. Make a culture for each AD array plate to be tested on a given day by transferring 5 ml from each well of the thawed culture to each well of a 96-well plate with 150 μl SD-w/D/g per well (see Note 10). Incubate 30°C for 2 days. This will be enough to mate with 30 different BD bait strains. If more is needed use a deepwell plate and a larger volume of medium.
3. Spot 5 μl of ADs from the 96-well plates onto YPD plates. Allow to dry completely at room temperature.
4. Pour entire 10 ml BD culture into a grooved reservoir. Spot 5 μl of BD culture on top of the AD spots on the YPD plates. Allow plates to dry completely at room temperature and then invert and wrap with PVC wrap. Incubate inverted for 2 days at 30°C.
5. Transfer diploids from the YPD plate first to SD-uhw/GRM/X plates and then to SD-uhwl/GRM plates using velvets (see Note 14). Wrap with PVC wrap and incubate for 5 days at 30°C.
6. Photograph plates using a photography stand (see Note 11).
7. Score each mating position for its reporter activity. We use the following scheme for phase 2: *LEU2*: 0 (no growth) through 3 (full growth). *lacZ*: 0 (white) through 5 (dark blue).
8. Enter scores for individual spots into a spreadsheet or database, subtract the background for each BD (as determined by mating the BD with the empty AD vector strain in Subheading 3.2.2, step 6) and extract all records sufficiently above background (see Note 12). All positives should be retested to confirm the interaction.

3.2.4. Confirmation Assays

Spots can register false positives for a variety of reasons including cross-contamination or mutations acquired as the cells were grown or mated. Much greater accuracy in a final data set can be obtained by retesting all positives from phase 2 in one-on-one assays using fresh yeast cultures. The fresh yeast cultures can be started from frozen cultures from the original array of BD and AD clones.

An efficient confirmation assay requires arranging the BD and AD cultures in separate 96-well plates to match the positions of specific pairs of BDs and ADs so that they can be mated 96 at a time with multichannel pipettors or a robot. For a large number of these binary assays, this is best achieved in two rearray steps with a robot (see Note 15). The first rearray consolidates each of the BDs and ADs to be tested from the master arrays into separate sets of BD and AD rearray plates, respectively. The second rearray distributes the BDs and ADs into new 96-well plates in matching positions. For example, if BD X is to be tested with AD Y, they will be put into position A01 of the respective BD and AD 96-well plates. In the mating step the two cultures will be pipetted onto the same YPD plate.

Rearray BD and AD Strains

1. Thaw 96-well plates (hereafter called source plates) containing BD or AD fusions to be assayed in the confirmation phase on ice.
2. Pipette 10 μl of culture from the 96-well source plates into deepwell "consolidation" rearray plates containing 1 ml of media in each well; for BD strains use SD-uh/D/g; for AD strains use SD-w/D/g. Incubate cultures in a 30°C shaker at 250 rpm until fully grown (A_{600} of ~2). For RFY206 BD strains this requires 4 days. For EGY48 or RFY231 AD strains this will take 2 days. Once grown make a copy before moving on to step 3. To do this, simply transfer 5 μm into 100 μm of media and grow shaking at 30°C to full density. Freeze the copies at –80°C.
3. Rearray a second time so that the BD for each assay is in the corresponding position to its AD interactor. If there are 960 confirmation assays, for example, you should have exactly ten plates each of BD and AD clones (see Note 15). Do this by pipetting 5 μl of culture from the deepwell into 100 ml/well of medium appropriate for the BD or AD strain. Grow in a 30°C shaker at 250 rpm until fully grown (A_{600} of ~2). Use immediately to set up matings or freeze at –80°C and subculture prior to setting up matings.

Perform Yeast Two-Hybrid Assay

1. Set up matings using the fresh AD and BD cultures. Transfer 5 ml of the rearrayed BDs onto YPD. Allow spots to dry completely. Spot 5 ml of the corresponding rearrayed ADs on top. Allow to dry and then invert and wrap with PVC wrap. Incubate inverted for 2 days at 30°C.
2. Transfer diploids to SD-uhwl/GRM and SD-uhw/GRMX plates using velvets, wrap with PVC wrap and incubate for 5 days at 30°C.
3. Photograph plates using a photography stand (see Note 11).
4. Score each mating position for its reporter activity. We use the following scheme: *LEU2*: 0 (no growth) to 3 (full growth), *lacZ*: 0 (white) to 5 (dark blue).
5. Enter scores for individual spots into a spreadsheet or database, subtract the background for each BD (as determined from mating the BD with the empty AD vector strain) (see Subheading 3.2.2). Extract all records sufficiently above background. These are your final set of interactions (see Note 16).

4. Notes

1. Pouring agar plates is the rate-limiting step in the two-phase protocol. To maximize efficiency, rapidly cool down two agar bottles at a time by wrapping them in wet c-fold towels, using

a squirt bottle to keep each bottle wet. Alternatively, place bottles in a 50°C water bath. A single lab worker can easily autoclave and pour 6 l per batch. Plates can be rapidly labeled using a color code indicating the amino acid(s) that are left out and the carbon source that is added.

2. Firm plates are essential for efficient transfer; inclusion of 2.5% agar (vs. the standard 2%) and one NaOH pellet (~110 mg) is imperative. Use Bacto Agar because yeast may not grow as well on granulated agar.
3. To dry plates, store them agar side up for 2–3 days after pouring and hardening. Failure to do so will cause cultures to run together or to dry very slowly after spotting. Alternately, plates can be dried for 15–30 min with the lid cracked open about one centimeter in a laminar flow hood. If plates have been kept at 4°C, take them out the night before and set them on the benchtop.
4. Adding 10× BU salts to medium which is too hot will lead to precipitation (apparent by cloudiness) which will decrease the buffering capacity of the medium. Neutral pH is optimal for *lacZ* reporter activity.
5. Temperatures above ~50°C will destroy X-gal. Indicator plates must be tested to ensure efficacy. It is useful to keep on hand a diploid culture with a known pair of interactors for testing each batch of plates before they are used in screening.
6. Hand-labeling is to be avoided throughout the entire procedure from array construction to screening. Standard Avery labels provide a cheap and rapid means of labeling the thousands of entities requiring labeling. Take care to use cryoresistant labeling for anything that must be frozen such as AD and BD arrays.
7. In library matings, it is useful to have an excess of the BD strain to increase the proportion of AD library members which are mated. The AD library is generally made by plating yeast transformed with the AD plasmid library and then scraping and mixing together all transformants into a single pool. Aliquots are then frozen at −80°C in glycerol. A sufficient number of aliquots should be made so that each is only thawed once, i.e., the aliquots are for a single use. It is also useful to have frozen single-use aliquots of the AD strain containing the empty AD vector, lacking any cDNA insert. These will be used as a control to establish the level of reporter activation for each BD bait strain in the absence of a particular AD fusion protein. A representative aliquot of each frozen culture should be thawed and used to determine the concentration of viable cells by plating dilutions on media to select for the AD library plasmid. A typical library will have ≥10^8 viable cells (colony

forming units, or CFUs) per ml. The BD culture, on the contrary, will be grown fresh and thus it will not be practical to test the plating efficiency before mating. A rough estimate of viable cells, therefore, can be used: a fresh culture of $A_{600} = 1$ has ~3×10^7 cells/ml, most of which are viable.

8. Mating is the crucial step in a library screen. Proper contact between the filter bottom of the filter plate and the agar surface is imperative. Yeast will only mate in contact with the surface of solid rich medium. To introduce as few bubbles as possible, place one side of the plate on the agar surface and smoothly bring the other side down to the surface. Look from directly above for air bubbles which appear white against the tan background of the YPD plate. Too much weight can break the agar surface or compromise the seal between the filter and well; it is much better to use a moderate amount of weight. Once weight is placed on lids, do not disturb for duration of incubation (2 days) so as not to introduce air bubbles (hence, the covered plates are placed in the incubator before adding weight to lids).
9. Library screens are particularly prone to false positives due to rare mutations. This is because the nature of a library screen is to search for a relatively rare event: the presence of one or a few interactors among a large background of noninteractors. One type of false positive can arise if the BD activates the reporters on its own, without an AD. This so-called "transactivation potential" of the BD can be determined by mating the BD with the AD strain containing the empty vector. The level of BD activation can be expressed as the number of colonies that grow on the –leu plate (Leu^+ colonies) per colony forming unit (CFU) plated. The number of CFU in an aliquot of mated cells is determined by plating on leucine-containing diploid selection plates (SD-uwh/D). If Leu^+/CFU is over ~10^{-4} for a particular BD, the library screen is unlikely to work.
10. Failure to include 15% glycerol in the first culture that is spotted to YPD agar leads to doughnut-shaped spots.
11. Securely tape a cutout made from a durable material such as foam board or wood to the base of the photography stand in the shape of the plate(s) to be photographed. Reproducible photographs are much more rapidly scored by hand and are essential if image analysis software is to be used (26).
12. Scoring and setting cutoffs should take into account several factors. Proteins with exceptionally large numbers of yeast two-hybrid interactors are not worthwhile to take into phase 2 because they are likely to constitute false positives; these proteins are sometimes referred to as "sticky" although the reasons for reporter activity may have nothing to do with a bait–prey protein

interaction. Bait–prey combinations that activate both reporters should be valued highly. However, since weaker interactions can lead to activation of only the growth reporter (e.g., *LEU2*), such positives may be worth pursuing in phase 2, particularly if they are the only positives available for a particular BD. A BD pool should not be considered positive unless it has a considerably higher score than the average score for that BD across all pools.

13. Although many BDs will be used several times during a day of phase 2 mating, making a separate BD culture for each mating avoids contamination issues inherent in reusing reservoirs. Furthermore, replicate cultures provide robustness, saving time by avoiding rethawing an AD array to redo matings for which a BD is contaminated or which did not grow.
14. Transferring first to SD-uhw/GRM/X reduces the number of yeast transferred to the nutritional reporter plates (SD-uhwl/GRM), thereby decreasing background.
15. The intermediate rearray (step 2) is necessary because some fusions will be used many times (so that some wells in the BD and AD array library would be depleted if used directly). Further, rearraying from the entire library to a smaller number of plates first reduces the complexity of the logistics required to get the required plates into the finite positions on the deck of a robot. Since about half of clones never test positive in a given two-hybrid screen, this step should at least decrease by half the number of plates.
16. A major difference between phase 2 and phase 1 is that in phase 2 BDs and ADs are mated one-on-one; technical true positives would thus be expected to show much more complete growth. Thus, separate scoring systems for phase 1 and phase 2 are highly recommended. A simple and effective rule for phase 2 is that scores greater than or equal to 1 above background (BD vs. empty AD vector only) are worthy of confirmation.

References

1. Fields, S. (2005) High-throughput two-hybrid analysis. The promise and the peril, *FEBS J* **272**, 5391–5399.
2. Parrish, J. R., Yu, J., Liu, G., Hines, J. A., Chan, J. E., Mangiola, B. A., Zhang, H., Pacifico, S., Fotouhi, F., DiRita, V. J., Ideker, T., Andrews, P., and Finley, R. L., Jr. (2007) A proteome-wide protein interaction map for Campylobacter jejuni, *Genome Biol* **8**, R130.
3. Bendixen, C., Gangloff, S., and Rothstein, R. (1994) A yeast mating-selection scheme for detection of protein-protein interactions, *Nucleic Acids Res* **22**, 1778–1779.
4. Finley, R. L., Jr., and Brent, R. (1994) Interaction mating reveals binary and ternary connections between Drosophila cell cycle regulators, *Proc Natl Acad Sci USA* **91**, 12980–12984.
5. Kolonin, M. G., Zhong, J., and Finley, R. L. (2000) Interaction mating methods in two-hybrid systems, *Methods Enzymol* **328**, 26–46.
6. Ito, T., Chiba, T., Ozawa, R., Yoshida, M., Hattori, M., and Sakaki, Y. (2001) A comprehensive two-hybrid analysis to explore the yeast protein interactome, *Proc Natl Acad Sci USA* **98**, 4569–4574.

7. Ito, T., Tashiro, K., Muta, S., Ozawa, R., Chiba, T., Nishizawa, M., Yamamoto, K., Kuhara, S., and Sakaki, Y. (2000) Toward a protein-protein interaction map of the budding yeast: A comprehensive system to examine two-hybrid interactions in all possible combinations between the yeast proteins, *Proc Natl Acad Sci USA* **97**, 1143–1147.
8. Uetz, P., Giot, L., Cagney, G., Mansfield, T. A., Judson, R. S., Knight, J. R., Lockshon, D., Narayan, V., Srinivasan, M., Pochart, P., Qureshi-Emili, A., Li, Y., Godwin, B., Conover, D., Kalbfleisch, T., Vijayadamodar, G., Yang, M., Johnston, M., Fields, S., and Rothberg, J. M. (2000) A comprehensive analysis of protein-protein interactions in Saccharomyces cerevisiae, *Nature* **403**, 623–627.
9. Walhout, A. J., Sordella, R., Lu, X., Hartley, J. L., Temple, G. F., Brasch, M. A., Thierry-Mieg, N., and Vidal, M. (2000) Protein interaction mapping in C. elegans using proteins involved in vulval development, *Science* **287**, 116–122.
10. Giot, L., Bader, J. S., Brouwer, C., Chaudhuri, A., Kuang, B., Li, Y., Hao, Y. L., Ooi, C. E., Godwin, B., Vitols, E., Vijayadamodar, G., Pochart, P., Machineni, H., Welsh, M., Kong, Y., Zerhusen, B., Malcolm, R., Varrone, Z., Collis, A., Minto, M., Burgess, S., McDaniel, L., Stimpson, E., Spriggs, F., Williams, J., Neurath, K., Ioime, N., Agee, M., Voss, E., Furtak, K., Renzulli, R., Aanensen, N., Carrolla, S., Bickelhaupt, E., Lazovatsky, Y., DaSilva, A., Zhong, J., Stanyon, C. A., Finley, R. L., Jr., White, K. P., Braverman, M., Jarvie, T., Gold, S., Leach, M., Knight, J., Shimkets, R. A., McKenna, M. P., Chant, J., and Rothberg, J. M. (2003) A protein interaction map of Drosophila melanogaster, *Science* **302**, 1727–1736.
11. Stanyon, C. A., Liu, G., Mangiola, B. A., Patel, N., Giot, L., Kuang, B., Zhang, H., Zhong, J., and Finley, R. L., Jr. (2004) A Drosophila protein-interaction map centered on cell-cycle regulators, *Genome Biol* **5**, R96.
12. Flajolet, M., Rotondo, G., Daviet, L., Bergametti, F., Inchauspe, G., Tiollais, P., Transy, C., and Legrain, P. (2000) A genomic approach of the hepatitis C virus generates a protein interaction map, *Gene* **242**, 369–379.
13. McCraith, S., Holtzman, T., Moss, B., and Fields, S. (2000) Genome-wide analysis of vaccinia virus protein-protein interactions, *Proc Natl Acad Sci USA* **97**, 4879–4884.
14. Uetz, P., Dong, Y. A., Zeretzke, C., Atzler, C., Baiker, A., Berger, B., Rajagopala, S. V., Roupelieva, M., Rose, D., Fossum, E., and Haas, J. (2006) Herpesviral protein networks and their interaction with the human proteome, *Science* **311**, 239–242.
15. Titz, B., Rajagopala, S. V., Goll, J., Hauser, R., McKevitt, M. T., Palzkill, T., and Uetz, P. (2008) The binary protein interactome of Treponema pallidum--the syphilis spirochete, *PLoS ONE* **3**, e2292.
16. Rual, J. F., Venkatesan, K., Hao, T., Hirozane-Kishikawa, T., Dricot, A., Li, N., Berriz, G. F., Gibbons, F. D., Dreze, M., Ayivi-Guedehoussou, N., Klitgord, N., Simon, C., Boxem, M., Milstein, S., Rosenberg, J., Goldberg, D. S., Zhang, L. V., Wong, S. L., Franklin, G., Li, S., Albala, J. S., Lim, J., Fraughton, C., Llamosas, E., Cevik, S., Bex, C., Lamesch, P., Sikorski, R. S., Vandenhaute, J., Zoghbi, H. Y., Smolyar, A., Bosak, S., Sequerra, R., Doucette-Stamm, L., Cusick, M. E., Hill, D. E., Roth, F. P., and Vidal, M. (2005) Towards a proteome-scale map of the human protein-protein interaction network, *Nature* **437**, 1173–1178.
17. Stelzl, U., Worm, U., Lalowski, M., Haenig, C., Brembeck, F. H., Goehler, H., Stroedicke, M., Zenkner, M., Schoenherr, A., Koeppen, S., Timm, J., Mintzlaff, S., Abraham, C., Bock, N., Kietzmann, S., Goedde, A., Toksoz, E., Droege, A., Krobitsch, S., Korn, B., Birchmeier, W., Lehrach, H., and Wanker, E. E. (2005) A human protein-protein interaction network: a resource for annotating the proteome, *Cell* **122**, 957–968.
18. Jin, F., Avramova, L., Huang, J., and Hazbun, T. (2007) A yeast two-hybrid smart-pool-array system for protein-interaction mapping, *Nat Methods* **4**, 405–407.
19. Thierry-Mieg, N. (2006) A new pooling strategy for high-throughput screening: the Shifted Transversal Design, *BMC Bioinformatics* **7**, 28.
20. Zhong, J., Zhang, H., Stanyon, C. A., Tromp, G., and Finley, R. L., Jr. (2003) A strategy for constructing large protein interaction maps using the yeast two-hybrid system: regulated expression arrays and two-phase mating, *Genome Res* **13**, 2691–2699.
21. Fields, S., and Song, O. (1989) A novel genetic system to detect protein-protein interactions, *Nature* **340**, 245–246.
22. Gyuris, J., Golemis, E., Chertkov, H., and Brent, R. (1993) Cdi1, a human G1 and S phase protein phosphatase that associates with Cdk2, *Cell* **75**, 791–803.
23. Finley, R. L., Jr., Zhang, H., Zhong, J., and Stanyon, C. A. (2002) Regulated expression of proteins in yeast using the MAL61-62 promoter and a mating scheme to increase dynamic range, *Gene* **285**, 49–57.

24. Golemis, E. A., Serebriiskii, I., Finley, R. L., Jr., Kolonin, M. G., Gyuris, J., and Brent, R. (2008) Interaction trap/two-hybrid system to identify interacting proteins, *Curr Protoc Mol Biol* **Chapter 20**, Unit 20 21.
25. Estojak, J., Brent, R., and Golemis, E. A. (1995) Correlation of two-hybrid affinity data with in vitro measurements, *Mol Cell Biol* **15**, 5820–5829.
26. Jafari-Khouzani, K., Soltanian-Zadeh, H., Fotouhi, F., Parrish, J. R., and Finley, R. L., Jr. (2007) Automated segmentation and classification of high throughput yeast assay spots, *IEEE Trans Med Imaging* **26**, 1401–1411.

Chapter 4

A Stringent Yeast Two-Hybrid Matrix Screening Approach for Protein–Protein Interaction Discovery

Josephine M. Worseck, Arndt Grossmann, Mareike Weimann, Anna Hegele, and Ulrich Stelzl

Abstract

The yeast two-hybrid (Y2H) system is currently one of the most important techniques for protein–protein interaction (PPI) discovery. Here, we describe a stringent three-step Y2H matrix interaction approach that is suitable for systematic PPI screening on a proteome scale. We start with the identification and elimination of autoactivating strains that would lead to false-positive signals and prevent the identification of interactions. Nonautoactivating strains are used for the primary PPI screen that is carried out in quadruplicate with arrayed preys. Interacting pairs of baits and preys are identified in a pairwise retest step. Only PPI pairs that pass the retest step are regarded as potentially biologically relevant interactions and are considered for further analysis.

Key words: Protein–protein interactions, Interactome mapping, Yeast-two hybrid, Large-scale screen, Network biology, STUB1/Chip, FKBP6/FKBP36, PRKACA/PKC alpha

1. Introduction

The yeast two-hybrid (Y2H) system (1) is a widely used tool for the discovery of protein–protein interactions (PPIs). Hundreds of laboratories successfully used the system to find novel proteins that prove to be important for the biological system under study. In the last 10 years, the Y2H system has been developed as a tool enabling the systematic, large-scale analysis of protein–protein interactions (2–15). At present, it is one of the most powerful methods for the generation of proteome-wide, binary protein–protein interaction maps (16–22) and will play a crucial role in whole-organism interactome mapping (23–27).

Bernhard Suter and Erich E. Wanker (eds.), *Two Hybrid Technologies: Methods and Protocols*, Methods in Molecular Biology, vol. 812, DOI 10.1007/978-1-61779-455-1_4, © Springer Science+Business Media, LLC 2012

Here, we describe our version of the Y2H system, which should be useful for experimentalists searching for protein–protein interactions as well as researchers who use the protein–protein interaction data generated in our lab for further computational and experimental analysis.

In contrast to Y2H library screens, matrix screens use ordered arrays of hybrid proteins, so-called baits and preys that are subcloned individually and characterized. The major advantage of array screens is that all protein pairs are tested with equal probabilities and interaction screens can be repeated. This allows quality-controlled, systematic generation of protein–protein interaction data on a large scale.

We describe a three-step matrix approach. First, after preparation of yeast strains in array format, we identify and eliminate autoactivating bait strains that would compromise the whole screen. Nonautoactivating strains are then used for the primary PPI screen and interacting pairs of baits and preys are identified in a third step using fresh yeast, a pairwise retest. Only PPI pairs that pass the third step are regarded as potentially biologically relevant interactions and are considered for further analysis.

1.1. Parameters for a Stringent Y2H System

The stringency of a Y2H screening experiment critically depends on three parameters. These parameters are of great importance for obtaining high quality interaction data in Y2H analyses; however, they can be addressed differently as exemplified by other versions of the Y2H system used in several other laboratories (28–31).

First, it is important to express bait and prey fusion proteins at very low levels. It is the rule rather than the exception that bait and prey proteins cannot be seen via western blotting of total yeast lysates even though the protein pair generates a genuine positive Y2H signal. In our system, low protein expression levels are achieved by using very weak promoters.

Second, reporter genes integrated in the genome of the Y2H yeast strains should not allow any activity in the absence of interacting baits and preys even after long incubation times. In particular for large-scale PPI experiments a background-free setup is essential as every single bait–prey combination has to be unambiguously classified as positive or negative. As the strength of the Y2H signal depends on the protein expression level, stability, localization and other parameters which are variable between different proteins, colony size is not a reliable measure for interaction detection and shows only weak correlation with affinity (32). Weak Y2H signals, which may result from high affinity interactions and give a strong signal in other interaction assays, can only be detected efficiently in a background-free assay. In our system, the *his3* reporter is the most stringent readout fully suppressing growth in the absence of an interaction.

Third, to detect interactions with a weak Y2H signal, the experiments have to be performed in several replicas using fresh yeast strains. Statistical analysis of the data helps identifying those interactions and largely improves data quality. Although established Y2H PPIs are highly reproducible, in large-scale screens only a subset of interactions is detected in every experiment. The sampling sensitivity of matrix screening approaches has recently been determined from repeated screening experiments (27). In our system, after four rounds of primary screen and retest approximately 42, 26, 18, and 14% of PPIs seen in the 1st, 2nd, 3rd, and 4th repeat of screening are new, respectively. Approximately 46% of the detected interactions are found once, 44% two times and 10% of the PPIs are detected more than two times. Thus, the number of interactions found only once decreases in successive screens and interactions that have been identified multiple times become the majority. Therefore, statistical data analysis can efficiently identify interactions, also those that give relatively weak Y2H signals, in a large-scale screen, even at low repeat numbers.

1.2. Performance of the Y2H System

Recent studies demonstrated that large-scale Y2H screens can result in high precision data. A study examining large-scale PPI data from yeast (22) has shown that data from Y2H and affinity purification coupled to mass spectrometry are of similar quality when the benchmarking sets used for quality measurements account for the different and complementary nature of the interactions. Quality estimates were also provided for large-scale *Caenorhabditis elegans* (33) and human PPI data (27) using empirical measurements with standard interaction sets and independent PPI detection methods. Venkatesan et al. measured the precision of two large single screen/single retest human Y2H datasets (18, 19) retesting random samples of 200 interactions from each with MAPPIT (34). These screens, each of which examined more than 25×10^6 protein pairs for interaction, showed an average precision of ~80% in independent experiments.

However, the sensitivity of the individual Y2H systems is 5–20%, only (24, 27). This holds true for other PPI-detection methods as well. Importantly, when a set of true interaction from the literature is being examined with different PPI methods, each method detects its own subset (24, 27). In the overlap we find exactly the number of interactions that is statistically expected for independent measures. This means that sensitivity can be increased by combining different PPI detection methods (35–37). Notably, this does also hold true for different versions of the Y2H system. Provided that different Y2H systems are producing high-precision data, parallel use of several Y2H setups will simply increase sensitivity and yield more complete interaction maps (24, 26, 30).

Clone selection is decisive for whether interactions are found with Y2H. Since a high fraction of baits is either autoactive and cannot be used in a Y2H experiment in the first place or simply does not yield interactions, the use of several different clones covering each protein increases the chance of finding PPIs substantially. Sometimes even a second full-length ORF for a protein that is inserted differently in the Y2H vector interacts while the first does not. To increase the chance of finding PPI partners, for large proteins in particular, it is advisable to screen with well characterized domains or protein fragments that ideally have been shown to be functional in other systems. In the end, the Y2H system is an extremely powerful PPI discovery tool. As such, it provides a wealth of high quality information for further experimental and computational analysis.

2. Materials

2.1. Labware

1. 96-Well MTPs with lid, PP, sterile, flat bottom (Greiner Bio-One GmbH).
2. 384-Well MTPs with lid, PP, sterile, flat bottom (Greiner Bio-One GmbH).
3. Omnitrays (Nunc GmbH & Co. KG).
4. Agar-plates (low profile bioassay dishes 241 × 241 × 20; Nunc GmbH & Co. KG).
5. 96-Well PCR plate (Costar).
6. 96-Well deepwell plates (2,000 μl/well; Eppendorf).
7. Pin tools with 96 and 384 pins. The steel pins are cylindrical with a diameter of 1.3 mm and the edge of the flat top that is touching the agar is beveled 45°at 0.2 mm. Sterilize by heating the pins until they glow red. Let them cool in a sterile environment (see Note 1).
8. Plastic tape for sealing PCR plates and MTPs (Costar or Thermo Scientific).
9. Sterile breathable sealing films (Aeraseal).

2.2. Solutions

1. Ampicillin stock (100 mg/ml): Dissolve 100 mg ampicillin sodium salt in 1 ml water. Store at –20°C. Dilute to a final concentration of 100 μg/ml (see Note 2).
2. Tetracycline stock (12.5 mg/ml): Dissolve 12.5 mg tetracycline hydrochloride in 1 ml 50% ethanol. Store at –20°C until use. Dilute to a final concentration of 20 μg/ml.
3. Tris/EDTA buffer pH 7.5 (10× TE): 100 mM Tris–HCl, pH 7.5, 10 mM EDTA, pH 8. Autoclave. Store at room temperature.

2.3. Media Stocks

1. LB medium: 10 g Bacto tryptone, 5 g yeast extract, 10 g NaCl, adjust to a final volume of 1 l with water and autoclave. Store at room temperature. Supplement with the appropriate antibiotics and mix before use.
2. LB agar: 10 g Bacto tryptone, 5 g yeast extract, 10 g NaCl, 20 g agar, adjust to a final volume of 1 l with water and autoclave. Store at room temperature. For agar plates, heat until the agar is dissolved, cool to 60°C, supplement with the appropriate antibiotics, stir and pour into sterile Petri dishes. Store plates at 4°C.
3. 1.25× YPD liquid medium: 5 g Yeast extract, 10 g peptone. Fill up to 400 ml with water and autoclave. Store at room temperature.
4. 1.25× YPD agar: 5 g Yeast extract, 10 g peptone, 10 g agar. Fill up to 400 ml with water and autoclave. Store at room temperature.
5. 2.5× Yeast medium (NB): 6.7 g Yeast nitrogen base. Fill up to 400 ml with water and autoclave. Store at room temperature.
6. 1.25× Yeast liquid medium (NB): 3.35 g Yeast nitrogen base. Fill up to 400 ml with water and autoclave. Store at room temperature.
7. 20× Glucose stock solution: 200 g Glucose monohydrate. Fill up to 500 ml with water and autoclave. Heating up helps dissolving the glucose before autoclaving. Store at room temperature.
8. 100× Amino acid/nucleoside stock solutions: Dissolve 4 g of leucine (L) in 400 ml water and autoclave. For histidine (H), adenine (A), uracil (U), and tryptophan (T) dissolve 0.8 g of the corresponding amino acid/nucleoside in 400 ml water and autoclave. Store at room temperature.
9. 2.5× Agar: For 500 ml of selective medium, autoclave 10 g agar in 200 ml water, store at room temperature.
10. 1.25× Yeast storage medium (NBG): 3.35 g Yeast nitrogen base, 250 ml glycerol (99%), and 29.44 g betain. Adjust to 400 ml with water and autoclave. Store at room temperature.
11. Sterile water (see Note 2).

2.4. Yeast Media Preparation

In this section, we describe the preparation of different media from the stock solutions. The media are named after the missing and required amino acids/nucleosides. Amino acids/nucleosides are abbreviated with a single letter as followed: "H" for histidine, "A" for adenine, "U" for uracil, "L" for leucine, and "T" for tryptophan. Anabolites omitted are marked by a minus sign and separated by a slash from the amino acids/nucleosides which are added to the media. The order on both sides of the slash is always HAULT.

1. Liquid medium: Add 25 ml 20× glucose stock solution and 5 ml of each required 100× amino acid/nucleoside stock solution to 400 ml 1.25× NB or 1.25× NBG and adjust to a final volume of 500 ml with sterile water. Pipette 100–120 μl into each well of 96-well MTPs and 35–45 μl into each well of 384-well MTPs.
2. Solid medium: Add 200 ml of 2.5× NB, 25 ml 20× glucose stock solution, and 5 ml of each required 100× amino acid/nucleoside stock solution to 200 ml 2.5× agar. Adjust to a final volume of 500 ml with sterile water. Heat up in a microwave until everything is dissolved. Cool the medium to 60°C and pour 200 ml into each agar plate under a sterile hood (see Note 3).
3. YPD liquid medium: Add 25 ml 20× glucose stock solution and 5 ml 100× adenine stock solution (optional) to 400 ml 1.25× YPD and adjust to a final volume of 500 ml with sterile water.
4. YPD solid medium: Add 25 ml 20× glucose stock solution and 5 ml 100× adenine stock solution (optional) to 400 ml 1.25× YPD agar, fill up to 500 ml with sterile water and heat in a microwave until dissolved. Cool to 60°C and pour 200 ml into each agar plate under a sterile hood.

2.5. Y2H Vectors

We use Gateway-compatible Y2H destination vectors: the DNA binding domain (BD)-containing pBTM116-D9 (baits) is a derivate of pBTM116 (Clontech); the activation domain (AD)-containing vector pACT4-DM (preys) is based on pACT2 (Clontech). As an alternative prey vector, pGAD426-D3 is used, which originates from pGAD426 (Clontech). The vectors contain a bacterial origin of replication and selectable antibiotic markers, the β-lactamase gene *Amp*^{R} (pACT4-DM, pGAD426-D3) or the tetracycline-resistance gene *Tet*^{R} (pBTM116-D9), respectively. All three yeast expression vectors are 2 μm vectors and contain a truncated ADH1 promoter and an ADH1 terminator.

Bait open reading frames (ORFs) are subcloned into pBTM116-D9 which contains an N-terminal LexA DNA-binding domain and a *TRP1* selection marker enabling growth selection of yeast transformants. The preys are generated by inserting ORFs into pACT4-DM or pGAD426-D3, respectively; both carry the *LEU2* marker gene and contain an N-terminal GAL4 transcription activation domain. An advantage of the N-terminal BD- and AD-fusions is that ORFs can be used irrespective of whether they contain a stop codon at the end. Only a few C-terminal amino acids are added to the open ORFs because all vectors contain a C-terminal stop codon after the attB2 Gateway recombination site.

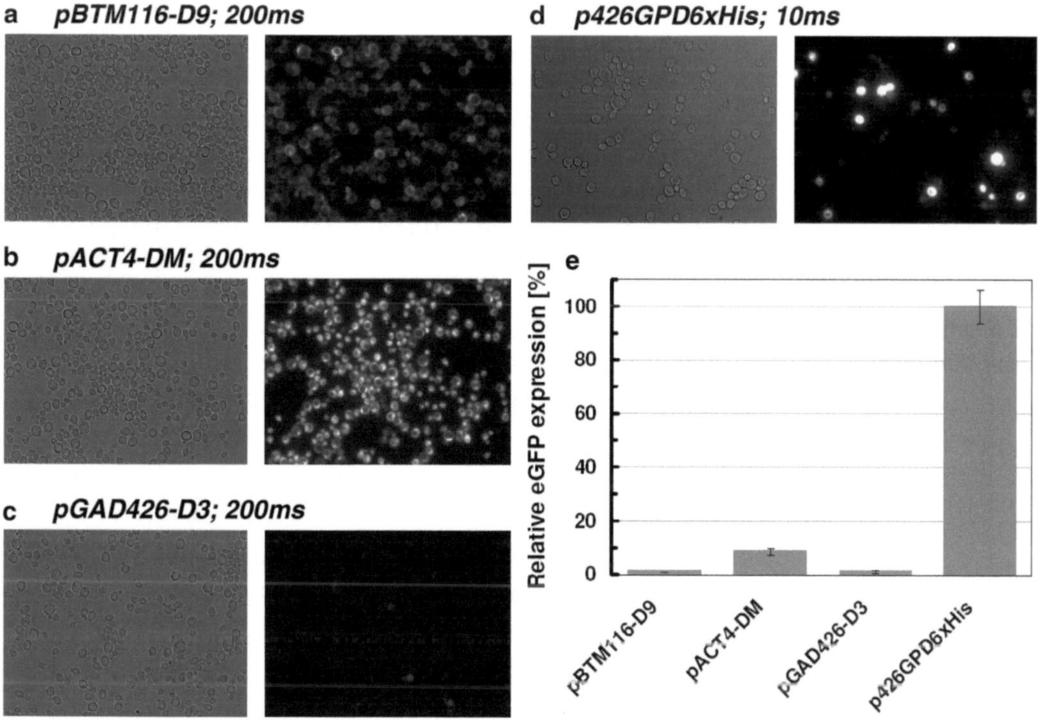

Fig. 1. Semiquantitative comparison of promoter activity of Y2H plasmids in *Saccharomyces cerevisiae*. Y2H MAT*a* strain was transformed with eGFP inserted in the Y2H pBTM116-D9 bait (**a**), pACT4-DM (**b**) and pGAD426-D3 (**c**) prey vectors. For comparison, eGFP expression was also assessed from a Gateway compatible p426GPD6xHis vector (**d**, 2 μ, ura auxotrophic marker). In this vector, eGFP is constitutively expressed from a glyceraldehyde-3-phosphate dehydrogenase (GPD) promoter, while in the Y2H vectors a truncated ADH1 promoter is used. Yeast strains were grown on selective media and then replicated (solid–liquid–solid) on YPD. After 30 h, bright light and fluorescence pictures of eGFP-expressing yeast cells were taken for 200 ms (Y2H plasmids) and 10 ms (PGPD6xHis). (**e**) Comparison of eGFP expression after normalization to cell density measured at 600 nm. Relative fluorescence was measured in an MTP reader (Biomek DTX 800/880).

Relative protein expression levels have been assessed with eGFP, demonstrating that the truncated ADH1 promoter on our vectors drives very low levels of gene expression (Fig. 1). Although the pACT4-DM and pGAD426-D3 contain the same truncated ADH1 promoter, they differ in protein expression levels.

2.6. Gateway-Cloning

1. Destination vector (150 ng/μl).
2. Entry clone (obtained via 96-well *Escherichia coli* mini-prep, see Subheading 3.3).
3. pENTR-gus for positive controls (Invitrogen).
4. LR Clonase Enzyme Mix II (Invitrogen).
5. Proteinase K Solution (Invitrogen).

2.7. Transformation of Competent *E. coli*

1. pUC19 DNA for positive controls (Invitrogen).
2. Chemically competent DH10B *E. coli* cells.

3. SOC Medium: for 100 ml SOC medium supplement 99 ml SOB medium with 1 ml 20× glucose stock solution. SOB medium (1 l): 20 g tryptone, 5 g yeast extract, 0.5 g NaCl. Add water to a final volume of 1 l and autoclave. Add 10 ml of filter-sterilized 1 M $MgCl_2$ and 10 ml of filter-sterilized 1 M $MgSO_4$.
4. LB Plates containing 100 μg/ml ampicillin or 20 μg/ml tetracycline.

2.8. Ninety Six-Well *E. coli* Mini-prep

1. 50% Glycerol: 581 ml 86% glycerol (Merck), bring up to 1 l with water and autoclave. Store at room temperature.
2. Buffer P1: 50 mM Tris, pH 8, 10 mM EDTA, pH 8. Autoclave, store at 4°C after addition of 50 mg/l RNAse A (f.c.).
3. Buffer P2: 200 mM NaOH, 1% SDS (w/v). Store at room temperature.
4. Buffer P3: 300 ml 5 M potassium acetate pH 5.5, 57.5 ml glacial acetic acid, 145.5 ml water. Store at room temperature.
5. Isopropanol (p.a.) (Merck).
6. 70% (v/v) ethanol: 700 ml absolute ethanol (Merck), bring up to 1 l with water.

2.9. Yeast Strains

We use L40ccu (MAT*a*) and L40ccα (MATα) (7), both of which have a *LacZ* ((*lexAop*)$_8$-GAL1TATA-*lacZ*) and a *HIS3* ((*lexAop*)$_4$-HIS3TATA-*HIS3*) reporter, additionally L40ccu has a *URA3* reporter ((*lexAop*)$_8$-GAL1TATA-*URA3*). L40ccu and L40ccα are auxotroph for leucine (*leu2-3,112*) and tryptophan (*trp1-901*), the *ADE2* gene is deleted in L40ccα (see Note 4). While choosing strains, keep in mind that the two strains, typically with the same genetic background, have to be of different mating type. Furthermore, the strains have to carry reporter genes with promoters compatible with the binding domain of your bait vector.

2.10. Yeast Transformation

1. YPDA liquid medium.
2. MAT*a* and MATα yeast strains to be transformed, e.g., L40ccα for preys cloned into pACT4-DM or pGAD426-D3, and L40ccu for baits in pBTM116-D9.
3. 1 M LiOAc, autoclave and store at room temperature.
4. 2 M Sorbitol, autoclave and store at room temperature.
5. 60% PEG-3350, autoclave and store at room temperature.
6. Expression vectors, e.g., pBTM116-D9 as the BD- and pACT4-DM or pGAD426-D3 as the AD-containing vectors, respectively.
7. Empty prey vector for the autoactivation test.

8. Carrier DNA: Dissolve 5 mg/ml salmon sperm DNA (Sigma) in 1× TE, heat up to 95°C for 5 min, put on ice, aliquot and store at −20°C.
9. Selective agar plates (-L/HAUT for preys and -AT/HUL for baits, see Subheading 2.4).

2.11. Beta-Galactosidase Assay

1. Diploid yeast containing putatively interacting bait–prey combinations.
2. Liquid nitrogen.
3. Sterile nylon membranes (MagnaCharge from MSI).
4. Z-buffer: 60 mM Na_2HPO_4, 40 mM NaH_2PO_4, 10 mM KCl, 1 mM $MgSO_4$.
5. 1 M Dithiothreitol (DTT).
6. Whatman 3MM Chromatography paper.
7. 5-Bromo-4-chloro-3-indolyl-β-D-galactopyranoside (X-Gal), 20 mg/ml in dimethylformamide.
8. Agar plates containing-HAULT medium.

3. Methods

We describe a three-step matrix screening protocol (Fig. 2), including a description of basic tools (see Subheadings 2.5 and 2.9) and yeast handling procedures (Subheading 3.4), the generation of the hybrid constructs (Subheadings 3.1–3.3), the transformation of yeast strains in 96-well format (Subheading 3.5), assaying for autoactivation (Subheading 3.6), screening of a prey matrix for primary interaction partners (Subheading 3.7) and the final identification of interacting protein pairs by retesting (Subheadings 3.8 and 3.9).

3.1. Gateway-Cloning

This method is used for the parallel site-specific recombination of ORFs from entry into Y2H destination vectors in 96 well format. Gateway-compatible entry plasmids carrying ORFs can be obtained from various sources and distributors (38–40).

1. Prepare a master mix of 0.5 μl destination vector (150 ng/μl, see Note 5 for 2-in-1-LR Clonase reaction), 2 μl TE-buffer and 1 μl LR Clonase Enzyme Mix II per reaction, mix and transfer 3.5 μl master mix per well into a PCR plate.
2. Add 1.5 μl entry vector per well (DNA obtained with the 96-well mini-prep can be used here; protocol described in Subheading 3.3). Include a negative control (i.e. only master mix and water or elution buffer instead of the entry vector)

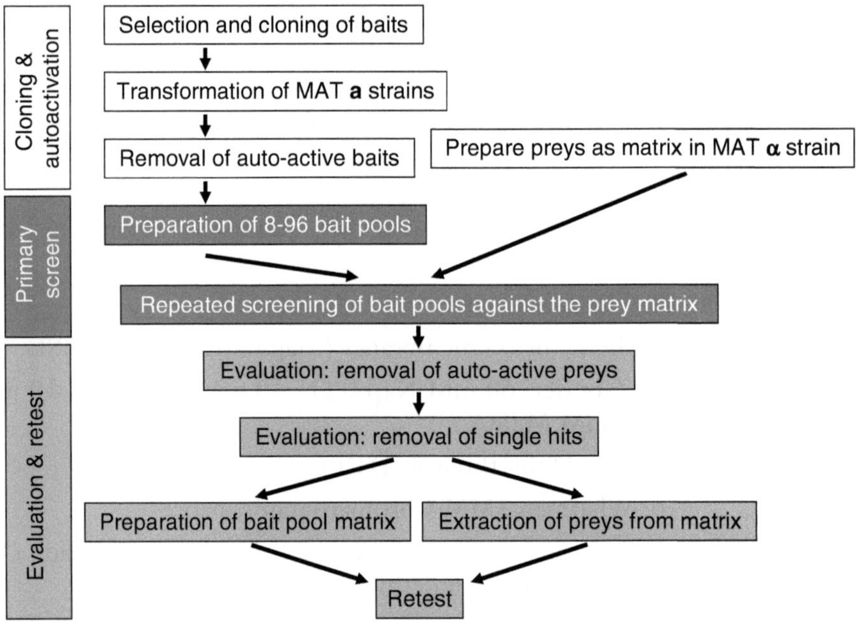

Fig. 2. Schematic depiction of the experimental flow. First, baits proteins are selected for cloning and MAT*a* strains are transformed with the bait plasmids and tested for autoactivity. After removal of autoactives, baits are pooled and screened against a previously prepared prey matrix. The resulting interacting pairs are filtered and tested again in an independent experiment using fresh yeast. *Shading* indicates the three parts of the protocol.

and a positive control (i.e. a well-tried entry vector or the pENTR-gus control vector provided by Invitrogen).

3. Seal the PCR plate with plastic tape, mix and spin down.
4. Incubate the plates at 25°C for 3–18 h.
5. Add 0.5 μl of Proteinase K solution to each reaction.
6. Seal the 96-well plate with plastic tape, vortex for 20 s and spin down.
7. Incubate the plates for 10 min at 37°C.
8. Use reaction directly for transformation or store at –20°C.

3.2. Transformation of Competent *E. coli* with LR Reaction Mixtures

This protocol is suitable for 96-well format transformations and yields colonies for 95–100% of the LR reactions if chemically competent DH10B are used (41). For missing clones, we use electroporation to obtain colonies, which is successful most of the time because of the higher transformation efficiency.

1. Prepare 200 μl SOC-Medium per well in a deepwell plate and warm up to 37°C.
2. Thaw the required amount of competent bacteria (30 μl per reaction) on ice and pipette into a PCR plate.
3. Add 2.5 μl (per well) of the Proteinase K-treated LR reaction prepared above (also include a positive control of 10 pg pUC19 DNA (Invitrogen) for transformation efficiency).

4. Seal the PCR plate with plastic tape, vortex softly and incubate on ice for 30 min.
5. Incubate at 42°C for 90 s.
6. Incubate on ice for 5 min.
7. Add 70 μl prewarmed SOC-Medium per well from the first column of the deepwell plate to the first column of the PCR plate and directly transfer the whole content of the wells back into the deepwell plate.
8. Repeat step 7 for the remaining 11 columns. Shake the deepwell plate at 37°C for 1 h.
9. Pipette a total of 50 μl from each well on selective LB plates by making rows of drops of 5 μl on an agar plate. Use LB plates containing 20 μg/ml tetracycline for pBTM116-D9 or 100 μg/ml ampicillin for pACT4-DM.
10. Incubate the LB plates for 16–20 h at 37°C.
11. Use colonies directly for the inoculation of LB-Medium to obtain DNA via 96-well mini-prep.
12. If you do not obtain clones in the first place repeat transformation of 1–2 μl of the remaining LR reaction using an electroporation protocol.

3.3. Ninety Six-Well *E. coli* Mini-prep

We developed a protocol that produces DNA of sufficient quantity (10–20 μg of plasmid DNA) and quality for LR-reactions, sequencing and yeast transformation. We use a pipetting robot to add buffers and to take off the supernatant but a pipette can also be used.

1. Pipette 1,000 μl LB medium containing the appropriate antibiotic into each well of a deepwell plate and inoculate with a single *E. coli* colony.
2. Seal the deepwell plate with breathable sealing film and grow for 15–18 h in a shaker at 37°C.
3. Transfer 50 μl of culture into a 96-well MTP that is filled with 50 μl 50% glycerol, mix and store at −80°C.
4. Spin the deepwell plate at 4°C, 1,258 × *g* for 30 min; pour off the supernatant and dry by tapping on a paper towel a couple of times.
5. Add 300 μl Buffer P1 per well (ensure that RNAse A has been added), seal the plates with plastic tape (see Note 6) and vortex vigorously for 2–3 min; make sure the pellet is completely dissolved.
6. Add 300 μl Buffer P2 per well, seal the plates with plastic tape, mix thoroughly by inverting the plate 3–4 times, and incubate for 5 min at room temperature.
7. Add 300 μl Buffer P3 per well, seal the plates with plastic tape, mix thoroughly by inverting the plate 3–4 times.

8. Centrifuge for 1 h at 3,220 × g at room temperature.
9. Transfer 750 μl of the supernatant from step 8 to a new deepwell plate and add 530 μl isopropanol to each well.
10. Seal the plates with plastic tape, mix thoroughly by inverting the plate 3–4 times.
11. Spin the deepwell plate at 3,220 × g for 45 min at room temperature; pour off the supernatant and dry by tapping on a paper towel a couple of times.
12. Add 1,000 μl 70% ethanol per well, seal the plates with plastic tape, and mix by inverting the plate 3–4 times.
13. Spin the deepwell plate at 3,220 × g for 30 min at 4°C; pour off the supernatant and dry by tapping on a paper towel a couple of times.
14. Allow for the pellet to dry. This might take up to 30 min.
15. Add 100 μl sterile water per well and dissolve the pellet by incubating over night at 4°C or by shaking for 1 h.
16. Control the success of the mini-prep in a BsrGI restriction analysis (see Note 7).

3.4. Yeast Handling

For Y2H experiments, yeast needs to be grown in and thus be transferred to different liquid and solid media. The yeast can be transferred from liquid to solid medium, solid to liquid, liquid to liquid and solid via liquid to solid. Instead of a solid–solid transfer, the solid–liquid–solid step is used to get a better selection of the yeast, because it prevents the yeast from clumping together. Correct growth and storage of yeast is important for successful screening. The yeast gets inoculated and streaked out in different formats as described below:

1. Liquid–solid: Streak (inoculation loop), drop (pipette) or stamp (pin tool) liquid yeast culture on agar plates. Take care that the liquid culture is mixed well. To grow the yeast on agar, streak it out, wrap the agar plate in foil and incubate it for 2–7 days at 30°C (see Note 8).
2. Solid–liquid: Scrape off yeast colonies from agar plates and transfer to liquid medium with an inoculation loop or with a pin tool. It is important to vortex the medium to separate the yeast clumps and get a uniform suspension of cells. Using a pin tool, scrape off yeast from the agar and transfer it to MTPs containing the liquid medium, repeat at least once and mix well. The liquid cultures obtained in the MTPs are quite dense and do not grow to much higher densities when incubated.
3. Solid–liquid–solid: Scrape off yeast colonies from agar plates like described above and transfer the resuspended yeast culture to solid medium with a pin tool, pipette or inoculation loop.

If using a pin tool stamp the yeast culture at least two times on the agar, wrap in foil and incubate at 30°C for 2–7 days.

4. Liquid–liquid: It is possible to inoculate liquid media with liquid yeast culture. To grow yeast in MTPs, inoculate each well with 5 μl of yeast culture, mix, wrap MTPs in foil and incubate at 30°C for 18–20 h. In contrast to 96-well MTPs, 384-well MTPs are incubated without shaking. To grow yeast in deepwell plates pipette 10 μl of yeast culture per well into deepwell plates containing 1 ml medium per well. Seal the deepwell plate with breathable sealing film and grow for 18–20 h in a shaker at 30°C.
5. Yeast storage: Yeast colonies can be stored for at least 1 month on foil-wrapped agar at 4°C. For longer storage scrape off yeast from agar, transfer to NBG medium (see solid–liquid replication), incubate for 18–20 h, mix and freeze at -80°C. Frozen yeast can be thawed two times only and needs to be replicated on agar (liquid–solid) for an extra generation before starting a Y2H experiment.

3.5. Ninety-Six-Well Yeast Transformation

With this 96-well format yeast transformation protocol, haploid MAT*a* and MATα yeast strains are transformed with the bait and prey plasmids respectively. Selection of the transformed yeast cells requires -AT/HUL and -L/HAUT media because in our system the bait vector pBTM116-D9 has a tryptophan and the prey vectors (pACT4-DM/pGAD429-D3) have a leucine selection marker (see Subheading 2.5). It is important that the MATα strain is also transformed with a prey vector without insert because this strain is needed for the autoactivation test. Colonies obtained from the transformation have to be solid–liquid–solid replicated at least once (see Subheading 3.4) before they are used in an autoactivation or mating experiment. We generally create four biological replicas (quadruplicates) and keep them separated throughout the autoactivation test, the primary screen and the retest.

The following protocol describes the transformation of eight 96-well plates in parallel:

1. Inoculate 12.5 ml of YPDA liquid medium with yeast strains freshly grown on YPD agar, vortex and grow for 15–18 h at 30°C with shaking.
2. Prepare PCR plates filled with plasmid DNA: pipette 5 μl plasmid DNA into each well (DNA from the 96 well mini-prep described in Subheading 3.3. can be used here), spin down, add 5 μl carrier DNA to each well, mix and spin down. Include one negative control (i.e. only carrier DNA) and a positive control (i.e. a well-tried vector preparation).
3. Inoculate 250 ml of YPDA to an OD_{600} of 0.10–0.15 with the over-night culture and grow at 30°C with shaking until an OD_{600} of 0.6–0.8 is reached.

4. Freshly prepare Mix 1 and Mix 2. Mix 1 (10 ml for eight plates): 1 ml 1 M LiOAc, 0.5 ml 10× TE, 5 ml 2 M sorbitol, 3.5 ml sterile water. Mix thoroughly. Mix 2 (60 ml for 8 plates): 6 ml 1 M LiOAc, 6 ml 10× TE, 40 ml 60% PEG-3350, 8 ml sterile water. Mix thoroughly.
5. Once the culture reaches the desired OD, split and transfer to five 50 ml screw-cap centrifuge tubes and centrifuge at 805 × *g* for 5 min.
6. Remove the supernatant and resuspend each pellet in 20 ml sterile 1× TE. Centrifuge at 805 × *g* for 5 min, remove the supernatant.
7. Resuspend each pellet in 2,000 μl Mix 1 and pool them (total volume of 10 ml) and incubate at room temperature for 10–60 min.
8. Pipette 11 μl of the yeast Mix 1 into each well of the PCR plate containing plasmid and carrier DNA.
9. Seal the plate with plastic tape and mix. Do not spin the plate.
10. Pipette 58 μl of Mix 2 into each well of the PCR plate.
11. Seal the plate with plastic tape and mix for 1 min.
12. Incubate the plates at 30°C for 30 min.
13. Add 8 μl DMSO to each well.
14. Seal the plate with plastic tape and mix for 1 min.
15. Incubate the plates at 42°C for 7 min in a thermocycler.
16. Create four biological replicas by transferring the cells to four selective agar plates. Transfer baits to -AT/HUL and preys to -L/HAUT to select for transformed yeast. Typically we use a pin tool to spot the cells five times on the same position of selective agar plates (e.g., Petri dishes or Omnitrays). Alternatively a pipette or a pipetting robot can be used to transfer 5 μl to selective agar plates. Allow the spots to dry on the plates (to speed up the process dry the open plates under laminar air flow).
17. Incubate at 30°C for 3 days.
18. Scrape off transformed yeast from the agar using a pin tool, transfer to 96-well MTPs containing NBG, mix well and stamp three times on selective agar (solid–liquid–solid). Incubate the MTP at 30°C for 12 h, mix and store at −80°C.
19. Incubate the agar plates at 30°C for 48–72 h.
20. This freshly grown yeast can be used for yeast two-hybrid experiments.

3.6. Autoactivation Test

It is important to remove autoactive bait strains before the matrix screen, because autoactivating baits will always grow after mating and mask any interaction signal of other baits in a pool.

Autoactivation is operationally defined as detectable bait-dependent reporter gene activation in the presence of any prey vector, even without insert. To detect autoactivity, the bait strains are mated with a prey strain carrying the prey plasmid without insert. Baits growing on -HAULT medium are autoactive and should not be used in a pooled matrix screen (see Note 9). Test all four replicas for autoactivation.

1. Prepare bait strains on selective agar in 96-well format and incubate for 3–4 days.
2. One day before mating inoculate freshly grown prey strain carrying the plasmid without insert in liquid -L/HAUT medium (solid–liquid), vortex and grow 18–20 h at 30°C in a shaker. Prepare 20 ml for each bait plate to be tested.
3. Mating: Pipette the prey strain into 96-well MTPs (100 μl per well). Transfer each of the baits from the agar into the MTPs, which contain the prey strain without insert and stamp the bait and prey strain mixture directly onto YPD agar. Incubate for 36–44 h at 30°C.
4. To select for diploid yeast cells take off the yeast from the YPD agar, resuspend in -ALT/HU MTPs and transfer to -ALT/HU agar (solid–liquid–solid). This is important because autoactivation can only be reliably assayed in diploid strains (see Note 10).
5. After four nights of incubation at 30°C transfer the yeast from -ALT/HU agar to -HAULT agar (solid–liquid–solid via -ALT/HU MTPs) to select for the growth reporter gene activity. Take pictures of the -ALT/HU agar plates. Incubate the -HAULT agar plates for 5–7 days at 30°C.
6. Take pictures of the -HAULT agar plates.

Remove all bait strains which do not grow on -ALT/HU agar plates as well as those growing on -HAULT agar plates. Usually, autoactive baits grow in all replicas, but occasionally single autoactive spots are detected and must not be used in further experiments (see Note 11).

3.7. Screening Bait Pools Against a Prey Matrix

Yeast strains expressing BD-fusion proteins are screened for primary protein–protein interactions with every strain in the prey matrix. Independent of the actual pool size we repeat each screen four times with distinct bait replicas. For large screens, when many prey MTPs have to be assayed, we create bait pools that contain between 8 and 96 different bait strains. Very efficient pooling strategies have been reported that can increase specificity and sensitivity of large screens without an additional deconvolution step (42, 43). However, we use a retest (Subheading 3.8) that shows which baits in a pool are interacting with the prey that was positive in the primary screen. Importantly, bait strains are grown separately and the baits belonging to one pool are combined directly before mating. Each prey is mated with each pool of baits and primary protein–protein interactions are identified after transfer to selective medium. For large prey matrixes automated screening is recommended. Prey vector controls are not required as protein–protein interactions are very rare events (27, 44) and thus a large majority of preys will not give any growth signal.

1. Three to four days before mating, prepare prey strains on selective agar in 384-well format and bait strains on selective agar in 96-well format (see Subheading 3.4).
2. Day 1: Prepare bait strains in liquid culture: Transfer (solid–liquid) the baits to -AT/HUL MTPs. Liquid–liquid inoculate -AT/HUL medium in flasks or deepwell plates with the bait strains (see Note 12) and grow them for 18–22 h at 30°C in a shaker to early stationary phase (OD_{600} = 1.5–3). Inoculate at least 20-ml liquid medium per prey MTP to be screened. Grow each bait strain separately to avoid growth competition.
3. Day 2: Mating: Make sure that the baits are grown to early stationary phase and that all the yeast is completely suspended before pooling. Combine all bait strains belonging to the same pool in one beaker and mix thoroughly, keeping replicas separated. Pipette the bait strains into 384-well MTPs (40 μl per well). Scrape off prey strains from the agar using a pin tool, transfer to the 384-well MTPs containing the corresponding bait pool cultures, mix well and stamp on YPD agar. This way in each position one prey strain is mated with all baits in one pool. Incubate for 36–44 h at 30°C to allow mating. In essence the mating step is a solid–liquid–solid replication step of the prey matrix, from -L/HAUT agar to YPD agar, except that the MTPs do not contain fresh medium but rather bait pools.
4. Day 4: Interaction selection: Transfer the colonies from YPD agar to 384-well MTPs containing -ALT/HU medium, then to -HAULT agar (solid–liquid–solid). Incubate the agar plates at 30°C for 5–7 days.
5. Mating control: Control the diploid recovery by taking some of the -ALT/HU 384-well MTPs with the yeast mixtures from step 4 and stamping onto -ALT/HU agar (liquid–solid). Grow for 3–4 days at 30°C.
6. Day 8: Take pictures of mating control plates. The diploid recovery should be close to 100% (see Note 13).
7. Day 11: Take pictures of the -HAULT agar plates.

3.8. Retesting

After the primary screen, a retest is necessary to verify and deconvolute the results. High confidence in the final interaction set is guaranteed by using fresh yeast with a low generation number and small culture volumes keeping the probability of acquiring mutation or recombination events very low (29).

In principle, all bait pool–prey combination that result a growth signal in the primary screen should be retested. However, there is a trade-off between the absolute number of interactions recovered and the retest success rate. We present simple rules designed to optimize this. We exclude autoactive/"sticky" preys and bait pool–prey combination that grow only once out of four times from the retest.

As in the autoactivation test, the bait and prey handling is opposite to the screen. The baits are prepared in matrix format on agar while the preys are grown in flasks. This way, each yeast spot corresponds to one bait–prey combination only. Also like in the autoactivation test, there is a diploid selection step between the mating on YPD and the interaction-specific selection.

1. Collect all primary PPIs in a relational database (SQL): For each yeast colony on the -HAULT agar plates, determine the matrix position, corresponding to the prey strain, as well as the bait pool. Make sure that agar numbers, plate numbers, row and column denominators are collected in separate fields and that each colony has a separate entry (see Note 14). All combinations of prey positions and bait pools collected constitute the primary retest list.
2. Removal of autoactive/"sticky" preys: For each prey position, determine the number of colonies and the number of bait pools that produce colonies (see Note 15). Remove any bait pool–prey combination with prey positions that score higher than 50% in both categories from the retest list (see Note 16).
3. Remove singletons: Remove all bait pool–prey combinations that produce yeast spots in only one out of four replica screens from the retest list.
4. Prepare prey cultures: Prepare an agar plate with freshly grown preys from the retest list by pipetting 5 μl of liquid culture on -L/HAUT agar and incubating for 24–36 h (see Note 17). Determine the number of bait pools for each prey. For retesting, prey spots are resuspended directly in 20–100 ml -L/HAUT liquid medium (20 ml per bait pool) and grown 18–22 h at 30°C. For preys with more than five bait pools, prepare 3 ml precultures 1 day in advance and inoculate cultures with 1% of preculture.
5. Prepare bait pool agars: For each bait pool, a 384-well MTP is prepared by combining the four 96-well MTPs containing the four replicas in the following manner. Replica A and B are put into the first (A1 of the 96-A replica in 384-A1) and fourth (A1 of the 96-B replica in 384-B2) quadrant, respectively. For replica C and D the 96-well MTPs are turned by 180° and put

into the second (H12 of the 96-C replica in 384-A2) and third (H12 of the 96-D replica in 384-B1) quadrant, respectively (see Note 18). One of these MTPs is sufficient for stamping 120 copies on agar (see Note 19).

6. Stamp (liquid–solid) the 384-well bait pool MTPs on -AT/HUL agar once for each associated prey on the retest list and incubate at 30°C for 60–72 h.
7. Mating: Transfer prey cultures from flasks to 384-well MTPs (40 μl/well). Scrape off bait colonies from the agar using a pin tool, transfer to the 384-well MTPs containing the corresponding prey solutions (one MTP per bait pool–prey combination), mix well, and stamp on YPD agar (solid–liquid–solid).
8. Recover diploids: After 38–44 h of incubation at 30°C, take off the yeast spots, resuspend in -ALT/HU MTPs, and stamp on -ALT/HU agar (solid–liquid–solid).
9. Interaction selection: After four nights of incubation at 30°C, take off the yeast spots, resuspend in -ALT/HU MTPs, and stamp on -HAULT agar (solid–liquid–solid). For assaying β-galactosidase activity (see Subheading 3.9) save the MTPs at this point.
10. After 5–7 days, take pictures of the agar plates.

Count the number of colonies for each interaction. Expect MTPs with more than one interaction at this point (see Note 20). Verified interactions show up as characteristic patterns, i.e., four yeast colonies that appear as two "anticorrelated" diagonal pairs of yeast colonies (Fig. 3). The Y2H interactions can be further validated applying other PPI detection methods (see Note 21). However, the Y2H PPI information as such is very useful for network analyses (45–48) and most promising starting points for functional studies (7, 12, 49, 50).

3.9. Beta-Galactosidase Assay

In our Y2H system the activity of the third reporter, the *E. coli LacZ* gene, is not tested via growth but in an enzymatic assay. The gene's product, beta-galactosidase, is a protein that cleaves 5-Bromo-4-chloro-3-indolyl-β-D-galactopyranoside (X-Gal) into galactose and 5-bromo-4-chloro-3-hydroxyindole, which oxidizes to 5,5′-dibromo-4,4′-dichloro indigo, resulting in a blue stain. The enzymatic assay is used in addition to growth reporter selection to further qualify Y2H interactions. The protocol for this assay is detailed here.

1. Place a nylon-membrane on a -HAULT agar by lifting two opposing corners with two pairs of forceps. Place the other two corners on an agar plate, then carefully let go of the first two corners. Remove any air bubbles that you may have created in the process. Prepare one agar for every six MTPs to be tested.

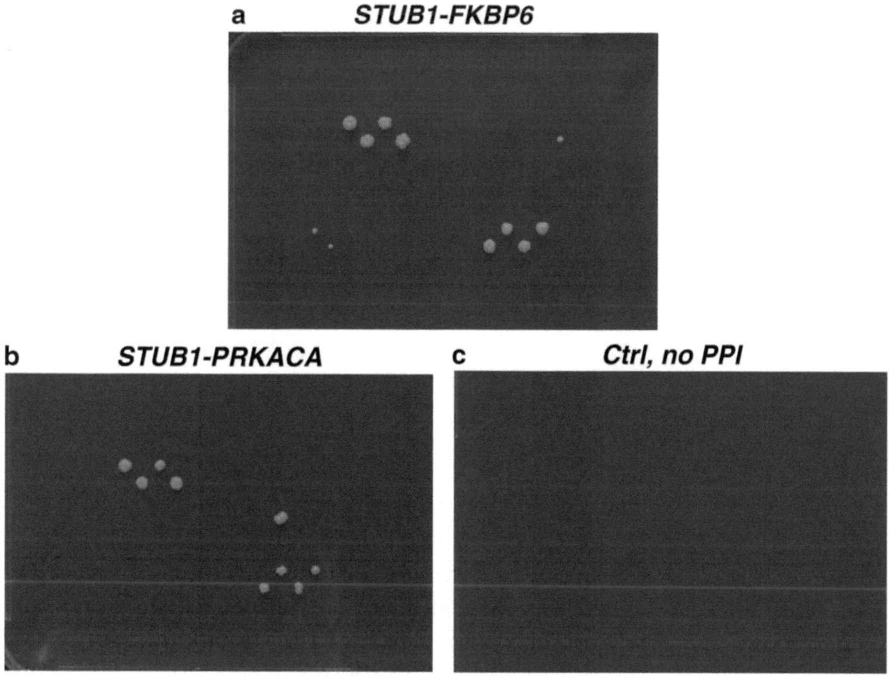

Fig. 3. Characteristic growth patterns indicate interactions in a retest experiment. The Y2H interactions between STUB1/Chip and FKBP6/FKBP36 (**a**) and STUB1/Chip and PRKACA/PKC alpha (**b**) are shown together with an experiment with a prey that does not interact with any of the baits (**c**). In this retest experiment, a 384 MTP that contains 96 baits in four replicas is tested against preys that were positive in the primary screen. Two full-length bait clones of STUB1/Chip (NP_005852, aa 1–303) are in position E7, F8, K18, L17 and E9, F10, K16, L15, respectively. Both constructs grow on selective media (–HAULT) when tested with FKBP6/FKBP36 (NP_003593, aa 1–322) or PRKACA/PKC alpha (NP_002721, aa 1–335) preys for interaction resulting a characteristic pattern of yeast colonies. STUB1/Chip is an E3 ubiquitin-protein ligase and HSPA8/HSC70 cochaperone for which quite a few protein interaction partners have been reported, e.g., ref. 51. STUB1/Chip function has been linked to neurodegenerative diseases (52). FKBP6/FKBP36, a peptidyl-prolyl *cis/trans* isomerase, is associated with Clathrin and Hsp72 in spermatocytes (53). FKBP6 deletions were shown to be associated with Williams–Beuren syndrome. The second potential STUB1/Chip interaction partner reported here, Protein Kinase C alpha (PRKACA/PKC alpha), is a major player in several well studied signaling cascades (54, 55).

2. Prepare diploid yeast containing bait and prey plasmids for each interaction to be tested like before. If you saved the -ALT/HU MTPs from the retest (step 10), these can be used.
3. Stamp the diploid yeast onto the -HAULT agars with membranes (liquid–solid, if using MTPs from retest). Use a solid–liquid–solid step if fresh diploids are grown on ALT/HU agar (see Note 22).
4. After 5–7 days, remove the membranes from the agar plates with forceps, freeze them in liquid nitrogen, let thaw at room temperature, freeze and thaw again (see Note 23).
5. Add 400 μl 1 M DTT and 624 μl 20 mg/ml X-Gal solution to 40 ml Z-buffer for each membrane.

6. Place two sheets of Whatman 3MM Chromatography paper in an empty agar dish and soak with Z-buffer.
7. Carefully place the membrane on the Whatman paper (again with forceps) and remove air bubbles.
8. Incubate the membrane for 3 h at 37°C or until blue staining can be seen (see Note 24).
9. Dry the membranes and take pictures. Blue staining signifies interactions.

4. Notes

1. The pins can also be sterilized by moving through a 70% ethanol bath equipped with a brush, moving into an ultrasonic bath for 10 s, moving through a second 70% ethanol bath with a brush and drying with hot air (300–400°C). This option is relevant when using a robot for automated stamping. On a related note, the tips of a pipetting robot can be sterilized by subsequently pipetting the maximal volume 70% ethanol, bleach (~1% free chlorine), 70% ethanol, and sterile water three times each. When working with robots, be sure to include sterility controls in every experiment.
2. In this publication, water refers to deionized water with a specific resistance of at least 18 MΩ.
3. It is possible to prepare plates with an agarclave. For 10 l selective agar, autoclave (15 min, 121°C) 67 g yeast nitrogen base and 300 g agar in 9.2–9.5 l water (depending on the required amino acids/nucleosides), cool the medium to 60°C, add 500 ml 20× glucose stock solution and 100 ml of each required amino acid/nucleoside stock solution. Note that you need 1.5 times the amount of agar when autoclaving agar and nitrogen base together. YPD agar plates can be prepared with the agarclave, too. For 10 l YPD agar, autoclave 100 g yeast extract, 200 g peptone and 200 g agar in 9.5 l water. Cool to 60°C and add 500 ml 20× glucose. Pour 200 ml into each agar plate under a sterile hood.
4. In contrast to L40ccu, L40ccα needs adenine-supplemented medium. If the adenine concentration is low the yeast turns red. This can be avoided by adding adenine to the YPD agar or doubling the amount of adenine to -L/HAUT plates.
5. It is possible to shuttle one ORF into two destination vectors in a single 2-in-1-LR Clonase reaction provided that the selection markers are different (e.g., ampicillin and tetracycline resistance in pACT4-DM and pBTM116-D9, respectively). Just add

0.25 µl of each destination vector instead of 0.5 µl of one destination vector. After the transformation of *E. coli*, plate one half of the SOC medium on LB agar containing ampicillin and the other half on LB-Agar containing tetracycline. Checking for cotransformation is very important because *E. coli* cotransformants containing both pACT4 and pBTM116 plasmids can lead to yeast cotransformed with both vectors. This results in growth on -HAULT agar if the protein in question is able to form homodimers, i.e., the protein will appear autoactive although it may not be. To check for cotransformation, plate 5 µl of the overnight culture on LB-Agar with ampicillin and on LB-Agar with tetracycline. If the overnight culture grows on both plates, dilute the plasmid-DNA 1:1,000 and transform again or pick a different colony.

6. Care should be taken to seal the plates properly in all steps because vigorous vortexing or inversion might cause cross-contamination.
7. Normally the 96-well *E. coli* mini-prep protocol yields 10-20 µg plasmid-DNA. BsrGI restriction analysis is necessary because free nucleotides are copurified making the measurement of the OD at 260 nm meaningless. To determine the DNA concentration, compare to known amounts of DNA on the same gel.
8. The yeast growth rate depends on the medium. Grow bait and prey strains for 3 days on -AT/HUL or -L/HAUT, respectively. To allow mating, grow yeast on YPD agar for 30–48 h. Select diploid strains by growing on -ALT/HU medium for 3–4 days. Select for protein–protein interactions by growing yeast on -HAULT medium for 5–7 days.
9. We do not test for prey autoactivity because autoactive preys occur in less than 1% of all cases and preys are not pooled. Autoactive preys are not removed from prey arrays as they are useful as mating controls and allow identification of prey plates from large collections at first glance once you gathered some experience.
10. Some degree of autoactivation can also be observed testing haploid bait strains for reporter activity, but only baits that do not autoactivate in a diploid context can be used for a pooled screen.
11. If three out of four replicas are autoactive, consider removing the fourth copy from the pooled approach and screening the bait separately.
12. Depending on the number of baits use flasks or deepwell plates. We use flasks for pools of 8 baits and deepwell plates for pools of 96 baits.
13. We differentiate diploid recovery and mating efficiency. Mating efficiency is defined as the number of diploids produced by an

equal mixture of MAT*a* and MATα haploid strains divided by half the number of haploid cells before mating and can be determined for individual mating reactions. The diploid recovery is defined as the fraction of positions of a MTP that grow on -ALT/HU agar after mating.

14. Since the growth rate of yeast colonies is only weakly correlated with the probability of successful retesting, be sure to count colonies regardless of size, but do not count very faint spots.
15. If you have information about primary hits from a previous screen with the same prey matrix, this information should be added to the current screens. Also, if all the baits are functionally related, the rules for removing preys should be relaxed.
16. This number is chosen to yield a success rate of about 80% in retesting. By adjusting the number of preys removed, the absolute number of interactions recovered and the success rate can be traded off.
17. Alternatively, preys can be stamped in 96-well format with a pin tool (liquid–solid). In that case, incubate the agar plate for 2–3 days.
18. Removal of autoactive baits is not necessary at this point.
19. Propagation of bait pool plate in liquid 384-well MTP format is not recommended.
20. At this point, it is important to exclude autoactive baits from the analysis. Autoactive preys can be recognized by stochastic distribution of a high number of colonies.
21. Keep in mind that protein–protein interaction assays are orthogonal. This means that while interactions validated by other methods do have a higher likelihood of being true, a large number of true interactions will inadvertently be lost.
22. The β-galactosidase activity can also be assayed from diploid strains gown on -ALT/HU, and the results can be compared to independent growth reporter readouts. However, we use this assay on top of the growth reporter readout as the most stringent way of assaying PPIs.
23. Before thawing the membranes the second time, they can be stored at −80°C for weeks.
24. Since the leuco form of the indigo reaction product is soluble, unspecific staining will occur once the DTT is oxidized, so make sure to stop the assay before this point.

Acknowledgments

J.M. Worseck, A. Grossmann, and M. Weimann contributed equally to the writing of this protocol. We would like to thank Erich Wanker (MDC-Berlin) and the members of his group for continuing support and for their contributions in developing the Y2H setup.

References

1. Fields S, Song O. (1989) A novel genetic system to detect protein-protein interactions. *Nature* **340**, 245–6.
2. Fromont-Racine M, Rain JC, Legrain P. (1997) Toward a functional analysis of the yeast genome through exhaustive two-hybrid screens. *Nat Genet* **16**, 277–82.
3. Uetz P, Giot L, Cagney G, et al. (2000) A comprehensive analysis of protein-protein interactions in *Saccharomyces cerevisiae*. *Nature* **403**, 623–7.
4. Walhout AJ, Sordella R, Lu X, et al. (2000) Protein interaction mapping in C. elegans using proteins involved in vulval development. *Science* **287**, 116–22.
5. Ito T, Chiba T, Ozawa R, Yoshida M, Hattori M, Sakaki Y. (2001) A comprehensive two-hybrid analysis to explore the yeast protein interactome. *Proc Natl Acad Sci USA* **98**, 4569–74.
6. Tong AH, Drees B, Nardelli G, et al. (2002) A combined experimental and computational strategy to define protein interaction networks for peptide recognition modules. *Science* **295**, 321–4.
7. Goehler H, Lalowski M, Stelzl U, et al. (2004) A protein interaction network links GIT1, an enhancer of Huntingtin aggregation, to Huntington's disease. *Mol Cell* **15**, 853–65.
8. Colland F, Jacq X, Trouplin V, et al. (2004) Functional proteomics mapping of a human signaling pathway. *Genome Res* **14**, 1324–32.
9. Formstecher E, Aresta S, Collura V, et al. (2005) Protein interaction mapping: a Drosophila case study. *Genome Res* **15**, 376–84.
10. Miller JP, Lo RS, Ben-Hur A, et al. (2005) Large-scale identification of yeast integral membrane protein interactions. *Proc Natl Acad Sci USA* **102**, 12123–8.
11. Lim J, Hao T, Shaw C, et al. (2006) A protein-protein interaction network for human inherited ataxias and disorders of Purkinje cell degeneration. *Cell* **125**, 801–14.
12. Kaltenbach LS, Romero E, Becklin RR, et al. (2007) Huntingtin interacting proteins are genetic modifiers of neurodegeneration. *PLoS Genet* **3**, e82.
13. Markson G, Kiel C, Hyde R, et al. (2009) Analysis of the human E2 ubiquitin conjugating enzyme protein interaction network. *Genome Res* **19**, 1905–11.
14. van Wijk SJ, de Vries SJ, Kemmeren P, et al. (2009) A comprehensive framework of E2-RING E3 interactions of the human ubiquitin-proteasome system. *Mol Syst Biol* **5**, 295.
15. Bell R, Hubbard A, Chettier R, et al. (2009) A human protein interaction network shows conservation of aging processes between human and invertebrate species. *PLoS Genet* **5**, e1000414.
16. Giot L, Bader JS, Brouwer C, et al. (2003) A protein interaction map of Drosophila melanogaster. *Science* **302**, 1727–36.
17. Li S, Armstrong CM, Bertin N, et al. (2004) A map of the interactome network of the metazoan C. elegans. *Science* **303**, 540–3.
18. Stelzl U, Worm U, Lalowski M, et al. (2005) A human protein-protein interaction network: a resource for annotating the proteome. *Cell* **122**, 957–68.
19. Rual JF, Venkatesan K, Hao T, et al. (2005) Towards a proteome-scale map of the human protein-protein interaction network. *Nature* **437**, 1173–8.
20. LaCount DJ, Vignali M, Chettier R, et al. (2005) A protein interaction network of the malaria parasite Plasmodium falciparum. *Nature* **438**, 103–7.
21. Parrish JR, Gulyas KD, Finley RL, Jr. (2006) Yeast two-hybrid contributions to interactome mapping. *Curr Opin Biotechnol* **17**, 387–93.
22. Yu H, Braun P, Yildirim MA, et al. (2008) High-quality binary protein interaction map of

the yeast interactome network. *Science* **322**, 104–10.
23. Parrish JR, Yu J, Liu G, et al. (2007) A proteome-wide protein interaction map for Campylobacter jejuni. *Genome Biol* **8**, R130.
24. Braun P, Tasan M, Dreze M, et al. (2009) An experimentally derived confidence score for binary protein-protein interactions. *Nat Methods* **6**, 91–7.
25. Figeys D. (2008) Mapping the human protein interactome. *Cell Res* **18**, 716–24.
26. Schwartz AS, Yu J, Gardenour KR, Finley RL, Jr., Ideker T. (2009) Cost-effective strategies for completing the interactome. *Nat Methods* **6**, 55–61.
27. Venkatesan K, Rual JF, Vazquez A, et al. (2009) An empirical framework for binary interactome mapping. *Nat Methods* **6**, 83–90.
28. Walhout AJ, Vidal M. (1999) A genetic strategy to eliminate self-activator baits prior to high-throughput yeast two-hybrid screens. *Genome Res* **9**, 1128–34.
29. Vidalain PO, Boxem M, Ge H, Li S, Vidal M. (2004) Increasing specificity in high-throughput yeast two-hybrid experiments. *Methods* **32**, 363–70.
30. Rajagopala SV, Hughes KT, Uetz P. (2009) Benchmarking yeast two-hybrid systems using the interactions of bacterial motility proteins. *Proteomics* **9**, 5296–302.
31. Golemis EA, Serebriiskii I, Finley RL, Jr., Kolonin MG, Gyuris J, Brent R. (2009) Interaction trap/two-hybrid system to identify interacting proteins. *Curr Protoc Protein Sci* Chapter **19**, Unit19 2.
32. Estojak J, Brent R, Golemis EA. (1995) Correlation of two-hybrid affinity data with in vitro measurements. *Mol Cell Biol* **15**, 5820–9.
33. Simonis N, Rual JF, Carvunis AR, et al. (2009) Empirically controlled mapping of the Caenorhabditis elegans protein-protein interactome network. *Nat Methods* **6**, 47–54.
34. Eyckerman S, Lemmens I, Lievens S, et al. (2002) Design and use of a mammalian protein-protein interaction trap (MAPPIT). *Sci STKE* **2002**, pl18.
35. Stelzl U, Wanker EE. (2006) The value of high quality protein-protein interaction networks for systems biology. *Curr Opin Chem Biol* **10**, 551–8.
36. Suter B, Kittanakom S, Stagljar I. (2008) Two-hybrid technologies in proteomics research. *Curr Opin Biotechnol* **19**, 316–23.
37. Sanderson CM. (2009) The Cartographers toolbox: building bigger and better human protein interaction networks. *Brief Funct Genomic Proteomic* **8**, 1–11.
38. Lamesch P, Li N, Milstein S, et al. (2007) hORFeome v3.1: a resource of human open reading frames representing over 10,000 human genes. *Genomics* **89**, 307–15.
39. Goshima N, Kawamura Y, Fukumoto A, et al. (2008) Human protein factory for converting the transcriptome into an in vitro-expressed proteome. *Nat Methods* **5**, 1011–7.
40. Temple G, Gerhard DS, Rasooly R, et al. (2009) The completion of the Mammalian Gene Collection (MGC). *Genome Res* **19**, 2324–33.
41. Chung CT, Niemela SL, Miller RH. (1989) One-step preparation of competent Escherichia coli: transformation and storage of bacterial cells in the same solution. *Proc Natl Acad Sci USA* **86**, 2172–5.
42. Jin F, Hazbun T, Michaud GA, et al. (2006) A pooling-deconvolution strategy for biological network elucidation. *Nat Methods* **3**, 183–9.
43. Xin X, Rual JF, Hirozane-Kishikawa T, et al. (2009) Shifted Transversal Design smart-pooling for high coverage interactome mapping. *Genome Res* **19**, 1262–9.
44. Jansen R, Gerstein M. (2004) Analyzing protein function on a genomic scale: the importance of gold-standard positives and negatives for network prediction. *Curr Opin Microbiol* **7**, 535–45.
45. Chuang HY, Lee E, Liu YT, Lee D, Ideker T. (2007) Network-based classification of breast cancer metastasis. *Mol Syst Biol* **3**, 140.
46. Lage K, Karlberg EO, Storling ZM, et al. (2007) A human phenome-interactome network of protein complexes implicated in genetic disorders. *Nat Biotechnol* **25**, 309–16.
47. Kohler S, Bauer S, Horn D, Robinson PN. (2008) Walking the interactome for prioritization of candidate disease genes. *Am J Hum Genet* **82**, 949–58.
48. Taylor IW, Linding R, Warde-Farley D, et al. (2009) Dynamic modularity in protein interaction networks predicts breast cancer outcome. *Nat Biotechnol* **27**, 199–204.
49. Pujana MA, Han JD, Starita LM, et al. (2007) Network modeling links breast cancer susceptibility and centrosome dysfunction. *Nat Genet* **39**, 1338–49.
50. Deribe YL, Wild P, Chandrashaker A, et al. (2009) Regulation of epidermal growth factor receptor trafficking by lysine deacetylase HDAC6. *Sci Signal* **2**, ra84.
51. Grelle G, Kostka S, Otto A, et al. (2006) Identification of VCP/p97, carboxyl terminus of Hsp70-interacting protein (CHIP), and

amphiphysin II interaction partners using membrane-based human proteome arrays. *Mol Cell Proteomics* **5**, 234–44.

52. Dickey CA, Patterson C, Dickson D, Petrucelli L. (2007) Brain CHIP: removing the culprits in neurodegenerative disease. *Trends Mol Med* **13**, 32–8.
53. Jarczowski F, Fischer G, Edlich F. (2008) FKBP36 forms complexes with clathrin and Hsp72 in spermatocytes. *Biochemistry* **47**, 6946–52.
54. Mackay HJ, Twelves CJ. (2007) Targeting the protein kinase C family: are we there yet? *Nat Rev Cancer* **7**, 554–62.
55. Roffey J, Rosse C, Linch M, Hibbert A, McDonald NQ, Parker PJ. (2009) Protein kinase C intervention: the state of play. *Curr Opin Cell Biol* **21**, 268–79.

Chapter 5

High-Throughput Yeast Two-Hybrid Screening of Complex cDNA Libraries

Kerstin Mohr and Manfred Koegl

Abstract

Yeast two-hybrid screening can be used to find cDNAs encoding proteins which bind to a given bait protein in large, pooled cDNA libraries. Screening of complex, pooled libraries is slower and more laborious than screening of arrayed collections of cDNAs, but has several advantages. First, the complexity of a pooled library can be orders of magnitude larger than the size of a typical arrayed library. Second, as long as available cDNA collections are incomplete and limited to full-length cDNAs, pooled cDNA libraries offer a more complete search space and are often the only way to do screens in organisms other than human and a few model organisms. We have streamlined and optimised the screening of pooled libraries in a format which uses micro-titre plates and produces quantitative signals for the selection of hits. This format has the advantage that automation of the process is straightforward and allows a throughput of up to at least 1,000 screens per year per person.

Key words: Two hybrid, Protein–protein interaction, Yeast mating, High-throughput screening, Functional proteomics

1. Introduction

In this chapter, we present an optimised and streamlined protocol for efficient screening of complex cDNA libraries. While small libraries, such as collections of all or most open reading frames (ORFs) of an organism (1), are of a complexity that is small enough to allow a one-by-one screening protocol, cDNA libraries that are based on reversely transcribed mRNA samples need to be of a high complexity, typically exceeding a million clones, which precludes their screening in an arrayed form. This high complexity is necessitated in part by the fact that the representation of cDNAs in the library depends primarily on the expression level of the mRNA in the tissue such that highly expressed mRNAs are overrepresented over

Bernhard Suter and Erich E. Wanker (eds.), *Two Hybrid Technologies: Methods and Protocols*, Methods in Molecular Biology, vol. 812, DOI 10.1007/978-1-61779-455-1_5, © Springer Science+Business Media, LLC 2012

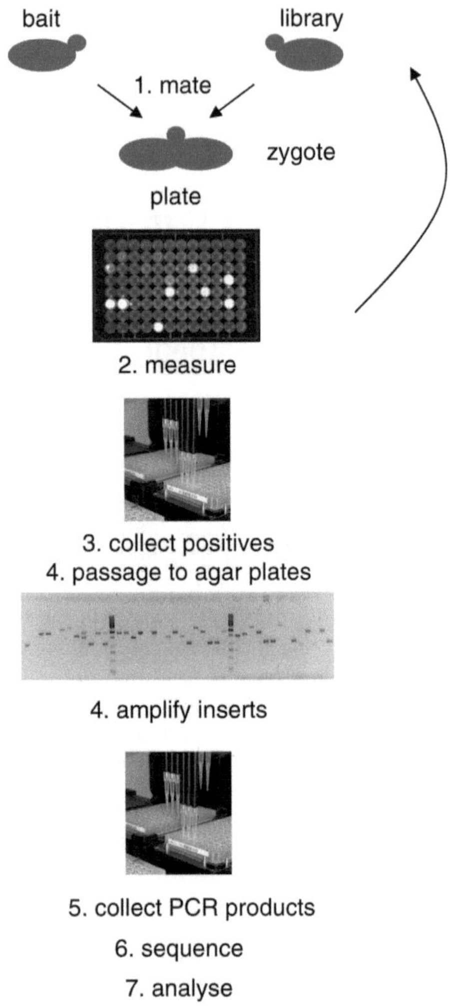

Fig. 1. Schematic of the screening process. (1) Bait and library cells are mated and plated on selective medium. (2) The activation of the two reporters, HIS3 and MEL1, is measured concomitantly via the fluorescent signal. (3) Yeasts from wells with interaction signals are collected and passaged once on a microtitre plate in liquid medium and once more onto an agar plate. (4) The library insert is amplified by PCR, and (5) PCR products of sufficient quality for sequencing are collected. (6) The PCR products are sequenced. (7) The preys' identity is determined by sequence comparison to a public database, and the results are summarised in a table. Reliable interactions are selected by filtering out single hits, non-natural peptides, as well as promiscuous preys. Note the two re-arraying steps at 3 and 5, which should be documented in a database. The curved arrow indicates where test screens or screens which have a too high frequency of wells with more than one colony are aborted to restart at adjusted conditions.

lowly expressed mRNAs. The presence of a majority of non-functional expression constructs in the library, owing to fusions in a wrong reading frame or within non-coding parts of the protein, adds to the need for a high complexity.

An overview of the protocol for screening of such complex libraries is given in Fig. 1. One of the key steps of the procedure is

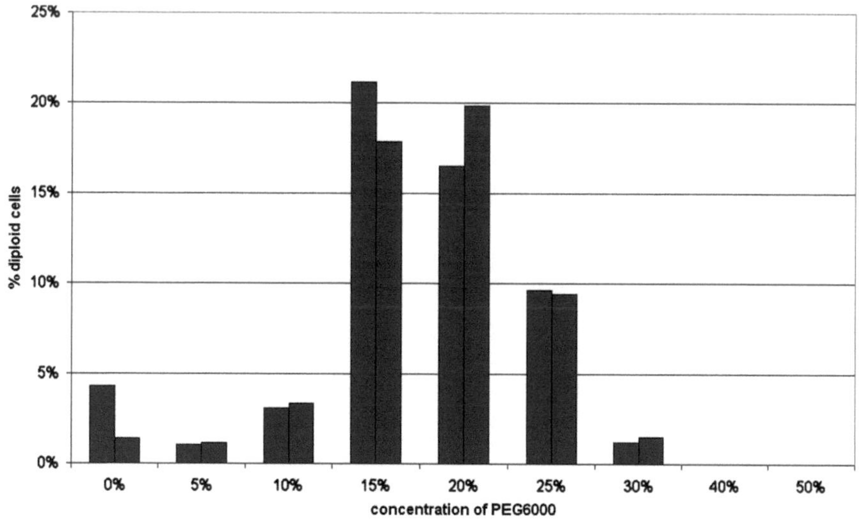

Fig. 2. Effect of PEG concentration on mating efficiency. Cells were mated as described in the protocol at the indicated concentrations of PEG6000 in YPDA. Results from duplicate experiments are depicted. Typical mating efficiencies in the high-throughput process range from 3 to 10%.

the use of a highly efficient mating reaction to bring the bait strain and the library together in the same cell. This requires the use of pre-transformed libraries, i.e., libraries which consist of yeast cells harbouring the library plasmids (see Note 1). Thus, the bait plasmid and the library plasmid pool are transformed independently into haploid yeast cells of compatible mating type. When these strains are brought into contact under appropriate conditions, the haploid cells fuse to form a zygote which harbours both the bait plasmid and a library plasmid. The addition of polyethylene glycol to the mating medium induces highly efficient mating in suspension cultures (shown in Fig. 2), avoiding the need to plate the cells on agar plates and to scrape them off for further processing.

The selection of positive yeast cells, i.e., yeast cells in which bait and prey interact and cause reporter gene activation, is done in micro-titre plates. Two reporters are used: (1) *HIS3* is used as an auxotrophy reporter which allows selective enrichment of positive cells in medium lacking histidine and (2) *MEL1* is used to induce the expression of alpha-galactosidase in positive cells (2). In contrast to beta-galactosidase, which is a located in the cytoplasm of positive cells, alpha-galactosidase is secreted into the medium. To detect alpha-galactosidase activity, a fluorigenic substrate for alpha-galactosidase is added to the medium. Only colonies that are positive for both reporters produce a strong fluorescent signal such that wells containing positive colonies can be easily and automatically identified by a plate reader. Note that this assay is homogeneous, i.e., after plating of the cells onto the micro-titre plates, it does not require any further steps, such as cell lysis or the addition of reagents before the plates can be read. This is a major advantage over protocols relying on the measurement of beta-galactosidase as a reporter, which typically involve several additional steps.

All presently commercially available yeast two-hybrid (Y2H) libraries from higher organisms are based on cDNA generated by reverse transcription of mRNA preparations. These libraries express a large fraction of non-natural peptides and fragmented proteins, since their cDNA inserts include often 5′ untranslated regions (UTRs), pieces of the coding regions, usually truncated at the 5′ end due to directional priming at the 3′ poly(A) tail, and fragments that solely consist of 3′ UTR. These non-natural peptides are a rich source for non-natural interactions with the bait protein. Since most of these interactions only produce weak interaction signals, they can be suppressed by adjusting the stringency of the screening conditions to a sufficiently high level. Some bait proteins are more prone to such interactions than others, which makes an individual adjustment of screening conditions for each bait necessary.

At a mating efficiency of 2.5–10%, each well of a micro-titre plate typically contains ~5,000 to ~20,000 different library clones. While there is no way to predict if a weakly positive clone is harbouring a true interaction or an artefact, it is reasonable to assume that a hit rate higher than 1 in 50,000 library clones is mainly due to artefacts. Higher screening stringency, in most cases, reduces the number of artificial interactions and increases the fraction of relevant interactions among the hits. In any case, screening conditions should be adjusted such that no more than 1 in 7 wells harbours a colony and chances of double colonies are not higher than 2% of all positive wells. This is best done by doing a test screen employing varying concentrations of 3-amino triazole (3-AT) per plate. 3-AT is an inhibitor of the *HIS3* reporter gene product. In the presence of 3-AT, a stronger activation of the *HIS3* reporter is required to overcome the inhibition. Test screens are discarded after reading the plate (step 2 in Fig. 1), and the screen is repeated on a larger scale using the optimal concentration of 3-AT (see Note 2).

This protocol can be performed manually, i.e., without the application of robotic systems. If the steps of plate reading and hit picking are automated, a throughput of at least 1,000 screens (i.e., combinations of baits and libraries) per person and year can be achieved. For an automated set-up, the process needs to be supported by a proper database. The database records (1) the input of the experimenter, e.g., the identity of the bait and the screening conditions used and (2) machine-read data, such as the fluorescent interaction signal; (3) keeps track of the re-array steps, which occur after the identification of positives and after the analysis of PCR products; and (4) summarises the results of the screens after analysis of the prey sequences. Lastly, the automated set-up demands automation of the analysis of results, including the comparison of hit sequences with the sequences in public databases, typically done by BLAST searches, determination of the hit fragments' start relative

Table 1
Shows the 20 preys that have been most frequently isolated in yeast two-hybrid screens with 635 baits

Gene symbol	# of baits	Library	Frequency (%)
CRX	137	ORF	31
COPS5	107	Testis	18
MEOX2	87	ORF	20
PCBP1	64	Bone marrow	13
OTX2	55	ORF	19
COX1	44	Heart	14
UBE2I	44	ORF	15
PGM1	34	HeLa	19
HBB	32	Bone marrow	25
TAF1	32	Brain	8
ACTB	31	Bone marrow	7
COX3	31	Kidney	12
PAX6	31	Brain	6
CREB3	30	Heart	6
MAP1B	29	Brain	9
C11orf17	28	Brain	5
CYB	28	Heart	14
COX2	27	Mouse day 17 embryo	32
BEND7	26	ORF	19
EEF1A1	26	Bone marrow	7

The preys are listed by their gene symbols. # of baits indicates the number of different baits which have led to the isolation of the prey in yeast two-hybrid screens, library is the library in which the prey is isolated with the highest frequency, and frequency is the percentage of screens in which this prey is isolated in this library

to the full-length mRNA, counting of the number of identical hits per bait, and other steps.

For further scientific analysis, only interactions should be considered which are observed at least twice. Single hits are unreliable since they can be due to a number of technical artefacts, including more than one library plasmid per diploid cell, more than one colony per well, or mutations in the bait. Furthermore, it is necessary to exclude promiscuous preys which are isolated with a very high number of baits, and are thus classified as non-specific binders (3). A list of the 15 most frequently isolated preys from our screens is given in Table 1. Arguably, the observed promiscuity need not be an artefact, but may be a reflection of a large natural interaction repertoire of the prey. In practice, however, preys, such as COPS5 which is isolated in libraries from testis in 18% of all screens, are very likely to be artefacts.

2. Materials

2.1. Yeast Strains

Baits:

CG1945 *MAT a, ura3-52, his3-200, ade2-101, lys2-801, trp1-901, leu2-3 112, gal4-542, gal80-538, cyh^r2, LYS2::GAL1*$_{UAS}$-*GAL1*$_{TATA}$-*HIS3, URA3::GAL4*$_{17mers(x3)}$-*CYC1*$_{TATA}$ -*lacZ*

or

AH109 *MAT a, ura3-52, his3-200, trp1-90, leu2-3 112, gal4 Δ; gal80Δ, LYS2::GAL1*$_{UAS}$-*GAL1*$_{TATA}$-*HIS3, GAL2*$_{UAS}$-*GAL2*$_{TATA}$-*ADE2, URA3::MEL1*$_{UAS}$-*MEL1*$_{TATA}$ -*lacZ, MEL1.*

Preys:

Y187 *MAT alpha, ura3-52, his3-200, ade2-101, trp1-901, leu2-3, 112, gal4 Δ, met-, gal80 Δ, URA3::GAL1*$_{UAS}$-*GAL1*$_{TATA}$ -*lacZ, MEL1.*

2.2. Yeast Stock Solutions

If not indicated otherwise, all media are prepared with deionised water, sterilised by sterile filtration (see Note 3), and stored at room temperature. All yeast media preparations (YNB, YPD, and CSM dropout mixes) are from MP Biomedicals, Solon, Ohio.

1. Sterilise 350 ml water per 500-ml bottle by autoclaving.
2. Sterilise 8 g agar agar in 350 ml water in 500-ml bottles by autoclaving.
3. 2× YPD: 100 g YPD mix, add water to 1 l.
4. 20% glucose: 200 g glucose, add water to 1 l.
5. 10× YNB: 67 g of yeast nitrogen base with ammonium sulphate, water to 1 l. Store at 4°C.
6. Dropout mixes: Prepare 500 ml of tenfold concentrated solutions of CSM-Leu, CSM-Trp, CSM-Leu-Trp, and CSM-His-Leu-Trp.
7. 100× adenine: 200 mg adenine, water ad 100 ml, store at 4°C.
8. Pen/Strep: Penicillin 10,000 U/ml, streptomycin 10,000 μg/ml.
9. 60 mM 4-MU: Dissolve 100 mg 4-methylumbelliferyl-alpha-D-galactoside (4Mu-X, Biosynth) in 5 ml dimethylformamide. No sterilisation required. Store at –20°C.
10. 3-AT: 3-Amino-triazole 1 M stock in SD-Leu-Trp-His.
11. 40% (w/v) PEG 6000 in water. Autoclave.

2.3. Yeast Media

1. YPDA: 250 ml sterile water, 250 ml 2× YPD, 5 ml 100× adenine, 2.5 ml Pen/Strep (penicillin 10,000 U/ml, streptomycin 10,000 μg/ml).
2. SD plates: Melt 350 ml agar agar (as prepared in 2.2, point 2), add 50 ml 20% glucose, 50 ml 10× YNB, 50 ml 10× CSM with appropriate amino acids, 2.5 ml Pen/Strep. Prepare 1 SD-Trp-Leu 10-cm plate per screen, and ~ 1 SD-Trp-Leu-His 15-cm plate per screen.
3. SD-liquid medium: 350 ml autoclaved water, 50 ml 20% glucose, 50 ml 10× YNB, 50 ml 10× CSM with appropriate amino acids, 2.5 ml Pen/Strep.
4. 20% PEG/YPDA: Mix 250 ml 2× YPD and 250 ml 40% PEG 6000, add 5 ml 100× adenine.

2.4. Yeast Transformation

1. Stocks: 1 M bicine, pH 8.35, 50% (w/v) PEG 1000, 1 M sorbitol, 5 M NaCl.
2. Solution I (500 ml): 500 mM sorbitol, 10 mM bicine, pH 8.35, 3% (v/v) ethylene glycol, 5% (v/v) DMSO.
3. Solution II (50 ml): 40% (w/v) PEG 1000, 200 mM bicine, pH 8.35.
4. Solution III (50 ml): 150 mM NaCl, 10 mM bicine, pH 8.35.
5. Carrier DNA: Salmon sperm DNA, 10 mg/ml.

2.5. PCR

1. Stocks: Red Taq (1 U/μl, Bioline, Luckenwalde), dNTPs 100 mM each, 10× buffer, 50 mM $MgCl_2$, oligos: 100 μM in water, 20% Triton X-100.
2. Oligos: PCR mix 1: o136: 5′ CTACAGGGATGTTTAATAC CACTACAATGG 3′, o137: 5′GGTTACATGGCCAAGATT GAAACTTAGAGG 3′.
   PCR mix 2: o0080: 5′ TGTTTAATACCACTACAATGGATG ATG 3′.
   o0081: 5′ CATAAAAGAAGGCAAAACGATG 3′.
   pACT2-FP sequencing primer: 5′ GATGATGAAGATACCC CAC 3′.
   pACT2-RP sequencing primer: 5′ CAGTTGAAGTGAACTT GC 3′.
   These primers work for the following prey vectors: pGADT7, pGAD424, pGADT7-Rec, and pACT2.
3. 0.25% SDS in water.
4. PCR mix 1 per well: 5 μl 10× buffer, 2 μl $MgCl_2$, 0.5 μl dNTPs, 0.3 μl oligo o136, 0.3 μl oligo o137, 2.5 μl 20% Triton X −100, 40 μl water. Mix by mild vortexing. Add 1 U Taq polymerase (= 1 μl). Mix by mild vortexing.

5. PCR mix 2 per well: 5 µl 10× buffer, 2 µl $MgCl_2$, 0.5 µl dNTPs, 0.3 µl oligo o0080, 0.3 µl oligo o0081, 40 µl water. Mix by mild vortexing. Add 1 U Taq polymerase (=1 µl). Mix by mild vortexing.

2.6. Disposable Plastic Ware

1. 96-tip replicator pins, long (Scinomix, Earth City).
2. Flat-bottom transparent micro-titre plates, 96 wells, without lid, not sterilised (see Note 4).
3. Adhesive plastic foil to cover the topmost micro-titre plates of a stack.
4. Saran wrap.

3. Methods

3.1. Preparation of Competent Yeast Cells

1. Inoculate a 5 ml culture of the appropriate yeast strain and grow overnight at 30°C in 1× YPDA.
2. Dilute the overnight culture cells in 500 ml YPDA (for 50 transformation reactions). Grow to an OD_{600} of 0.6–1.0.
3. Pellet cells at 1,000 g for 5 min in disposable 50-ml plastic tubes.
4. Wash cells in 250 ml solution I, and pellet at 1,000 g for 5 min.
5. Resuspend the cells in 10 ml of solution I.
6. Slowly freeze as 0.2-ml aliquots (in 2-ml plastic tubes) by placing them in a −80°C freezer (see Note 5).
7. Store at −80°C.

3.2. Transformation of the Bait

1. Denature 5 µl of carrier DNA for 5 min at 95°C, and place on ice immediately afterwards.
2. Add carrier DNA and plasmid DNA (<10 µl, ~200 ng/µl) (see Note 6) on top of the frozen competent yeast (CG1945 or AH109) cell suspension.
3. Incubate yeast at 30°C and mix by hand every 10–15 s until solutions are completely melted.
4. Add 1.4 ml of solution II and mix by vortexing for 1 min.
5. Shake at 800 rpm for 1 h at 30°C in a micro-tube shaker.
6. Centrifuge at 3,000 × *g* for 5 s.
7. Discard supernatant, and re-suspend cells in 100 µl solution III.
8. Plate cell suspension on appropriate SD plates and incubate for 40 h at 30°C.

3.3. Mating of the Bait to the Library

1. Inoculate a single yeast colony from the bait transformation in 5 ml SD-TRP, and grow overnight at 30°C shaking at 180 rpm (see Note 7).
2. Per screen, inoculate 12 ml of the bait and 12 ml of the library to an OD_{600} of 0.05 from the overnight culture. Grow at 30°C shaking at 180 rpm for 16 h in a closed, disposable 50-ml plastic tube (see Notes 8 and 9).
3. The next day, cultures should have an OD_{600} of 0.8–1.2 when they are used for mating. Mix 12 ml of bait culture and 12 ml of the library culture in disposable 50-ml plastic tube.
4. Spin down the cells for 2 min, 1,000 g, at room temperature, and discard supernatant (see Note 10).
5. Add 24 ml (see Note 11) 20% PEG/YPDA (see Note 12), and re-suspend by vortexing (see Note 13).
6. To allow mating, incubate the cells at 30°C with gentle agitation (100 rpm) for 3 h (see Note 14).
7. Spin down the cells for 2 min at 1,000 g at room temperature, and discard supernatant.
8. Resuspend in 10 ml of SD-Leu-Trp-His, spin 2 min at 1,000 g, and discard supernatant.
9. Resuspend in 200 ml SD-Leu-Trp-His, and add 75 µl 60 mM 4-MU.

3.4. Test Screens (see Note 2)

1. Split the suspension into aliquots of 20 ml.
2. To each aliquot, add varying amounts 3-AT from a 1 M stock. Include the following concentrations: 0, 0.5, 1, 2, 6, 20, and 60 mM (see Note 2).
3. Distribute the cell suspension into seven (unless additional concentrations of 3-AT are tested) 96-well micro-titre plates, 200 µl per well.
4. To determine the mating efficiency, plate an aliquot on SD-Leu-Trp agar plates to measure the number of viable diploid bait- and prey-expressing cells. Do a 1:100 dilution in SD-Leu-Trp-His plate 100 µl.
   Per ten micro-titre plates, screening depth in millions = colony number ×0.2.
5. Put an adhesive plastic seal (or a lid, see Note 14) on the topmost plate and pack the plate stacks in Saran wrap to avoid desicca tion. Incubate for 6 days at 30°C (see Note 15).
6. Measure the fluorescence of the micro-titre plate wells at an excitation of 360 nm and an emission of 465 nm (see Note 16).
7. Analyse the fluorimetric readout in a graphical representation. Determine the number of positive wells (wells which have a fluorescent signal which is significantly higher than the plate

mean, see Note 17). To determine a concentration of 3-AT suitable for screening, choose conditions in which the fraction of positive wells per plate is less than 15. Visually inspect the plates for the presence of colonies in the wells (see Note 18).

3.5. Screens

1. Do the protocol 3.3 of this procedure ("Mating of the bait to the library").
2. Split the suspension into aliquots of 20 ml.
3. To each aliquot, add an optimal amount 3-AT as determined in a test screen from a 1 M stock (see Note 2).
4. Distribute the cell suspension into ten 96-well micro-titre plates, 200 μl per well.
5. To determine the mating efficiency, plate an aliquot on SD-Leu-Trp agar plates to measure the number of viable diploid bait- and prey-expressing cells. Do a 1:100 dilution in SD-Leu-Trp-His plate 100 μl.
   Per ten micro-titre plates, screening depth in millions = colony number ×0.2.
6. Put an adhesive plastic seal (or a lid, see Note 15) on the top-most plate and pack the plate stacks in Saran wrap to avoid desiccation. Incubate for 6 days at 30°C.
7. Measure fluorescence of the micro-titre plate at an excitation of 360 nm and an emission of 465 nm (see Note 16).
8. Visually inspect the plates and analyse the fluorimetric readout in a graphical representation (see Note 18). For screens with an acceptable number of hits, collect all hits whose signals are twofold higher than the plate mean (see Note 17) by re-suspending the cells and transferring 10 μl to a new micro-titre plate with 200 μl SD-Trp-Leu-His per well (see Note 19).
9. Grow the passaged yeast cells for 48 h at 30°. Transfer to a 15-cm SD-Trp-Leu-His agar plate using a plastic 96-pin replicator, and grow for 48 h (see Note 20).

3.6. Amplification of the Library Insert by PCR

1. Dispense 25 μl of 0.25% SDS to each well of a PCR plate (=PCR plate 1).
2. Prepare PCR mix 1 and dispense 50 μl to each well of a second PCR plate (=PCR plate 2).
3. Use 96-pin plastic replicators to pick the colonies from the 15-cm agar plate and re-suspend in PCR plate 1 (see Note 21).
4. Heat to 95°C for 5 min.
5. Transfer 5 μl from PCR plate 1 to PCR plate 2 (see Note 22).
6. Heat to 95°C for 5 min.
7. Do 27 cycles of the following PCR cycle:
   30 s 95°C; 90 s 55°C; 210 s 72°C

8. Prepare PCR mix 2 and dispense 50 µl to each well of a PCR plate (=PCR plate 3).
9. Transfer 1 µl from PCR plate 2 to PCR plate 3.
10. Heat to 95°C for 5 min.
11. Do 20 cycles of the following PCR cycle:

    30 s 95°C; 90 s 58°C; 210 s 72°C

    End by 5 min at 72°C.
12. Analyse 10 µl of the PCR products by 1.5% agarose gel electrophoresis.
13. Discard all PCR reactions which have more than one product (see Note 23) or no product (see Note 24).
14. Collect (see Note 25) and sequence the remaining clones from the 5′ end using pACT2-FP sequencing primer.

3.7. Filtering of Interaction Data

1. Determine the identity of the encoded protein and the 5′ end of the cDNA fragment by comparison to a public sequence database, e.g., RefSeq (see Note 26).
2. Generate a table that displays the identity of the bait, screening conditions, library used, interaction signal, size of the PCR product, and 5′ end of the isolated fragment.
3. Disregard all hits that correspond to 3′ UTRs, that are known as frequently isolated promiscuous preys, or that have been isolated only once with a given bait.
4. Select all remaining protein pairs that have been observed multiple times for further analysis (see Notes 26 and 27).

4. Notes

1. This method works well with libraries in pGAD424 and pGBKT7, with GATEWAY-compatible version of these, as well as with libraries in pACT2 and pGADT7-rec, such as the MATCHMAKER libraries (Clontech, Palo Alto). All libraries have to be pre-transformed in yeast strain Y187.
2. For libraries generated from pools of full-length ORFs, which have a complexity at least 100-fold lower than classical cDNA libraries and which do not contain non-natural peptides, it is more economical to skip the test screens and do all screens at a fixed concentration of 3-AT at first, and only apply test screening if necessary. For baits in pGBT9 and full-length ORF libraries in pGAD424 or related vectors, most screens can be done at 0.4 mM 3-AT.

3. Avoid autoclaving of yeast media, since it deteriorates media quality and can introduce variations in media quality owing to fluctuation in the duration of the heating and cooling times.
4. In our experience, non-sterilised micro-titre plates and lids can be used. We have never encountered any problems of contaminating germs with such plates. For large projects, the use of barcode-labelled micro-titre plates for automatic tracking is a prerequisite.
5. Avoid shock freezing in liquid nitrogen when preparing competent yeast cells. It is best to simply put the yeast at –80°C.
6. This method works with a large variety of bait plasmids. We have done successful screens with pGBT9 and pGBKT7, as well as with pAS2, pGBDU, pGBKCg, and pDB-Leu.
7. When screens are repeated, the initial 5 ml culture of the bait can be stored at 4°C for up to 4 weeks and used to inoculate the overnight cultures for mating.
8. In high-glucose media, yeast cells do not need intensive aeration for growth. Therefore, the overnight cultures of baits and libraries can be conveniently grown in disposable 50-ml plastic tubes with the lid closed. Shaking is only needed to keep the cells in suspension.
9. Regeneration and limited expansion of the library overnight, involving 4–5 cell divisions, increases mating efficiency. Library complexity suffers if this expansion is extended.
10. For optimal mating efficiency and efficient outgrowth of positive cells, avoid cooling of the cells and do not leave cells unsuspended after centrifugation.
11. Cell density at the mating reaction should not be increased, since this negatively affects mating efficiency.
12. Twenty per cent PEG 6000 is optimal for the combination of yeast strains we use. At concentrations higher than 20%, a loss of viability is observed.
13. After re-suspension of the cells in 20% PEG/YPDA, small clumps of cell aggregates should be apparent.
14. Do not extend the mating time, since zygotes divide upon prolonged incubation. Independent isolation of interaction pairs in multiple independent clones is an important criterion for the selection of reliable interactions. Clones that are based on the same zygote and do not represent truly independent isolates artificially increase the number of multiple isolates.
15. The topmost micro-titre plates of a stack are best covered with adhesive plastic seals. Lids can be used instead, but are more expensive. Seals do not need to be permissive for air. Seals for PCR plates are inconvenient, since they stick more tightly than necessary and are hard to remove.

16. For the fluorescence measurement, we strongly recommend to fix the gain of the instrument such that the interaction signals are comparable between different screens. Reading of the plates is an obvious step for acceleration of the method by the application of automated readers equipped with stackers, automated loading systems, and bar-code readers.
17. The plate mean is used as a measure of background fluorescence, i.e., the level of florescence which is generated in a well which does not contain a positive colony. In some screens, all or a very large fraction of wells contain cells which activate the MEL1 reporter and produce a fluorescent signal. For obvious reasons, the plate mean is not a useful measure of negative wells on such plates. If plates are read at a fixed gain, plates in which all wells are positive can be easily distinguished from plates in which none are positive. Also, visual inspection of plates helps here.
18. After the mating reaction, the plates are incubated for 6 days without agitation. Due to the lack of current inside the wells, positive clones grow into colonies similar to colonies on agar plates. This allows for visual judgment of the fraction of wells that contain more than one colony. Note that many contaminating micro-organisms also express alpha-galactosidases. Inspect some of the positive wells for the presence of yeast colonies to exclude artificial positives due to contaminations.
19. In some screens, a large number of colonies are observed that do not activate the *MEL1* reporter, and therefore do not generate a fluorescent signal. These colonies usually do not contain any meaningful interactions, and can be ignored.
20. PCRs from yeast cells do not work well when suspension cultures are used. The passage on agar plates serves to grow the clones into large colonies, which provide a richer source of yeast cells for the PCR reaction.
21. To inoculate the PCR reaction, try to pick as much of the yeast colony as possible with the 96-tip replicator. The more, the better!
22. After heating of the yeast in 0.25% SDS, it is advisable to transfer the clear supernatant and not the slurry of yeast, but do not worry if a bit of the yeast suspension is transferred along.
23. A faint band which runs about 500 bp behind the major product is often observed in these PCR reactions. This band does not pose problems for sequencing.
24. On average, 50% of all colonies yield useful PCR products ($n = 63{,}000$ colonies). However, this varies dramatically from screen to screen. One reason for this is that in some screens a large number of hits can arise from a particular interactor with a long insert, which is amplified less efficiently than short inserts.

25. The second re-arraying step in which the useful PCR products are collected can be omitted. However, the major cost factor of this protocol is the sequencing of the preys. Since only in 50% of cases the PCR reactions give rise to a single DNA fragment that can be sequenced, this second re-arraying step saves money.
26. On average, 80% of PCR fragments yield sequences that allow identification of the prey by BLAST searches in public sequence databases. Overall, for 40% of the collected positive clones, the identity of the prey can be determined.
27. The reading frame of the prey with respect to the GAL4 activation domain is not a straightforward criterion for the selection of relevant interactors (4, 5). For example, we have observed interactions from fragments encoding the same protein expressed in different reading frames. It appears that a shifted reading frame can in some cases still give rise to the natural protein. We speculate that such phenomena are due to yeast's ability to translate through a stop codon with a concomitant frame shift (6).

Acknowledgments

We are grateful to Sabine Asser-Kaiser for critical reading of the manuscript and helpful comments.

References

1. Lamesch, P., Li, N., Milstein, S., Fan, C., Hao, T., Szabo, G., Hu, Z., Venkatesan, K., Bethel, G., Martin, P., Rogers, J., Lawlor, S., McLaren, S., Dricot, A., Borick, H., Cusick, M. E., Vandenhaute, J., Dunham, I., Hill, D. E., and Vidal, M. (2007) hORFeome v3.1: a resource of human open reading frames representing over 10,000 human genes, *Genomics* **89**, 307–315.
2. Aho, S., Arffman, A., Pummi, T., and Uitto, J. (1997) A novel reporter gene MEL1 for the yeast two-hybrid system, *Analytical biochemistry* **253**, 270–272.
3. Albers, M., Kranz, H., Kober, I., Kaiser, C., Klink, M., Suckow, J., Kern, R., and Koegl, M. (2005) Automated yeast two-hybrid screening for nuclear receptor-interacting proteins, *Mol Cell Proteomics* **4**, 205–213.
4. Fromont-Racine, M., Rain, J. C., and Legrain, P. (1997) Toward a functional analysis of the yeast genome through exhaustive two-hybrid screens, *Nature genetics* **16**, 277–282.
5. Giot, L., Bader, J. S., Brouwer, C., Chaudhuri, A., Kuang, B., Li, Y., Hao, Y. L., Ooi, C. E., Godwin, B., Vitols, E., Vijayadamodar, G., Pochart, P., Machineni, H., Welsh, M., Kong, Y., Zerhusen, B., Malcolm, R., Varrone, Z., Collis, A., Minto, M., Burgess, S., McDaniel, L., Stimpson, E., Spriggs, F., Williams, J., Neurath, K., Ioime, N., Agee, M., Voss, E., Furtak, K., Renzulli, R., Aanensen, N., Carrolla, S., Bickelhaupt, E., Lazovatsky, Y., DaSilva, A., Zhong, J., Stanyon, C. A., Finley, R. L., Jr., White, K. P., Braverman, M., Jarvie, T., Gold, S., Leach, M., Knight, J., Shimkets, R. A., McKenna, M. P., Chant, J., and Rothberg, J. M. (2003) A protein interaction map of Drosophila melanogaster, *Science* **302**, 1727–1736.
6. Ivanov, I. P., and Atkins, J. F. (2007) Ribosomal frameshifting in decoding antizyme mRNAs from yeast and protists to humans: close to 300 cases reveal remarkable diversity despite underlying conservation, *Nucleic acids research* **35**, 1842–1858.

Chapter 6

Virus–Human Cell Interactomes

Lionel Tafforeau, Chantal Rabourdin-Combe, and Vincent Lotteau

Abstract

Using global approaches and high-throughput technologies in virology brings a new vision of the infections physiology and allows the identification of cellular factors, mandatory for viral life cycle, that could be targeted by original therapeutic agents. It opens perspectives for the treatment of viral infections by acting on cellular pathways that the virus must use for its own replication. Combining these new molecules with classical antiviral drugs and immunomodulators diversifies and enlarges the antiviral arsenal and contributes to fight drug resistance.

Our laboratory and others are constructing virus–human interactomes to propose a comprehensive analysis of viral infection at the cellular level. Studying these infection maps, where the viral infection can be visualized as perturbation of the human protein–protein interaction network, and identifying the biological functions that are impaired by these perturbations may lead to discovery of new therapeutic targets. These virus–human interaction maps are constructed in a stringent yeast two-hybrid system by screening human cDNA libraries with viral proteins as bait and integrating interactions mined from literature and public databases.

Key words: Yeast two-hybrid screen, Mating, Yeast two-hybrid array, Protein–protein interaction

1. Introduction

The rapidly growing knowledge of protein–protein interaction networks (interactomes) for model organisms and human provides a network-based model to understand molecular and cellular biology. Recently, virus–host relationships also began to be studied at the proteome level by identifying interactions between viral and host-cell proteins (1–5), as reviewed in ref. 6. These virus–host interactomes compose a repertoire of interactions between viral and human proteins that can be analyzed in a network approach. By this mean, viral infections can be viewed as the expression of new constraints imposed by the virus on the cellular interactome.

Bernhard Suter and Erich E. Wanker (eds.), *Two Hybrid Technologies: Methods and Protocols*, Methods in Molecular Biology, vol. 812, DOI 10.1007/978-1-61779-455-1_6, © Springer Science+Business Media, LLC 2012

Study of topological and functional properties that are lost, deregulated, or that emerged in the "infection network" leads to the identification of cellular functions that are mandatory for the virus life cycle. By developing therapeutic agents that act directly on these cellular factors, new ways for treatment of viral infections may emerge.

There are several methods to detect protein–protein interactions. The yeast two-hybrid system is widely used because it requires the manipulation of DNA exclusively (allowing standardization and automation) and because of its efficacy. However, the two-hybrid system is not amenable to detect all protein–protein interactions, since it is based on the nuclear localization of a transcriptional reporter system. Additional approaches need to be used to complement and improve coverage of protein–protein interaction maps generated by yeast two hybrid.

The yeast two hybrid is based on the observation that a transcription factor consists of two separate functional domains: a DNA-binding (DB) domain and a transactivation domain (AD). These two domains are separated and each is fused to a protein of interest (a bait X and a prey Y). Physical interaction between X and Y reconstitutes a transcription factor that binds to responsive elements upstream of a reporter gene and, thus, can activate its transcription (7). In the screening procedure we developed, we use DB and AD from the yeast transcription factor Gal4, *HIS3* as reporter gene, and the mating protocol, in which pretransformed haploid yeast cells form diploids that carry both bait and prey vectors (8).

Here, we describe yeast two-hybrid screens, DNA isolation protocols, and analysis strategies (Fig. 1) that were recently developed to generate the HCV infection map (4). For this interactome approach, 27 HCV ORFs (encoding full-length proteins or domains) were used as baits to screen 2 human cDNA libraries and generate a virus–host cell interaction map composed of 314 virus–host protein interactions. We added interactions retrieved from literature, constructing an HCV–human interaction network composed of 481 interactions involving 11 viral proteins and 421 cellular proteins. For this purpose, we developed VirHostNet, a database gathering virus–host protein interactions from literature and public databases, allowing a comprehensive analysis of virus–human protein interactions through a network view (9).

2. Materials

2.1. Plasmids

1. pGBKT7-gw: pGBKT7 is commercialized by Clontech (10). It is a bait vector, encoding Gal4DB domain (aa1-147) under the control of *ADH1* promoter (see Note 1). A Gateway® cassette was introduced downstream and in frame with Gal4DB

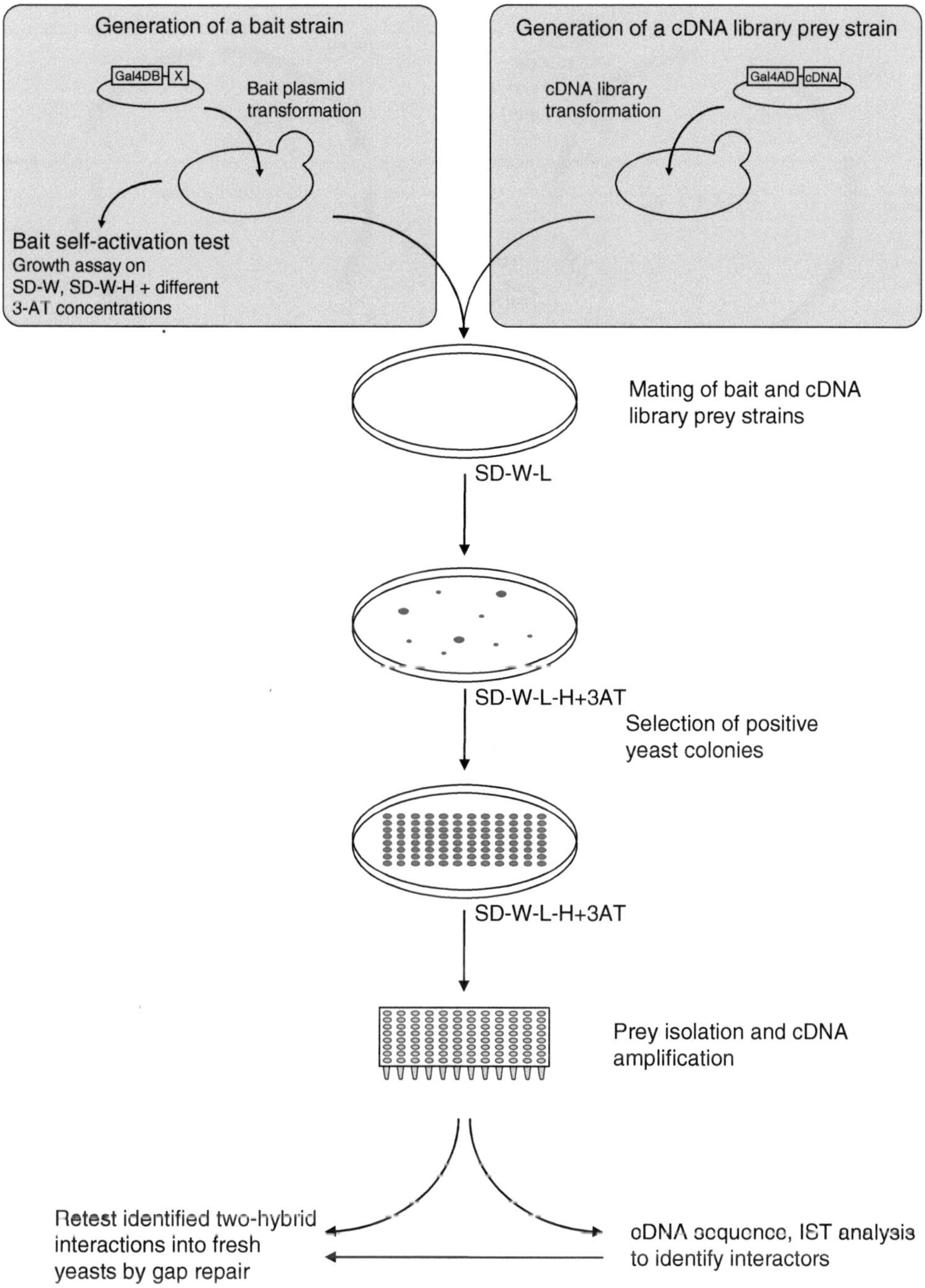

Fig. 1. Flowchart of the two-hybrid screening using the mating strategy. See text for details

to render it a destination vector (kindly provided by Dr. Yves Jacob, Institut Pasteur (11), Fig. 2a, see Notes 2 and 3).

2. pACT2-gw: pACT2 is commercialized by Clontech (12). It is a prey vector, encoding Gal4AD domain (aa 768–881) under the control of *ADH1* promoter (see Note 4). A Gateway® cassette

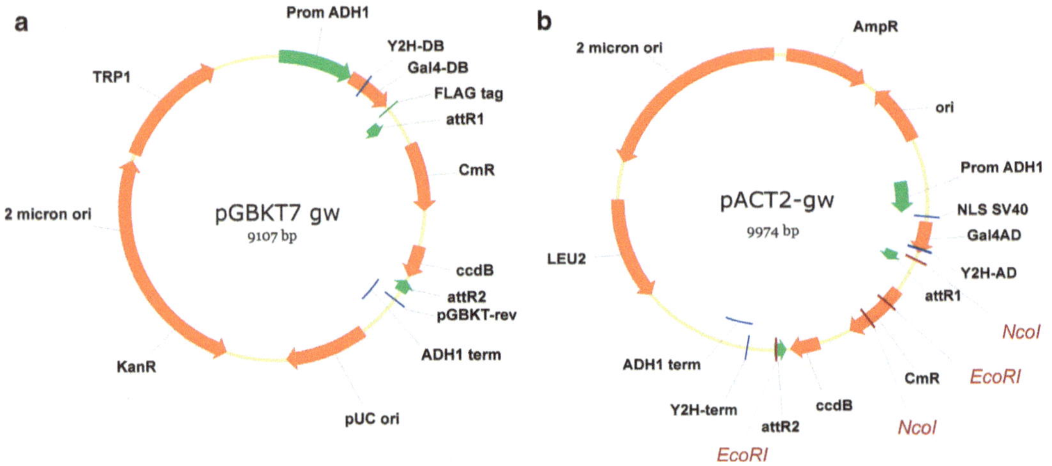

Fig. 2. Bait and prey plasmid maps used for yeast two hybrid. (**a**) pGBKT7-gw, bait vector: it encodes Gal4DB–bait protein of interest. See text for details. (**b**) pACT2-gw, prey vector: it encodes Gal4AD–prey protein. See text for details. Human cDNA libraries are cloned in pACT2-gw using the Invitrogen CloneMiner technique (see Note 5). In red are represented restriction sites used for gap repair.

was introduced downstream and in frame with Gal4AD to render it a destination vector (kindly provided by Dr. Yves Jacob, Institut Pasteur (11), Fig. 2b). We introduced human cDNA libraries into pACT2-gw (see Note 5).

2.2. Yeast Strains

The yeast strains used in our lab are from Clontech. AH109 is *MATa* and Y187 *MATα* (13, 14). The bait vectors are transformed in AH109 and the cDNA libraries (encoded in prey vector, pACT2-gw) are transformed in Y187. By simplicity, AH109 is called the yeast bait strain and Y187 the prey strain.

Genotypes (see Note 6):

AH109: *MATa, trp1-901, leu2-3 112, ura3-52, his3-200, gal4∅, gal80∅, LYS2::UAS*$_{GAL1}$*- TATA*$_{GAL1}$*- HIS3, UAS*$_{GAL2}$*- TATA*$_{GAL2}$*- ADE2, URA3::UAS*$_{MEL1}$*- TATA*$_{MEL1}$*- lacZ.*

Y187: MATα, trp1-901, leu2-3 112, ura3-52, his3-200, ade2-101, met–, gal4∅, gal80∅, URA3::UAS$_{GAL1}$- TATA$_{GAL1}$- lacZ.

2.3. Yeast Media

1. Rich medium – YPAD: Allowing propagation of AH109 and Y187, also used for the mating step during the yeast two-hybrid screen. YPAD (1% yeast extract, 2% peptone, 2% dextrose, 0.004% adenine) is dissolved in deionized water and autoclaved for 20 min at 120°C. Solid media are made by adding 2% agar.
2. Selective medium – SD: Used for selection of yeast transformants and detection of DB-X and AD-Y interactions. SD (8 g/l SD mix, 2% dextrose) is dissolved in deionized water and autoclaved for 20 min at 120°C. This medium does not contain

histidine, leucine, and tryptophan. Add 4 ml of stock solution of each required amino acids after autoclave; see ref. 5.

3. SD mix powder for 20 l: 34 g YNB (w/o amino acids and $(NH_4)_2SO_4$), 26 g amino acids mix, 100 g $(NH_4)_2SO_4$. Mix well to homogenize the powder.
4. Amino acids mix: Mix 6 g of each amino acid: alanine, arginine, asparagine, cysteine, glutamate, glutamine, glycine, isoleucine, lysine, methionine, phenylalanine, proline, serine, threonine, tyrosine, and valine. Add 6 g of adenine sulfate. Mix well to homogenize.
5. Stock solutions of histidine: 100 mM, leucine: 100 mM and tryptophan: 40 mM. Autoclave all, but tryptophan solution (heat labile): filter sterilize using 0.22 μm filter. Histidine and leucine solution is stored at RT. Tryptophan solution should be stored in the dark at 4°C.
6. 3-AT: 2 M stock solution. Filter sterilize using 0.22 μm filter. Store in the dark at 4°C.

2.4. Transformation Solutions

1. LiAc/TE: 100 mM LiAc, 10 mM Tris, pH 7.5, 1 mM EDTA from sterile stock solutions (1 M LiAc and 10× TE (100 mM Tris, pH 7.5, 10 mM EDTA); see Note 7).
2. LiAc/TE/PEG: 100 mM LiAc, 10 mM Tris, pH 7.5, 1 mM EDTA from sterile stock solutions (as above) and 40% PEG from a sterile 50% solution (see Note 8).
3. ssDNA: Deoxyribonucleic acid, single stranded from salmon testes (Sigma D-9156). This DNA is single-stranded denaturated at a concentration of 10 mg/ml.

2.5. Primers

Y2H-AD: 5′-CGATGATGAAGATACCCCACCAAA.
Y2H-Term: 5′-ACCAAACCTCTGGCGAAGAA.
Y2H-DB: 5′-GCACATCTGACAGAAGTGGA.
pGKBT7-rev: 5′-GAAATTCGCCCGGAATTAGC.

3. Methods

3.1. Yeast Two-Hybrid Library Screening Using the Mating Strategy

A bait strain is generated and self-activation is tested. Then, a cDNA library is transformed in a prey yeast strain, which allows interaction screening by mating with the bait strain. Selection of positive interactors leads to their identification by sequence analysis. These positive interactors are then retested into fresh yeasts by gap repair (schemed in Fig. 1).

A prerequisite step is to transform AH109 yeast strain with the bait vector (i.e., pGBKT7 bait of interest). The bait-encoding sequence, already cloned into a pDONR, is transferred into pGBKT7-gw by an LR reaction (Invitrogen, Gateway® recombination system). This highly efficient cloning method eliminates the need for gene-specific manipulations, such as restriction enzymes. The bait is tested for basal self-activation of *HIS3* reporter gene (see below) and toxicity in yeast. If the bait is not toxic for yeast and if it does not self-activate the expression of the reporter gene, it can be further used in a two-hybrid screen.

3.1.1. Bait Plasmid Transformation

1. Patch AH109 on a YPAD plate and incubate overnight at 30°C.
2. Inoculate 5 ml of YPAD with this fresh yeast culture and incubate overnight at 30°C under agitation (see Note 9).
3. Inoculate the yeast strain at O.D.$_{600\ \mathrm{nm}}$ 0.2 in 5 ml YPAD (5 ml of culture is needed for each plasmid to be transformed, inoculate a minimum of 20 ml). Incubate at 30°C under agitation until it reaches an O.D. of 0.6.
4. Transfer in a 50 ml conical tube and centrifuge for 5 min at 3,000 × *g* at RT. Discard the supernatant. Wash the pellet with an equal amount of sterile water. Discard the supernatant.
5. Resuspend yeast cells with 1 ml LiAc/TE solution (for each transformation to be done) and transfer in 1.5 ml tubes. Centrifuge for 30 s at 6,000 × *g* at RT and discard the supernatant.
6. Resuspend each pellet in 100 μl LiAc/TE.
7. Add to each pellet 2 μl of ssDNA and 1 μg of DNA bait vector to be transformed.
8. Mix by pipetting two times and incubate for 10 min at RT.
9. Add 230 μl LiAc/TE/PEG, and mix by inverting two times the tube. Incubate for 30 min at 30°C.
10. Add 43 μl DMSO and heat shock for 5 min at 42°C (mix by inversion from time to time).
11. Pellet the cells (30 s at 6,000 × *g*), discard the supernatant, and resuspend in 150 μl water.
12. Plate yeast cells on a 10-cm SD-W plate using sterile glass beads, and incubate at 30°C for 2–3 days (see Note 10).

3.1.2. Bait Basal Self-Activation

Bait strains should be examined for the bait self-activation prior to two-hybrid analysis. Self-activation is defined as a detectable DB-X-dependent reporter gene activation in the absence of any AD-Y prey protein, allowing the bait strain to grow on medium lacking histidine. Weak to intermediate self-activator DB-X can be used in two-hybrid experiments by titrating the basal yeast growth by using 3-AT, whereas strong self-activators must be discarded. Self-activation is examined on plates containing different concentrations of 3-AT (see Note 11).

1. Prepare selective plates lacking tryptophan and histidine and containing 0, 2.5, 5, 10, 15, 20, and 30 mM 3-AT.
2. Inoculate each bait strain in 5 ml SD-W and incubate overnight at 30°C under agitation.
3. Dilute each bait strain to a O.D.≈0.2 in SD-W and drop 5 µl of each bait strain on SD-W, SD-W-H, SD-W-H + increasing 3-AT concentration. Let the drops dry, and then incubate the plates for 2–5 days at 30°C. Leave the SD-W plate at day 1 and check others regularly.
4. Score the 3-AT plates. For bait strains that confer growth on SD-W-H + 30 mM 3-AT, one can test higher concentrations of 3-AT (up to 100 mM). Above this concentration, bait should be discarded for yeast two-hybrid screen: it is not able to discriminate between self-activation and positive interactions with an AD-Y partner during a screen. For bait strains that show a lower level of self-activation, i.e., no growth at 30 mM but at lower 3-AT concentrations, two-hybrid readouts are performed albeit at a slightly higher concentration of 3-AT. For bait strains that do not show any self-activation, e.g., that do not grow even on SD-W-H, we used to screen them on medium containing 5 mM 3-AT to select only positive interactions.

3.1.3. Transformation of the cDNA Library

1. Inoculate 50 ml of YPAD with a patch of freshly grown Y187 prey strain, and incubate overnight at 30°C under agitation.
2. Inoculate the yeast strain at O.D. 0.2 in 200 ml YPAD. Incubate at 30°C under agitation until it reaches an O.D. of 0.6.
3. Transfer in 50 ml conical tubes and centrifuge for 5 min at 3,000 × *g* at RT. Discard the supernatant. Wash the pellet with an equal amount of sterile water. Discard the supernatant. Wash the yeast cells in 50 ml LiAc/TE solution. Discard the supernatant.
4. Resuspend the yeasts in 10 ml LiAc/TE and transfer in six 15 ml conical tubes. Centrifuge for 5 min at 3,000 × *g* at RT and discard the supernatant.
5. Resuspend each pellet in 100 µl LiAc/TE.
6. Add sequentially in each tube 160 µl ssDNA, 40 µg of pACT2-cDNA library, 200 µl DMSO, and 10 ml LiAc/TE/PEG.
7. Mix by inverting the tubes 5–6 times, and incubate for 30 min at 30°C.
8. Heat shock for 15 min at 42°C (in a water bath, mix by inversion every 5 min).
9. Pellet the yeasts (5 min at 3,000 g), discard the supernatant, and resuspend the cells in 10 ml sterile water (for each 15 ml tube).
10. Plate 300 µl on large SD-L plates (200 plates). Incubate at 30°C for 4 days (see Note 12).

3.1.4. Harvesting and Pooling the Transformed Library

The transformed library is pooled in slurry, then aliquoted, and frozen. Each aliquot is a representative of the set of primary transformants and is used in subsequent mating.

1. Prepare 400 ml YPAD containing 20% glycerol, and autoclave it (see Note 13).
2. Pour 2 ml YPAD + glycerol per plate and scrap the yeast colonies using a spreader (see Notes 14 and 15). Collect the yeast slurry from each plate and pool in 50 ml conical tubes.
3. Resuspend the cells and aliquot by 1.8 ml in 2 ml sterile tubes, and freeze at −80°C.
4. Determine the transformed library titer by plating several dilutions (10^6, 10^7, and 10^8) of a thawed aliquot on SD-L plates. Count the CFU for each dilution and determine the number of yeast cells/milliliter of the transformed library. It should be around 10^9 cells/ml.

3.1.5. Two-Hybrid Screening by Mating Bait and Prey Yeast Strains

As bait and prey strains are ready, the next step is to mate both strains. For this purpose, the bait strain is mixed with a thawed aliquot of the prey strain. The cells are plated on rich media for 4–5 h to induce mating of cells of opposed mating type and forming diploids (see Note 16). The diploids cells, along with haploids (unmated yeasts), are collected and plated on interaction selection medium (SD-W-L-H + 3 AT). On this medium, only diploids, where an interaction between bait and prey takes place, are able to grow.

1. Inoculate 50 ml of SD-W with a patch of freshly grown bait strain, and incubate overnight at 30°C under agitation (see Note 17).
2. Thaw an aliquot of the prey strain, and take 60×10^7 cells that are incubated in 20 ml YPAD for 10 min at 30°C (in a 50 ml conical tube).
3. Measure the O.D. of the bait culture (a $O.D._{600\ nm}$ of 1 corresponds to 10^7 cells/ml), and collect the equivalent of 72×10^7 cells. Add this volume to the prey culture, and mix by inverting.
4. Centrifuge at 3,000 × *g* for 5 min at RT. Discard the supernatant and resuspend the yeasts in 1.1 ml YPAD (the yeast pellet corresponds approximatively to 400 μl).
5. Plate on three large YPAD plates (500 μl per plate), and incubate at 30°C for 4–5 h.
6. Scrap the yeasts using a spreader with 2 ml SD-W-L-H per plate. Collect the slurry and plate on ten selective medium (SD-W-L-H + appropriate concentration of 3-AT, according to the made basal self-activation test), thus 500 μl per plate (see Note 18).

7. Incubate for 6–10 days at 30°C. From day 6, inspect the plates daily and mark the location of colonies using a given color and each day mark the arising colonies with different colors. At day 10, collect the colonies (following the order of appearance, generally the first appeared are the biggest) using toothpicks and patch them in a 96-well format on a selective medium plate. Incubate for 48 h at 30°C and replica plate twice on selective medium (see Note 19).

3.1.6. Prey Isolation and cDNA Identification

AD prey-encoding sequences can be retrieved by PCR amplification directly from yeast colonies. For this purpose, we use primers that hybridize in Gal4-AD sequence and in *ADH1* Term (Y2H-AD and Y2H-Term, respectively; see Subheading 2). We perform yeast colony PCR directly on yeast colonies arrayed on a plate in a 96-well format.

1. Replica patch the yeasts, and incubate the plate during one day at 30°C (see Note 20).
2. Prepare on ice a PCR mix for a plate ($\approx$ 100 PCR): 1,000 μl GoTaq (Promega) 5× buffer, 300 μl $MgCl_2$ (supplied with GoTaq), 20 μl dNTP's (solution stock 20 mM), 15 μl Y2H-AD primer 100 μM, 15 μl Y2H-Term primer 100 μM, 3,130 μl distilled water, and 20 μl GoTaq (5 U/μl).
3. Transfer 45 μl aliquots of the PCR mix into each well of a 96-well PCR plate.
4. Using a sterilized pin-tool replicator, pick and resuspend 96 yeasts into 50 μl of water (dispensed in a 96-well PCR plate) and incubate at 100°C for 5 min.
5. Add 5 μl of boiled yeasts to the PCR mix and run the PCR with the following program: (a) 94°C, 5 min; (b) 94°C, 30 s; (c) 54°C, 45 s; (d) 72°C, 5 min (see Note 21); (e) 72°C, 3 min by repeating steps b–d 34 more times.
6. Load 5 μl on a 1% agarose gel (see Note 22). If there is an amplification band, we proceed for its sequencing.

3.2. Yeast Two-Hybrid Array Screening

In a two-hybrid array experiment or also called "two-hybrid matrix," pairs of interactions are examined individually by mating bait and prey yeast strains. After generation of diploids, yeasts representing single potential interaction pairs are dropped onto selective media and examined for two-hybrid phenotype. Such matrix experiments are limited by the number of pairwise combinations that can be reasonably tested (from a few hundreds when working manually to a few thousands when having a multichannel pipetting robotic platform). Since the baits' and preys' identity is known throughout the experiment, no sequencing step is needed to identify them, making this method rapid and cheap. But it requires that all bait and prey protein-encoding sequences are already cloned.

Another advantage of the array screen is its parallel character, allowing to compare baits' as well as preys' global responses. If different preys have different affinities to a given bait, it will be indicated by different colony growth on selective plates.

3.2.1. Retesting Identified Two-Hybrid Interactions

An array strategy can be used for the confirmation of interactions identified during a screen. In this case, preys are retested against the bait that trapped them during the two-hybrid screen. This array protocol is specific because prey plasmids are reconstituted by "gap repair" into a yeast strain already containing the bait plasmid, and the two-hybrid interaction phenotype scoring is conducted into a haploid strain. The gap repair is based on the homologous recombination that occurs in vivo in yeast and allows the reconstitution of the prey plasmid due to the identity between the PCR fragment and the pACT2-gw at the priming sites (15). The yeast strain is transformed by linearized prey plasmid and AD prey sequences obtained by PCR.

All interactors found (after identification of interaction sequence tag (IST) by BLAST, see below) are retested against the bait protein as well as the empty bait vector using this array screening method. Recombinant plasmids are selected onto SD-W-L plates.

DNA preparation

1. Linearize pACT2-gw using *NcoI* and *EcoRI* (plasmid purification is not necessary); 2.5 μg of vector are necessary for 96 gap repairs.
2. Per reaction, transform ≈ 25 ng linearized pACT2-gw and 3–5 μl PCR product in a 96-well plate format (see Notes 23 and 24).

Transformation

1. Inoculate 5 ml SD-W with fresh bait strain and incubate overnight at 30°C under agitation.
2. Inoculate the yeast strain at O.D. 0.2 in 200 ml SD-W. Incubate at 30°C under agitation until it reaches an O.D. of 0.6.
3. Transfer in 50-ml conical tubes and centrifuge for 5 min at 3,000 × *g* at RT. Discard the supernatant. Wash the pellet with an equal amount of sterile water. Discard the supernatant.
4. Wash the yeast cells in 50 ml LiAc/TE solution and pool in a single 50 ml conical tube. Discard the supernatant.
5. Wash the yeasts in 10 ml LiAc/TE solution. Discard the supernatant.
6. Resuspend the pellet in 2.5 ml LiAc/TE. Add 300 μl ssDNA. Transfer 30 μl of the suspension in each well of a 96-well PCR plate containing DNAs (linearized pACT2 and PCR product).

7. Add 100 μl of LiAc/TE/PEG solution in each well and resuspend carefully.
8. Incubate for 30 min at 30°C.
9. Heat shock for 15 min at 42°C.
10. Centrifuge the 96-well plate for 5 min at 1,500 × *g* at RT. Remove the supernatant using a multichannel pipette. Add 110 μl of water to each well and directly remove 100 μl (see Note 25).
11. Resuspend the yeasts in the remaining volume and spot the cells on an SD-W-L plate. Incubate overnight at 30°C.
12. Replicate softly the plate, and incubate for 2 days at 30°C (see Note 26).

    Compare the number of transformants with and without PCR products (linearized pACT2-gw without PCR product). The gap repair is considered successful when at least a minimum fivefold induction is observed.

Two-hybrid interaction phenotype scoring

1. From the master plate containing all the yeast transformants, inoculate a 96-well deep-well plate (each well containing 1.2 ml SD-W-L) using a pin-tool replicator. Incubate overnight at 30°C under agitation.
2. Dilute five times the yeast cultures and drop 5 μl using a multichannel pipette on large plates in a 96-well format: SD-W-L, SD-W-L-H, SD-W-L-H containing increasing amounts of 3-AT (classically: 2.5, 5, 10 mM 3-AT) (see Note 27).
3. Let the drops dry, and then incubate the plates for 2–5 days at 30°C. Leave the SD-W-L plate at day 1 and check others regularly.
4. Score the 3-AT plates and compare to controls (see Note 28).

3.2.2. Array Screening

Arrays can also be used as a general tool to screen a defined subset of proteins as a family of proteins or the members of a given cellular pathway. A prerequisite step is to clone the protein-encoding sequences to be tested. In our lab, this procedure has been used to screen whole-viral proteomes against a specific cellular pathway (for example, autophagy or innate immunity pathway that are both functionally related to viruses).

Generation of a prey strains collection.

Clone a set of preys into the pACT2-gw from entry clones using the Gateway® LR recombination and transform the Y187 yeast strain by these prey plasmids. In our particular case, preys are cellular proteins and baits viral proteins. Following the number of transformation, one can use the simple transformation protocol, as described in Subheading 3.1. (Bait plasmid transformation) or the 96-well transformation protocol described in Subheading 3.2.

(Preparing a prey array using gap repair). In this case, yeasts are transformed by prey plasmids without the need of a gap repair.

Dispose the prey yeasts collection in a 96-array format on SD-L plates. Include per plate at least a yeast strain, including empty prey vector.

Generation of a bait strains collection.

By the same way, clone a set of baits into the pGBKT7-gw from entry clones using the Gateway® LR recombination and transform the AH109 yeast strain by these bait plasmids. In our case, the baits are viral proteins and many of them are already cloned in pDONR vectors (see viral ORFeome database, http://www.viralorfeome.com (16)).

Dispose the bait yeasts collection in a 96-array format on SD-W plates. Include at least a yeast strain, including empty bait vector.

Screening the prey array with different baits.

When both prey and bait arrays are consolidated, the goal is to mate each bait against each prey strain.

Mating

1. Incubate the yeast prey array overnight in SD-L using a 96-well deep-well plate (each well containing 1.2 ml SD-L) inoculated by a pin-tool replicator under agitation at 30°C.
2. Grow each yeast bait colony overnight in 2 ml SD-W under agitation at 30°C (see Note 29).
3. Dilute five times the bait culture in a sterile reservoir and fill each well of a 96-well microtiter plate with 25 μl of the yeast bait suspension.
4. Transfer 5 μl of the prey array to the bait microtiter, mix the yeast suspension, and drop 5 μl of each mix on a large YPAD plate. Let the drops dry and then incubate overnight at 30°C. The bait and prey strains conjugate on this medium, forming diploids.
5. Using a velvet, replica plate the mated yeasts on a large SD-W-L plate. Incubate for 2 days at 30°C. This step selects diploids and ensures that all colonies have mated properly.

Screening the prey array

1. Inoculate the diploid colonies in a 96-well deep-well plate filled of SD-W-L (1.2 ml per well) using a pin-tool replicator, and incubate overnight at 30°C under agitation.
2. Dilute five times the cultures (in a 96-well microtiter plate) and drop 5 μl on selective plates: SD-W-L, SD-W-L-H, SD-W-L-H + increasing amounts of 3-AT (2.5, 5, and 10 mM 3-AT).
3. Let the drops dry, and then incubate the plates for 2–6 days at 30°C. Leave the SD-W-L plate at day 1 and check others regularly.

Score the 3-AT plate and compare to the controls.

3.3. IST Analysis

The two-hybrid screens generate a large amount of ISTs obtained by sequencing all AD cDNA isolated from yeast colonies (17). To establish a standardized system to analyze these ISTs, we developed pISTil, a bioinformatics pipeline combined to a user-friendly Web interface (see Note 30) (18).

The pISTil system is highly flexible and allows (a) systematic and fast assignation of ISTs to a unique protein accession number; (b) annotation of "in-frame" or not-in-frame ISTs; (c) manual checking and visualization of annotated ISTs through a user-friendly Web interface; and (d) export of ppi in multiple formats, such as MIMIx standard format (19).

IST chromatogram files, in Applied Biosystems INC (ABI) or Standard Chromatogram Format (SCF) formats, are filtered by Phred-pregap4 software (20) to extract nucleic sequences and their associated quality values. The resulting nucleic sequence of each IST is then translated into three frames and aligned against a protein sequence database (as defined in the configuration file by users, typically Ensembl or NCBI databases) by using BLASTX alignment software (21). Only alignment information for the best hit is subsequently retained. In addition, identification of Gal4AD on ISTs allows the true delineation of in-frame and not-in-frame ISTs. Thus, this filter can discard putative not-in-frame ISTs that may lead to false-positive interaction. All information generated at each step of the IST pipeline is stored into the pISTil database, such as sequence quality, percent identity, e-value, alignment position, frame, and protein sequence database source. Other data supplied by users, such as bait protein used for the screen (GenBank accession number) or description of cDNA libraries (host organism, tissue origin, cell type), can be integrated to the pISTil database. A demonstration of the Web site capabilities is available at http://www.pbildb1.univ-lyon1.fr/pistil.

4. Notes

1. The *ADH1* promoter of pGBKT7 is a truncated form (700 bp), leading to relatively high expression level of Gal4DB–bait fusion protein (22). This plasmid harbors a *TRP1* yeast marker, allowing its selection on medium lacking tryptophan, and the *KanR* marker, for its selection in bacteria. It possesses the 2 μ yeast replication origin (multicopy plasmid).
2. We are using the Gateway® recombinational cloning system (Invitrogen, (23, 24)) in order to clone viral ORFs and to construct viral ORFeomes (see our LIMS, viralORFeome at http://www.viralorfeome.com (16)). Using these viral ORFs entry clones, we can recombine them by an LR reaction into pGBKT7-gw.

3. We swapped the original c-myc epitope for a FLAG epitope between Gal4DB and the Gateway® cassette.
4. The *ADH1* promoter of pACT2 is a truncated form (410 bp), leading to medium expression level of Gal4AD–prey fusion protein (25). It harbors the *LEU2* yeast marker, allowing transformed yeasts to grow on medium lacking leucin, and the *AmpR* marker, for its selection in bacteria. It also possesses the 2 μ yeast replication origin (multicopy plasmid).
5. cDNA are obtained from mRNA extracted from tissues or cells by using a polyA primer fused to *attB2* sequence and a linker containing *attB1* sequence. Then, they are cloned into a pDONR vector by a BP recombination (Invitrogen, Gateway® CloneMiner cDNA library construction kit, ref: 18249–029). Next, the whole cDNA library is swapped into the pACT2-gw destination vector using an LR recombination. Using the Gateway® recombination system to clone cDNA avoids the use of restriction enzymes, and thus avoids cutting cDNA into fragments.
6. The genes encoding Gal4 and Gal80 (its inhibitor) are deleted in both yeast strains. AH109 possesses three reporter genes: *HIS3*, *ADE2*, and *LacZ* under control of different promoters (UAS*GAL TATAGAL*), whereas Y187 possesses one reporter gene, *LacZ*. We decided to only analyze the transactivation phenotype of the *HIS3* reporter gene, which encodes imidazole glycerolphosphate dehydratase or, more simply, His3 enzyme. A competitive inhibitor has been described: 3-amino 1,2,4 triazole (3-AT) that can be used to inhibit low levels of *HIS3* expression, and thus to suppress background growth (26, 27). It can also be used when baits show basal self-activation and titrate His3 that is produced by this basal activation.
7. For better transformation efficiency (for instance, for high-throughput transformation as a library transformation), this solution should be freshly prepared.
8. For 100 ml solution, weigh 50 g of polyethylene glycol, MW 3350 (Sigma, P-3640), in a 150 ml glass beaker and add 35 ml of distilled water. Stir with a magnetic bar until dissolved (can be heated to accelerate dissolution). Transfer the liquid to a 100-ml graduated cylinder. Rinse the beaker with a small amount of distilled water, add this to the graduated cylinder containing the PEG solution, and bring the volume to exactly 100 ml. Mix well by inversion. Autoclave for 20 min at 120°C.
9. Yeast culture and selection are not subjected to antibiotics resistance, as for bacteria. All the experiments should be done under sterile conditions.
10. Glass beads are used to plate yeast cells, as well as bacteria. Glass beads, 5 mm diameter, 1 kg (Fisher Scientific, ref: W0130M).

11. It is useful to have controls for this self-activation test. As negative control, use the bait strain transformed by pGBKT7 empty. As positive control, one can use a bait strain containing a couple of proteins that are known to interact (DB-X and AD-Y).
12. Plate 10^4, 10^5, and 10^6 dilutions on SD-L plates to obtain the transformation efficiency. A good transformation efficiency should yield $1–5 \times 10^4$ yeast colonies/µg DNA, thus $1–5 \times 10^4$ colonies per plate.
13. We add 10 µg/ml tetracyclin in YPAD in order to avoid microbial contamination of the library.
14. Spreaders are easily made by heating and bending a 20-cm length of 3 mm-diameter glass rod. Using a Bunsen burner, both ends are flame polished, and then the rod is heated and bent approximately 3 cm from one end to form an angle. Alternatively, one can also make spreaders from Pasteur pipettes. The spreader is sterilized by briefly dipping it in alcohol and passing it through the flame of a Bunsen burner.
15. More material is collected if the scrapping is repeated twice but, in that case, pour 2×1 ml of YPAD + 20% glycerol per plate.
16. This incubation period is an optimized time for mating and for limiting further growth of diploid cells (28).
17. We used to preinoculate 5 ml SD-W with a patch of bait strain, incubate it overday, and then dilute this preculture in 50 ml for an overnight culture.
18. At this step, it is important to evaluate the number of diploids that were formed and thus screened. For this purpose, plate different dilutions of the diploids mix (10^4, 10^5, and 10^6) on SD-W-L plates (unmated yeasts do not grow on this medium). Count the colonies that grow after 2–3 days at 30°C and determine the titer of the mated cells.
19. It is of note that yeast cells can maintain multiple distinct plasmids. When transformed by a cDNA library, the prey strain could be transformed by more than one prey plasmid. Then, when a positive colony is selected during the two hybrid screen, it could in fact possess more than one prey plasmid, where one giving the positive interaction and the others are simply contaminating. The double round of growth on selection medium maintains positive selection of DB-bait and AD-prey, whereas eventual contaminating prey plasmids are not selected and thus, can be lost (29).
20. Fresh cells give rise to the highest level of success in yeast colony PCR.
21. The elongation time is critical to obtain amplification bands.
22. We use precast E-gel 48.1% agarose (Invitrogen), highly convenient for high-throughput DNA manipulations (compatible

with the utilization of multichannel pipettes). E-gels 96 are also convenient, but the migration distance is less than for E-gel 48.

23. Controls should include "no DNA," circular pACT2-gw, and linearized pACT2-gw without PCR product.
24. The 96-well transformation protocol is very useful when dealing with hundreds to thousands of yeast transformations. It can easily be achieved using a multichannel pipetting robotic platform, except for the centrifugation step. In our lab, we use a Tecan® Freedom Evo platform.
25. This step allows elimination of most of the transformation mix containing PEG.
26. This replica-cleaning step is necessary to dilute out the non-transformed cells present in each spot.
27. When dealing with hundreds of tests, we used our multichannel pipetting robotic platform to drop yeast cultures onto selective plates, with special labware able to support three large petri dishes.
28. Controls should include prey proteins against the empty bait vector and bait proteins against the empty prey vector (circular pACT2-gw, from the "gap repair" transformation). Some prey proteins may show self-activation: transformed yeasts grow on selective reporter media, even those containing empty bait vector. In this case, one should discard these interactions.
29. If using dozens of bait strains, incubate them in 96-well deep-well plates.
30. pISTil is publicly available for download at http://www.sourceforge.net/projects/pISTil.

References

1. Uetz, P., Dong, Y. A., Zeretzke, C., Atzler, C., Baiker, A., Berger, B., Rajagopala, S. V., Roupelieva, M., Rose, D., Fossum, E., and Haas, J. (2006) Herpesviral protein networks and their interaction with the human proteome, *Science* 311, 239–242.
2. Calderwood, M. A., Venkatesan, K., Xing, L., Chase, M. R., Vazquez, A., Holthaus, A. M., Ewence, A. E., Li, N., Hirozane-Kishikawa, T., Hill, D. E., Vidal, M., Kieff, E., and Johannsen, E. (2007) Epstein-Barr virus and virus human protein interaction maps, *Proc Natl Acad Sci USA* 104, 7606–7611.
3. Konig, R., Zhou, Y., Elleder, D., Diamond, T. L., Bonamy, G. M., Irelan, J. T., Chiang, C. Y., Tu, B. P., De Jesus, P. D., Lilley, C. E., Seidel, S., Opaluch, A. M., Caldwell, J. S., Weitzman, M. D., Kuhen, K. L., Bandyopadhyay, S., Ideker, T., Orth, A. P., Miraglia, L. J., Bushman, F. D., Young, J. A., and Chanda, S. K. (2008) Global analysis of host-pathogen interactions that regulate early-stage HIV-1 replication, *Cell* 135, 49–60.
4. de Chassey, B., Navratil, V., Tafforeau, L., Hiet, M. S., Aublin-Gex, A., Agaugue, S., Meiffren, G., Pradezynski, F., Faria, B. F., Chantier, T., Le Breton, M., Pellet, J., Davoust, N., Mangeot, P. E., Chaboud, A., Penin, F., Jacob, Y., Vidalain, P. O., Vidal, M., Andre, P., Rabourdin-Combe, C., and Lotteau, V. (2008) Hepatitis C virus infection protein network, *Mol Syst Biol* 4, 230.

5. Zhang, L., Villa, N. Y., Rahman, M. M., Smallwood, S., Shattuck, D., Neff, C., Dufford, M., Lanchbury, J. S., Labaer, J., and McFadden, G. (2009) Analysis of vaccinia virus-host protein-protein interactions: validations of yeast two-hybrid screenings, *J Proteome Res* 8, 4311–4318.
6. Bailer, S. M., and Haas, J. (2009) Connecting viral with cellular interactomes, *Curr Opin Microbiol* 12, 453–459.
7. Fields, S., and Song, O. (1989) A novel genetic system to detect protein-protein interactions, *Nature* 340, 245–246.
8. Fromont-Racine, M., Rain, J. C., and Legrain, P. (1997) Toward a functional analysis of the yeast genome through exhaustive two- hybrid screens, *Nat Genet* 16, 277–282.
9. Navratil, V., de Chassey, B., Meyniel, L., Delmotte, S., Gautier, C., Andre, P., Lotteau, V., and Rabourdin-Combe, C. (2009) VirHostNet: a knowledge base for the management and the analysis of proteome-wide virus-host interaction networks, *Nucleic Acids Res* 37, D661–D668.
10. Louvet, O., Doignon, F., and Crouzet, M. (1997) Stable DNA-binding yeast vector allowing high-bait expression for use in the two-hybrid system, *Biotechniques* 23, 816–818, 820.
11. Gholami, A., Kassis, R., Real, E., Delmas, O., Guadagnini, S., Larrous, F., Obach, D., Prevost, M. C., Jacob, Y., and Bourhy, H. (2008) Mitochondrial dysfunction in lyssavirus-induced apoptosis, *J Virol* 82, 4774–4784.
12. Li, L., Elledge, S. J., Peterson, C. A., Bales, E. S., and Legerski, R. J. (1994) Specific association between the human DNA repair proteins XPA and ERCC1, *Proc Natl Acad Sci USA* 91, 5012–5016.
13. James, P., Halladay, J., and Craig, E. A. (1996) Genomic libraries and a host strain designed for highly efficient two-hybrid selection in yeast, *Genetics* 144, 1425–1436.
14. Harper, J. W., Adami, G. R., Wei, N., Keyomarsi, K., and Elledge, S. J. (1993) The p21 Cdk-interacting protein Cip1 is a potent inhibitor of G1 cyclin-dependent kinases, *Cell* 75, 805–816.
15. Petermann, R., Mossier, B. M., Aryee, D. N., and Kovar, H. (1998) A recombination based method to rapidly assess specificity of two-hybrid clones in yeast, *Nucleic Acids Res* 26, 2252–2253.
16. Pellet, J., Tafforeau, L., Lucas-Hourani, M., Navratil, V., Meyniel, L., Achaz, G., Guironnet-Paquet, A., Aublin-Gex, A., Caignard, G., Cassonnet, P., Chaboud, A., Chantier, T., Deloire, A., Demeret, C., Le Breton, M., Neveu, G., Jacotot, L., Vaglio, P., Delmotte, S., Gautier, C., Combet, C., Deleage, G., Favre, M., Tangy, F., Jacob, Y., Andre, P., Lotteau, V., Rabourdin-Combe, C., and Vidalain, P. O. (2010) ViralORFeome: an integrated database to generate a versatile collection of viral ORFs, *Nucleic Acids Res* 38, D371–D378.
17. Li, S., Armstrong, C. M., Bertin, N., Ge, H., Milstein, S., Boxem, M., Vidalain, P. O., Han, J. D., Chesneau, A., Hao, T., Goldberg, D. S., Li, N., Martinez, M., Rual, J. F., Lamesch, P., Xu, L., Tewari, M., Wong, S. L., Zhang, L. V., Berriz, G. F., Jacotot, L., Vaglio, P., Reboul, J., Hirozane-Kishikawa, T., Li, Q., Gabel, H. W., Elewa, A., Baumgartner, B., Rose, D. J., Yu, H., Bosak, S., Sequerra, R., Fraser, A., Mango, S. E., Saxton, W. M., Strome, S., Van Den Heuvel, S., Piano, F., Vandenhaute, J., Sardet, C., Gerstein, M., Doucette-Stamm, L., Gunsalus, K. C., Harper, J. W., Cusick, M. E., Roth, F. P., Hill, D. E., and Vidal, M. (2004) A map of the interactome network of the metazoan C. elegans, *Science* 303, 540–543.
18. Pellet, J., Meyniel, L., Vidalain, P. O., de Chassey, B., Tafforeau, L., Lotteau, V., Rabourdin-Combe, C., and Navratil, V. (2009) pISTil: a pipeline for yeast two-hybrid Interaction Sequence Tags identification and analysis, *BMC research notes* 2, 220.
19. Orchard, S., Salwinski, L., Kerrien, S., Montecchi-Palazzi, L., Oesterheld, M., Stumpflen, V., Ceol, A., Chatr-aryamontri, A., Armstrong, J., Woollard, P., Salama, J. J., Moore, S., Wojcik, J., Bader, G. D., Vidal, M., Cusick, M. E., Gerstein, M., Gavin, A. C., Superti-Furga, G., Greenblatt, J., Bader, J., Uetz, P., Tyers, M., Legrain, P., Fields, S., Mulder, N., Gilson, M., Niepmann, M., Burgoon, L., De Las Rivas, J., Prieto, C., Perreau, V. M., Hogue, C., Mewes, H. W., Apweiler, R., Xenarios, I., Eisenberg, D., Cesareni, G., and Hermjakob, H. (2007) The minimum information required for reporting a molecular interaction experiment (MIMIx), *Nat Biotechnol* 25, 894–898.
20. Staden, R., Beal, K. F., and Bonfield, J. K. (2000) The Staden package, 1998, *Methods Mol Biol* 132, 115–130.
21. Altschul, S. F., Madden, T. L., Schaffer, A. A., Zhang, J., Zhang, Z., Miller, W., and Lipman, D. J. (1997) Gapped BLAST and PSI-BLAST: a new generation of protein database search programs, *Nucleic Acids Res* 25, 3389–3402.
22. Ruohonen, L., Aalto, M. K., and Keranen, S. (1995) Modifications to the ADH1 promoter of Saccharomyces cerevisiae for efficient production of heterologous proteins, *J Biotechnol* 39, 193–203.

23. Hartley, J. L., Temple, G. F., and Brasch, M. A. (2000) DNA cloning using in vitro site-specific recombination, *Genome Res* 10, 1788–1795.
24. Walhout, A. J., Temple, G. F., Brasch, M. A., Hartley, J. L., Lorson, M. A., van den Heuvel, S., and Vidal, M. (2000) GATEWAY recombinational cloning: application to the cloning of large numbers of open reading frames or ORFeomes, *Methods Enzymol* 328, 575–592.
25. Tornow, J., and Santangelo, G. M. (1990) Efficient expression of the Saccharomyces cerevisiae glycolytic gene ADH1 is dependent upon a cis-acting regulatory element (UASRPG) found initially in genes encoding ribosomal proteins, *Gene* 90, 79–85.
26. Fields, S., and Sternglanz, R. (1994) The two-hybrid system: an assay for protein-protein interactions, *Trends Genet* 10, 286–292.
27. Durfee, T., Becherer, K., Chen, P. L., Yeh, S. H., Yang, Y., Kilburn, A. E., Lee, W. H., and Elledge, S. J. (1993) The retinoblastoma protein associates with the protein phosphatase type 1 catalytic subunit, *Genes Dev* 7, 555–569.
28. Soellick, T. R., and Uhrig, J. F. (2001) Development of an optimized interaction-mating protocol for large-scale yeast two-hybrid analyses, *Genome Biol* 2, RESEARCH0052.
29. Vidalain, P. O., Boxem, M., Ge, H., Li, S., and Vidal, M. (2004) Increasing specificity in high-throughput yeast two-hybrid experiments, *Methods* 32, 363–370.

Chapter 7

Interactome Mapping in Malaria Parasites: Challenges and Opportunities

Douglas J. LaCount

Abstract

Nearly two-thirds of the proteins encoded by *Plasmodium falciparum,* the parasite that causes the most deadly form of malaria, are annotated as "hypothetical." The yeast two-hybrid assay, which requires no prior knowledge about the target protein, has great potential to provide functional information about these uncharacterized proteins. However, *P. falciparum* yeast two-hybrid screens are hampered by the poor expression of *P. falciparum* genes in yeast. AU-rich sequences in nascent *P. falciparum* transcripts resemble the 3′ end processing sites in yeast mRNAs, and are prematurely cleaved and polyadenylated. In most cases, these aberrant messages are degraded and yield no protein. To overcome this limitation, we have developed methods to extensively fragment *P. falciparum* genes. Novel yeast two-hybrid vectors, in which auxotrophic markers are fused to the 3′ ends of the cloned inserts, are employed to identify those gene fragments that are expressed in yeast. In this chapter, we provide detailed protocols for fragmenting *P. falciparum* genes, creating *P. falciparum* activation domain libraries, and performing *P. falciparum* yeast two-hybrid screens. Though focused on *P. falciparum*, the approaches described here are applicable to other organisms and are likely to be especially useful for those with AT-rich genomes, which are also likely to be poorly expressed in yeast.

Key words: Plasmodium, Malaria, Yeast two-hybrid, Protein–protein interaction, Interactome

1. Introduction

The yeast two-hybrid assay is a powerful tool to obtain functional information about uncharacterized proteins (1). This approach has great potential for under-studied organisms like the malaria parasite *Plasmodium falciparum*, in which more than half the proteins are annotated as "hypothetical" and even fewer have been experimentally characterized. However, application of the yeast two-hybrid assay to *P. falciparum* is hampered by the high AT content of the

Bernhard Suter and Erich E. Wanker (eds.), *Two Hybrid Technologies: Methods and Protocols*, Methods in Molecular Biology, vol. 812, DOI 10.1007/978-1-61779-455-1_7,

P. falciparum genome (80% overall, 74% in coding regions) (2). AT-rich *P. falciparum* genes are transcribed into AU-rich RNAs that contain uninterrupted stretches of A and U. These sequences resemble the mRNA 3′ end processing sites that specify where nascent yeast RNAs are cleaved and polyadenylated (3). Thus, *P. falciparum* transcripts are frequently truncated in yeast because the yeast 3′ end processing machinery interprets the AU-rich sequences as processing signals (4–6). The improperly processed RNAs encode an open reading frame but have no stop codon and are targeted for degradation by the yeast mRNA surveillance pathway (4, 7). In most cases, no protein is produced.

Although most full-length *P. falciparum* genes are poorly expressed in yeast, most fragments of ~600 bp or less are expressed to detectable levels (our unpublished observations and (8)). Thus, to screen *P. falciparum* genes in the yeast two-hybrid assay, we use complex collections of overlapping fragments (see Subheading 3.5). To identify those *P. falciparum* gene fragments that are expressed, novel yeast two-hybrid vectors in which auxotrophic marker genes are placed downstream of the *GAL4* activation and DNA-binding domains (AD and DBD, respectively) are employed (Fig. 1) (8). The auxotrophic markers are out-of-frame with the *GAL4* domains in the parental vectors pOAD.102 and pOBD.111, so yeast harboring these plasmids are unable to grow on media lacking methionine and uracil, respectively. However, if a gene fragment is cloned in-frame with both the *GAL4* domain and the auxotrophic marker, ***and*** that gene fragment is expressed, a tripartite fusion protein consisting of the Gal4 domain, the insert and the auxotrophic marker is produced, which enables yeast to grow on media lacking methionine or uracil. Although these vectors do not solve the expression problem, they allow those *P. falciparum* gene fragments that have the potential to be functional in the yeast two-hybrid assay to be identified and prevent wasted effort on fragments that are not expressed.

This approach has several additional advantages that impact yeast two-hybrid screening. AD libraries generated in these vectors are smaller (have fewer independent clones) than traditional libraries since only clones with fragments that are in-frame with the *GAL4* AD are included in the libraries. (Two thirds of the inserts in a "normal" yeast two-hybrid library are not in-frame with the AD). Because the libraries are smaller, sampling all of the clones in the library during a yeast two-hybrid screen is easier, which in turn enables library screens to be performed in high-throughput format. Eliminating out-of-frame clones reduces an important source of false-positives – e.g., clones that produce nonfunctional peptides that are not normally expressed in cells. The presence of the auxotrophic markers also allows constant selection pressure to maintain the correct reading frame during the yeast two-hybrid screen, which eliminates frame shifting, another significant source of false-positives (9). Finally, the use of gene fragments has been reported

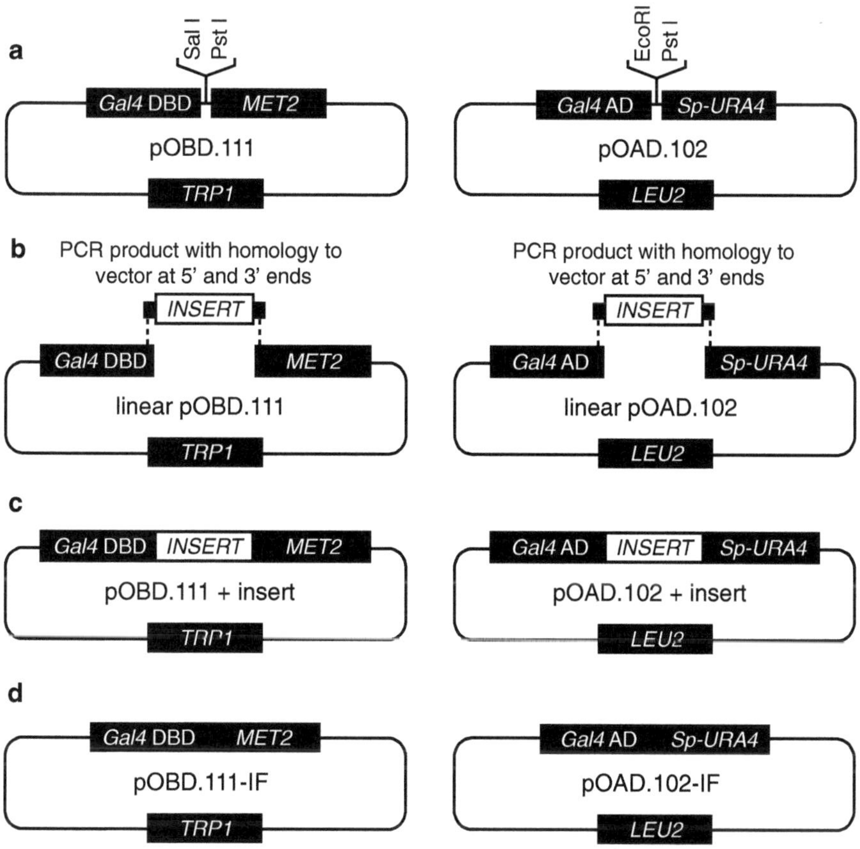

Fig. 1. Yeast two-hybrid vectors for the analysis of *P. falciparum* protein–protein interactions. (**a**) Parental vectors pOBD.111 and pOAD.102. Restriction enzyme sites used to linearize the vectors are shown. The parental pOBD.111 plasmid confers growth on medium lacking tryptophan, but not growth on medium lacking methionine because the *MET2* gene is out-of-frame with the *GAL4* DBD. Similarly, the parental pOAD.102 plasmid confers growth on medium lacking leucine, but not growth on medium lacking uracil because the *Sp-URA4* gene is out-of-frame with the *GAL4* AD. (**b**) PCR products with homology to the yeast two-hybrid vectors at the 5′ and 3′ ends can be cloned into pOBD.111 and pOAD.102 by homologous recombination in yeast. (**c**) Gene fragments that are cloned in-frame with the *GAL4* domain and the 3′ auxotrophic marker, and that are expressed in yeast, enable expression of fusion proteins that contain Met2 and Sp-Ura4. Clones with in-frame, expressed inserts can be selected by growth on medium lacking methionine (pOBD.111) or uracil (pOAD.102). (**d**) Negative control plasmids in which the auxotrophic markers are in-frame with the GAL4 domain. No insert is present. Yeast harboring these plasmids are able to grow on medium lacking tryptophan and methionine (pOBD.111-IF) or leucine and uracil (pOAD.102-IF).

to increase the number of true yeast two-hybrid interactions identified without increasing nonspecific interactions (10, 11). Thus, the modified yeast two-hybrid system described here enables identification of *P. falciparum* gene fragments that are expressed in yeast, improves throughput, reduces false positives, and may increase true positives.

The disadvantages of this system stem from the presence of the 3′ auxotrophic marker. These libraries are under-represented in 3′ end sequences because any clone with a stop codon will not express the auxotrophic marker and will not be selected. In addition, the presence of a C-terminal fusion alters the normal C terminus and will likely prevent interactions that depend on having a particular C-terminal sequence. Similarly, the auxotrophic marker may disrupt folding or may occlude some interactions. Despite these limitations, the substantial advantages of this approach for *P. falciparum* yeast two-hybrid assays far outweigh the potential disadvantages.

In this chapter, we describe methods to produce an AD library in pOAD.102 and DBD fusions from complex collections of *P. falciparum* gene fragments. Our approaches make extensive use of cloning by homologous recombination (gap repair) in yeast (12), a powerful, efficient, and cost-effective method that generates clones directly in the yeast strains that are used for the yeast two-hybrid assay (Fig. 1b). We then describe a detailed protocol for performing yeast two-hybrid library screens by mating and outline steps for retesting yeast two-hybrid positives. Confirmation of protein–protein interactions in *P. falciparum* and evaluation of their biological significance are beyond the scope of this article.

2. Materials

2.1. Yeast Two-Hybrid Vectors

1. pOAD.102: 8.9 kb, *GAL4* AD; truncated *ADH1* promoter; *Schizosaccharomyces pombe URA4* (an ortholog of *Saccharomyces cerevisiae URA3*) for selection of in-frame fusions to *GAL4* AD; MCS: *Eco*RI, *Pst*I; ampicillin resistance gene.
2. pOAD.102-IF. Negative control vector in which *Sp-URA4* is in-frame with the *GAL4* AD; used to test for self-activation of DBD fusions.
3. pOBD.111: 8.4 kb; *GAL4* DBD; truncated *ADH1* promoter; *cen* origin of replication; *TRP1*; *MET2* for selection of in-frame fusions to *GAL4* DBD. MCS: *Sal*I, *Pst*I; kanamycin resistance gene.
4. pOBD.111-IF. Negative control vector in which *MET2* is in-frame with the *GAL4* DBD; used to test specificity of AD fusions isolated in yeast two-hybrid screens.

2.2. Yeast Two-Hybrid Strains

1. DBD strain: R2HMet [*MATα ura3-52 ade2-101 trp1-901 leu2-3,112 his3-200 met2Δ::hisG gal4Δ gal80Δ*] (8).
2. Activation domain strain: BK100 [*MATa ura3-52 ade2-101 trp1-901 leu2-3,112 his3-200 gal4Δ gal80Δ GAL2-ADE2 LYS2::GAL1-HIS3 met2::GAL7-lacZ*]; a derivative of PJ69-4A. (8, 13).

2.3. Preparation of an AD cDNA Library

1. 2.0 μg PolyA + RNA from *P. falciparum*.
2. SuperScript™ III First-Strand Synthesis System (Invitrogen).

3. 10 mM dNTP mix.
4. 5× second strand buffer (Invitrogen).
5. *Escherichia coli* DNA ligase.
6. *E. coli* DNA polymerase.
7. T4 DNA polymerase.
8. T4 DNA ligase.
9. 0.5 M ethylenediaminetetraacetic acid (EDTA), pH 8.0.
10. Phenol:Chloroform:Isoamyl Alcohol (25:24:1, v/v).
11. 3 M sodium acetate (NaOAc), pH 5.2.
12. Absolute ethanol (EtOH), chilled to −20°C.
13. 70% EtOH, chilled to −20°C.
14. 0.1 M dithiothreitol (DTT).
15. High-fidelity thermostable DNA polymerase.
16. QIAEX II gel extraction kit (Qiagen).
17. Sephacryl S500 (GE Healthcare).
18. STE buffer: 150 mM NaCL, 10 mM Tris, 1.0 mM EDTA, pH 7.5.
19. Sterile plastic 2-ml pipette.
20. Sterile disposable 60-ml syringe.
21. Tygon tubing.
22. 1× TE buffer: 10 mM Tris–HCl, 1 mM EDTA, pH 8.0.
23. 10× GE buffer: 50% glycerol, 1 mM EDTA, pH 8.0.
24. DNA molecular size markers.
25. TBE electrophoresis buffer: 1 L of 10× contains 53 g of Tris base, 27.5 g of boric acid, 20 ml of 0.5 M EDTA pH 8.0.
26. Cheesecloth, for the removal of debris during AD library preparation; wrap in aluminum foil and autoclave to sterilize.
27. 96-well microtiter plates.
28. 96-well plate sealing film.
29. SeqPrep™ 96 HP Plasmid Prep System (EdgeBio).
30. Zymolyase.
31. 3′ Random primer: 5′-CTCCACCTCGAGAACGCGTTTGTC GNNNNNNNNNN-3′
32. 5′ Adapter-top oligo: 5′-CAGGAATTCAGCACATACTCATT GTTCAGCCG-3′
33. 5′ Adapter-bottom oligo: 5′-CGGCTGAACAATGAGTATGT GCTGAATTCCTGTGT-3′
34. Annealed 5′ adapter top and bottom oligos: Mix 40 μl of 100 μM top oligo with 40 μl of 100 μM bottom oligo, 1.0 μl of 5 M NaCl, and 19 μl dH_2O in a 1.5-ml microfuge tube.

Place tube in a heat block at 95°C for 5 min. Place the heat block on the bench and allow to cool to room temperature.

35. 5′ PCR primer to amplify cDNA: 5′-CTCCACCTCGAGAA CGCGTTT-3′
36. 3′ PCR primer to amplify cDNA: 5′-CAGGAATTCAGCA CATACTCATTGTTC-3′
37. cDNA-AF1: 5′-CTAGTTCTGGTAGCTCTAAAGGGTCCC TCGGTAGACAGGAATTCAGCACATACTCATTGTTC-3′; PAGE purified
38. cDNA-AR1: 5′-TGTGACCACTTCCCGAACTACTGCC TAACGAACTCCCTCCACCTCGAGAACGCGTTT-3′; PAGE purified
39. AD102-F1: PCR primer to amplify AD inserts from pOAD.102, 5′-CGACGACGAGGATACGCCACCGAAC-3′
40. AD102-F2: 5′ primer to sequence AD inserts from pOAD.102, 5′-ATACGCCACCGAACCCTAAGAAA-3′
41. AD102-R1: PCR primer to amplify AD inserts from pOAD.102, 5′-TGTGACCACTTCCCGAACTACTGCCTAA CGAACTCC-3′
42. AD102-R2: 3′ primer to sequence AD inserts from pOAD.102, 5′-TCGCCGAATAGCTCTGGAA-3′
43. Sephacryl S500 (GE Healthcare).
44. STE buffer: 150 mM NaCL, 10 mM Tris, 1.0 mM EDTA, pH 7.5.
45. Sterile plastic 2-ml pipette.
46. Sterile disposable 60-ml syringe.
47. Tygon tubing.
48. 1× TE buffer: 10 mM Tris–HCl, 1 mM EDTA, pH 8.0.
49. 10× GE buffer: 50% glycerol, 1 mM EDTA, pH 8.0.
50. DNA molecular size markers.
51. TBE electrophoresis buffer: 1 L of 10× contains 53 g of Tris base, 27.5 g of boric acid, 20 ml of 0.5 M EDTA pH 8.0.

2.4. Preparation of DBD Clones

2.4.1. Random Fragmentation with DNAse I

1. DNAse I (Novagen, $Mg2^{+}$ free).
2. 1 M $MnCl_2$, filter sterilized. Dilute 1:10 for 100 mM.
3. 10× DNAse I Buffer, 500 mM Tris–Cl pH7.5, 1 mg/ml BSA.
4. T4 DNA polymerase.
5. T4 DNA ligase.
6. 0.5 M EDTA, pH 8.0.
7. MinElute Reaction Cleanup kit (Qiagen).

8. 5′ BD RecTail Top oligo: 5′-GTTGAGCAGTCTGGGAGGA TTGAGCGGTAGCGGA-3′
9. 5′ BD RecTail Bottom oligo: 5′-TCCGCTACCGCTCAAT CCTCCCAGACTGCTCAACCCG-3′
10. 3′ BD RecTail Top oligo: 5′-CTAGGGTAGGCTCATCGC GTGAAGGTGGCTCTACCCG-3′
11. 3′ BD RecTail Bottom oligo: 5′-GTAGAGCCACCTTCACG CGATGAGCCTACCCTAG-3′
12. Annealed 5′ BD RecTail Top and Bottom oligos: See item 34 in Subheading 2.3.
13. Annealed 3′ BD RecTail Top and Bottom oligos: See item 34 in Subheading 2.3.
14. BD111-F1: PCR primer to amplify DBD inserts from pOBD.111, 5′-GCAACGACAGCTCACGGTATCTCAGG-3′
15. BD111-F2: 5′ primer to sequence DBD inserts from pOBD.111, 5′-CAGCTCACGGTATCTCAGGAAA-3′
16. BD111-R1: PCR primer to amplify DBD inserts from pOBD.111, 5′-CTCAATGTCCAGCTCTTGGAGC-3′
17. BD111-R2: 3′ primer to sequence DBD inserts from pOBD.111, 5′-TCCAGCTCTTGGAGCGTTTT-3′

2.4.2. Defined Fragments with Gene-Specific Primers

1. Gene-specific primers to amplify each fragment; append the sequence 5′-GAGGATTGAGCGGTAGCGGA-3′ to the 5′ end of the forward oligos; append the sequence 5′-TCACGCGATGAGCCTACCCT-3′ to the 5′ end of the reverse oligos.
2. High-fidelity thermostable DNA polymerase.

2.5. Yeast Growth, Transformation, and Yeast Two-Hybrid Screening

1. YEP (Yeast extract-peptone medium): 1% yeast extract, 2% peptone. Autoclave 20 min at 121°C.
2. YPDA (Yeast extract-peptone-dextrose plus adenine medium): 1% yeast extract, 2% peptone, 2% glucose, 0.003% adenine. For plates, add 1.6% agar. Autoclave 20 min at 121°C.
3. YPDA, pH 3.5.
4. YPDA, pH 4.5.
5. Synthetic drop-out medium (SD): 0.67% yeast nitrogen base (without amino acids), 2% glucose, 1× drop-out mix, 1.5× adenine; for plates, add 1.6% agar; autoclave at 121°C for 20 min.
6. 10× drop out mix: For 1 L, 0.2 g adenine, 0.2 g arginine, 0.2 g histidine, 0.3 g isoleucine 1 g leucine, 0.3 g lysine, 0.2 g methionine, 0.5 g phenylalanine, 2 g threonine, 0.2 g tryptophan, 0.3 g tyrosine, 0.2 g uracil, 1.5 g valine. Leave out appropriate components for drop-out medium (SD−TRP, SD−LEU, SD−TRP−LEU, SD−LEU−URA, SD−TRP−MET, SD−TR

P–LEU–MET–URA–HIS, SD–TRP–LEU–MET–URA–ADE). Autoclave 20 min at 121°C.

7. 100× adenine: 2.0 g/L. Add 5 ml of 4N NaOH to dissolve. Autoclave 20 min at 121°C.
8. 1 M 3-amino-1,2,4-triazole (3-AT); store at −20°C.
9. 1 M (10×) Lithium acetate (LiOAC).
10. 50% poly(ethylene glycol) average mol wt 3,350 (PEG); filter sterilize.
11. Salmon sperm DNA (used as carrier for yeast transformations; dissolve in TE buffer at a concentration of 2 mg/ml with vigorous mixing. Pass the solution through a 10 ml pipet several times to disperse clumps. Aliquot and store at −20°C. Prior to use, heat the DNA for 5 min at 95°C, then snap cool in an ice water slurry).
12. Dimethyl sulfoxide (DMSO), use fresh or store in 1 ml aliquots at −20°C.
13. Falcon 2059 tubes.
14. Glass beads, 4 mm, for distributing yeast cell on plates.

3. Methods

3.1. Yeast Strains

The strains R2HMet and BK100 are utilized for the *P. falciparum* yeast two-hybrid screens (8). DBD fusions are expressed in R2HMet, which does not encode any yeast two-hybrid reporters, whereas the AD library is expressed in BK100. A derivative of PJ69-4a (13), BK100 encodes two yeast two-hybrid reporter genes (*HIS3* and *ADE2*) under control of two different Gal4 responsive promoters. *HIS3* is the more sensitive reporter and is therefore used as the selection during library screens. This reporter has the added benefit that its protein product, His3, is competitively inhibited by 3-amino-1,2,4-triazole (3-AT). By increasing the amount of 3-AT in the culture medium, it is possible to suppress the background His3 activity produced by self-activating DBD fusions, which enables these fusions to be screened. *ADE2* expression is used to confirm yeast two-hybrid positives.

3.2. PCR Amplification of *P. falciparum* Genes from cDNA

Three critical changes are necessary for efficient amplification of *P. falciparum* genes. First the extension temperature must be reduced to 60 or 65°C (14). For amplification of rare targets from cDNA, 60°C extensions are optimal. Amplification of fragments from a plasmid template can usually be achieved with a 65°C extension. Second, reducing the ramp speed (the rate the thermal cycle changes temperature) between the denaturation and annealing temperatures and between the annealing and extension temperatures can significantly improve yields. A ramp speed of 0.2–0.5 degrees per second is optimal. Finally, to amplify fragments using oligos that have a 5′ extension (as for cloning by recombination), adding 5 cycles with a lower annealing temperatures (42–45°C) followed by 25 cycles with a higher annealing temperature (55°C) works well.

3.3. Preparation of Linear pOAD.102 and pOBD.111

1. Digest 50 μg of pOAD.102 with *Eco*R I and *Pst* I.
2. Digest 50 μg of pOBD.111 with *Sal* I and *Pst* I.
3. Purify the DNA using the Qiagen Qiaex II kit. Follow the manufacturer's protocol for desalting and concentrating DNA solutions.
4. Measure the DNA concentration. Load a sample of the digested DNA on an agarose gel to check the quality.
5. Add EDTA to 1 mM final concentration and store at −20°C.

3.4. Preparation of an AD Library in pOAD.102

P. falciparum yeast two-hybrid libraries are not commercially available, nor are kits to create libraries in pOAD.102. This section describes the generation of an AD library in pOAD.102 using individually purchased reagents. Beginning with polyA + RNA, first strand cDNA is reverse transcribed with random primers and converted to double-stranded cDNA. The cDNA fragments are then amplified by PCR with primers that add extensions to the 5′ and 3′ ends of the fragments. These extensions are homologous to sequences in the AD vector that flank the cloning site for the cDNA (Fig. 1b). The AD library is created by homologous recombination in the yeast two-hybrid strain BK100, which eliminates the need to first generate the library in *E. coli* or bacteriophage and then transform it into yeast. As described above, pOAD.102 enables selection of cDNA fragments that are cloned in-frame with *Gal4* AD and *Sp-URA4*, and that are expressed in yeast (i.e., those fragments that have the potential to be functional in the yeast two-hybrid assay). The key step in creating a high-quality cDNA library in pOAD.102 is the complete removal of primer dimers and small fragments from the cDNA preparation. These small DNAs are less likely to encode stop codons in the undesired reading frames than larger fragments. Thus, even if the small DNAs constitute only a tiny fraction of the total DNA, they can be selected at a much higher rate and will make up a high percentage of the clones in the AD library. To date, the only method that consistently eliminates all of the small fragments is fractionation on Sephacryl S400 or S500 columns, which is described here in detail.

3.4.1. Preparation of Double-Stranded cDNA

1. Prepare first strand cDNA from *P. falciparum* polyA + RNA using the Invitrogen SuperScript™ III First-Strand Synthesis System and the 3′ random primer. Do not treat with RNAse H.

2. To the 20 µl first strand reaction from **step 1**, add the following reagents in order: 91 µl dH_2O, 30 µl 5× second strand buffer, 3 µl 10 mM dNTP mix, 1 µl *E. coli* DNA ligase (10 U/µl), 4 µl *E. coli* DNA polymerase I (10 U/µl), and 1 µl *E. coli* RNAse H (2 U/µl).
3. Mix gently. Incubate 2 h at 30°C.
4. Add 2 µl T4 DNA polymerase. Incubate 5 min at 16°C.
5. Add 10 µl of 500 mM EDTA pH 8.0 to stop the reaction.
6. Extract once with an equal volume of Phenol:Chloroform: Isoamyl Alcohol (25:24:1). Transfer the aqueous phase to a new tube.
7. Add 1/10 volume 3 M NaOAc, pH 5.2, and two volumes of 100% ethanol. Mix vigorously and store overnight at −20°C.
8. Collect DNA by centrifugation for 20 min at 14,000×*g*.
9. Remove supernatant and wash the pellet with 70% ethanol. Centrifuge 5 min at 14,000×*g*. Carefully remove supernatant. Centrifuge briefly to collect any wash solution that remained in the tube. Carefully remove supernatant. Air dry pellet ~10–15 min.
10. Resuspend pellet in 18 µl dH_2O.
11. Ligate adapters to double-stranded cDNA by adding 10 µl 5× T4 DNA ligase buffer, 10 µl 40 µM annealed 5′ adapter primer, 7 µl 0.1 M DTT, 5 µl T4 DNA ligase (1 U/µl). Mix gently. Incubate at 14°C for >16 h.
12. Remove 1 µl and set up a PCR reaction with primers 35 and 36 (Subheading 2.3). If the synthesis of the double-stranded cDNA and ligation of the 5′ adapter was successful, a smear extending from ~100 bp to >2 kb will be observed.

3.4.2. Size Fractionation of cDNA on a Sephacryl S500 Column

1. Partially remove the cotton plug from a 2-ml disposal pipette and cut off half of it. Using a syringe, blow air through the pipette to force the plug to the bottom of the pipette. If it does not move to the bottom, wet the plug with a small amount of STE buffer and repeat.
2. Remove the plunger from the barrel of a 60-ml syringe. Attach the barrel to the 2-ml disposable pipette prepared in step 1 with a short piece of Tygon tubing. This will serve as the buffer reservoir for the column. Mount the syringe and pipette on a ring stand with at least two clamps. Make sure the syringe and the pipette are aligned and are vertical.
3. Attach an ~5 cm piece of tubing to the bottom of the pipette. Place the end of the tubing in a 150-ml beaker.
4. Add ~10 ml of STE in the syringe and clamp the tubing while the buffer is flowing through.
5. Add ~5 ml STE to the reservoir and then 5 ml Sephacryl S500. The Sephacryl S500 will gradually settle to form the column.

If necessary, add more Sephacryl S500 until the matrix fills the column nearly to the top. If there is excess Sephacryl, carefully pipette up and down to dislodge the matrix from the reservoir and remove.

6. Wash the column with ≥5 column volumes of STE buffer. Washing can be accomplished by filling the reservoir with STE and allowing the buffer to flow through the column overnight. The flow rate will be about 2 drops/min (~100 μl/min). Filling the reservoir with 70 ml of STE will allow the column to flow at least 11 h. Before allowing the column to run overnight, check the flow rate to ensure that the column will not dry out.
7. Calibrate the column by loading 10 μg of DNA molecular size markers. Mix the DNA ladder with 12.5 μl TE and 2.5 μl 10× GE gel loading buffer. Remove almost all of the STE from the reservoir, leaving only enough to ensure that the column does not run dry. Carefully pipette the DNA ladder onto the column without disturbing the surface. Carefully add STE to the column. Collect 2 drop fractions. Load 10 μl from fractions 9 through 30 on a 1% gel. The largest DNA bands are typically first observed in drop 21 (±2 drops). Note the fractions in which DNA is first detected and in which fractions the 400-, 300-, and 200-bp bands first appear.
8. Wash the column with ≥ 5 column volumes of STE.
9. Ethanol precipitate the double-stranded cDNA from Subheading 3.4.1 step 11 and resuspend in 22.5 μl of TE. Add 2.5 μl 10× GE.
10. Load double-stranded cDNA on the Sephacryl S500 column as in step 7. Collect 2-drop fractions.
11. Verify the presence of DNA by PCR using 1 μl of each fraction as template and with primers 35 and 36 (Subheading 2.3). Expect to detect a smooth smear of PCR products in the same fractions in which the DNA ladder was observed. The size of the smear should decrease as fraction number increases. The size range of the smear should approximately correspond to the sizes from the fractionation of the DNA size markers.

3.4.3. Optimization of Conditions to Amplify cDNA

The cDNA produced in the previous section is amplified with primers cDNA-AF1 and cDNA-AR1, which provide 5′ and 3′ extensions that are homologous to the sequences flanking the cloning site in pOAD.102. In this section, we optimize the amount of cDNA to use as template and the number of PCR cycles to perform in order to amplify the cDNA without skewing the representation.

1. Mix 5 μl from the fractions containing cDNAs ≥300 bp.
2. Prepare 6 threefold dilutions of the pooled cDNA fractions in STE.

3. Assemble 50 μl PCR reactions using 1 μl of each dilution as template, 1 μl of 10 μM cDNA-AF1 and 1 μl of 10 μM cDNA-AR1 as primers, and a high-fidelity DNA polymerase. These PCR primers will enable selection of one out of nine possible reading frames when the fragments are cloned into pOAD.102. See Note 1 for ways to increase the diversity of the AD library.
4. Amplify the cDNA using the following conditions: 3 min at 94°C; 25 cycles of 94°C for 30 s and 60°C for 3 min, 30 s; 10 min at 60°C; hold at 4°C. Reduce ramp rates to 0.2°C/s.
5. Electrophorese 5 μl of each reaction on a 1.2% agarose/TBE gel. Note the reaction with the greatest yield (as judged by brightness) and the largest size distribution of the DNA smear.
6. Using the dilution noted in step 5, set up six more PCR reactions. Remove one reaction after 15, 17, 19, 21, 23, and 25 cycles.
7. Electrophorese 5 μl of each reaction on a 1.2% agarose/TBE gel. Note the reaction with the greatest yield and the largest size distribution of the smear. No DNA is expected to be detected in the 15-cycle sample. A smear should be detectable in the 17 or 19 cycle lanes and should become more intense in the next sample (Fig. 2). With greater number of cycles the intensity of the smear may increase, but the size range will

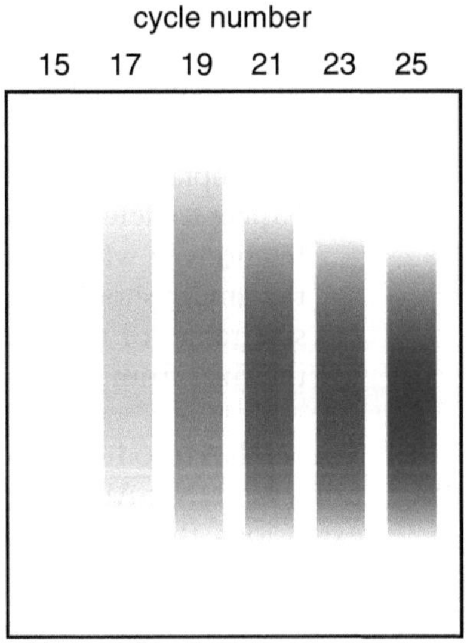

Fig. 2. Optimization of the number of cycles for amplification of cDNA. A schematic representation of the expected results from the cycle optimization experiment is shown. In this example, a smear of amplified cDNA is first detected after 17 cycles and increases in size and intensity in the 19-cycle sample. Additional PCR cycles cause a gradual shift toward smaller fragments. The optimal number of cycles to amplify this cDNA preparation is 19.

decrease as smaller PCR products are preferentially amplified. Choose the cycle number with the greatest yield of the smear with the largest size range. If a bright smear is detected in the 15-cycle sample, repeat the cycle titration using less cDNA and/or fewer cycles.

3.4.4. Test Transformation to Evaluate the Quality of the cDNA

1. Amplify cDNA using the optimal conditions determined in Subheading 3.4.3.
2. Grow a 50 ml culture of BK100 in YPDA to an OD_{600} of 1.5.
3. Collect cells by centrifugation for 3 min at 1,400 × *g*. Discard supernatant and wash once with dH_2O.
4. Resuspend cells in 300 μl 100 mM LiOAc, 10 mM Tris, 1 mM EDTA, pH 8.0 (1× LiOAc/1× TE).
5. Prepare transformation master mix consisting of 55 μl cells, 200 μl 50% PEG, 25 μl 1 M LiOAc, 15 μl heat denatured carrier DNA (2 mg/ml) per transformation. Distribute 290 μl to each tube.
6. Add the following DNAs and vortex to mix.

   Transformation #1: 10 ng circular pOAD.102 (positive control and test for transformation efficiency)

   Transformation #2: 250 μg linear pOAD.102 (negative control)

   Transformation #3: 250 μg linearized pOAD.102 plus 5 μl AD PCR product

   Transformation #4: 250 μg linearized pOAD.102 plus 10 μl AD PCR product

   Transformation #5: 250 μg linearized pOAD.102 plus 15 μl AD PCR product
7. Incubate 30 min at 30°C.
8. Add 23 μl fresh DMSO.
9. Heat shock at 42°C for 15 min.
10. Centrifuge 1 min at 1,400 × *g*. Discard supernatant. Resuspend cells in 500 μl dH_2O.
11. Plate 50 μl of transformations 1 and 2 on SD − LEU − URA + ADE and SD − LEU + ADE. For transformations 3–5, plate 50 μl on four SD − LEU − URA + ADE plates and 50 μl of 10^{-1} and 10^{-2} diluted transformation mix on SD − LEU + ADE. Incubate 3 days at 30°C.
12. Count the number of colonies on each plate. Calculate the number of colonies per ml of each transformation. Be sure to take into account the dilution factor.
13. Transformation 1 should yield 30–100 colonies on SD − LEU + ADE and zero on SD − LEU − URA + ADE. If pOAD.102 has been properly linearized, transformation 2 should yield very few colonies (ideally zero) on

SD−LEU+ADE and SD−LEU−URA+ADE; if colonies are observed, the ratio tends to be ~3:1 on SD−LEU+ADE compared to SD−LEU−URA+ADE. If the number of colonies from this transformation approaches that of the transformations 3–5, prepare a new batch of linear pOAD.102. For transformations 3–5, approximately 50- to 100-times more colonies should be present on SD−LEU+ADE than on SD−LEU−URA+ADE plates, since only 1–2% of the clones are typically in-frame and expressed.

14. Determine the amount of PCR product that yielded the largest number of colonies on SD−LEU−URA+ADE. A good co-transformation of cDNA with linear pOAD.102 will yield > 1,500 colonies on SD−LEU−URA+ADE. If the number of colonies is significantly lower, repeat the transformations with increasing amounts of linear pOAD.102 and the PCR product.
15. Pick 96 colonies from the SD−LEU−URA+ADE plates and inoculate one colony per well of a 96-well plate containing 110 μl YPDA per well. Cover with film and incubate overnight at 30°C.
16. Vortex gently to resuspend yeast. Inoculate 90 μl YPDA in a second 96-well microtiter plate with 10 μl of yeast culture. Cover plate with film and incubate overnight at 30°C. Add 5 μl DMSO, vortex gently to resuspend, and freeze at −80°C.
17. Collect remaining cells by centrifugation at 1,400 × *g* for 5 min. Remove supernatant by inverting plate over a beaker.
18. Resuspend cell pellet in 100 μl of Edge SeqPrep Lysis Solution supplemented with the manufacturer's Enzyme Mix plus 150 U/ml zymolyase. Transfer sample to SeqPrep™ 96 HP Plate. Incubate at 30°C for 30 min.
19. Remove lysate by decanting and blot plate vigorously on paper toweling (DNA should be bound to plate).
20. Add 100 μl of wash solution to samples. Vortex for 30 s.
21. Decant and blot plate on paper towels to remove residual wash solution.
22. Add 100 μl of 70% isopropanol to wells. Vortex for 30 s.
23. Decant and blot plate to remove isopropanol.
24. Repeat steps 21 and 22. Air dry plate face up for 30 min at room temperature. Do not leave plate for extended period.
25. Add 40 μl 10 mM Tris–HCl pH 8.0 to elute DNA. Incubate 5 min at room temperature.
26. Amplify the AD inserts by PCR using oligos AD102-F1 and AD102-R1 and 1 μl of DNA from step 25. Reduce the extension temperature to 65°C for optimal amplification. Check the size of the inserts on a 1.2% agarose/TBE gel.

27. Count the number of PCR products of 500 bp or greater, which corresponds to inserts of ≥300 bp (vector sequences account for ~180 bp of the PCR products). More than 90% of PCR products should have inserts ≥300 bp. If the percentage is significantly lower, small molecular weight DNA fragments remain in the cDNA prep and the fractionation must be repeated.

3.4.5. Large-Scale Transformation to Generate the AD Library

1. Using the optimal conditions determined in Subheading 3.4.3, set up 48 PCR reactions to amplify cDNA for a large-scale transformation.
2. Grow a culture of BK100 in 250 ml of YPDA overnight at 30°C.
3. Collect cells by centrifugation at 1,400×*g* for 10 min and resuspend in 20 ml fresh YPDA.
4. Inoculate 500-ml of YPDA with sufficient BK100 from step 2 to achieve an OD_{600} = 0.25. Incubate at 30°C with shaking until the OD_{600} = 1.5.
5. Collect cells by centrifugation, discard the supernatant, and wash the pellet once with water. Resuspend cells in sufficient dH_2O to give a total volume of 11.25 ml.
6. Add 1.25 ml 1 M LiOAc to the cells and mix well. Store at room temperature.
7. Prepare transformation mix: 41.2 ml 50% PEG, 5.0 ml 1 M LiOAc, 1.5 ml heat denatured carrier DNA, 35 μg linear pOAD.102, 1.4 ml amplified cDNA (see Note 2), 12.5 ml of yeast, and dH_2O to a final volume of 62 ml. Mix well by vortexing.
8. Distribute 3.1 ml to 20 Falcon 2059 tubes. Incubate 30 min at 30°C.
9. Add 230 μl DMSO to each tube. Vortex to mix.
10. Incubate 15 min at 42°C in a water bath.
11. Centrifuge 3 min at 1,400×*g*. Discard supernatant.
12. Resuspend pellet in 1 ml dH_2O per tube and distribute cells to four 150 mm SD−LEU−URA+ADE plates (80 total). Distribute the cells by adding 10-20 glass beads and shaking. Allow the plates to air dry and then remove the glass beads.
13. Incubate 2–3 days at 30°C until the colonies are ~1 mm in diameter. There should be ~2,000 colonies per plate.
14. Chill the plates at 4°C.
15. Scrape the cells from the plate in 8.0 ml YEP using a bent glass rod. Transfer cells to a sterile 1-L flask. Wash plates with 2.0 ml YEP. Add the wash to the flask.
16. Filter the yeast through several layers of sterile cheesecloth to remove any debris.

17. Measure the volume of yeast recovered in a sterile graduated cylinder.
18. Add 1/20th volume of DMSO and mix well. Distribute to sterile, pre-labeled tubes. It is useful to make ~100 1.0 ml aliquots and store the remaining library in 10 and 25 ml aliquots. Freeze at −80°C.

3.4.6. Evaluation of the Library

1. Thaw one 1.0 ml aliquot of the AD library on ice.
2. Dilute 10^{-4}, 10^{-5}, and 10^{-6} in dH_2O. Plate 100 ml of each dilution on two SD−LEU+ADE plates. Incubate 3 days at 30°C. Count colonies and determine the number of colony forming units (cfu)/ml. Expect a titer of $2–4 \times 10^8$ cfu/ml.
3. Plate the remaining portion of the library aliquot on SD−TRP to check for contamination. Incubate at least 7 days at 30°C. Any colonies or microbial growth indicates contamination since no yeast from the library should be able to grow on SD–TRP medium.
4. From the SD−LEU+ADE plates in step 2, pick at least 96 colonies and process them as described in Subheading 3.4.4, steps 15–26. Note the number of clones that have inserts ≥300 bp. Sequence the PCR products to determine the diversity of gene fragments in the library.
5. Perform test yeast two-hybrid screens of the new library. The *P. falciparum* genes *PFE1350c* and *PFC0255c* are good candidates for test screens because they strongly, specifically and reproducibly interact with each other. Due to their small size and relatively low AT content, full-length *PFE1350c* and *PFC0255c* can be expressed in the DBD vector pOBD.111. See Subheading 3.5 for DBD plasmids. It is also useful to screen the library with pOBD.111-IF to identify promiscuous AD clones. If the same AD clone is identified in multiple screens with unrelated DBD fusions, it may be necessary to generate a new library (see Note 3).

3.5. Preparation of DBD Clones in Plasmid pOBD.111

As described in Subheading 1, most full-length *P. falciparum* genes are poorly expressed in yeast. In order to circumvent this problem, it is necessary to use smaller fragments of *P. falciparum* genes. However, it is not currently possible to predict which gene fragments are likely to be expressed in yeast or which fragments will be functional in the yeast two-hybrid assay. Approaches that yield many overlapping fragments address both problems. Here, we describe two complementary methods to generate fragments of *P. falciparum* genes (Fig. 3). In Subheading 3.5.1, DNAse I is used to generate random fragments of *P. falciparum* genes. This approach yields the greatest diversity of fragments, but some regions of the target gene may be missed due to sampling errors.

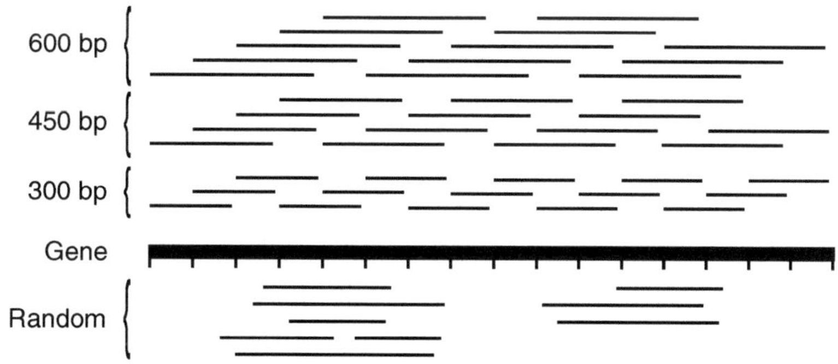

Fig. 3. Fragmentation of *P. falciparum* genes. The *black bar* represents a 2,400-bp *P. falciparum* gene. Tick marks are indicated at 150-bp intervals. *Horizontal lines* below the gene indicate random gene fragments produced by partial DNAse I digestion that were cloned into pOBD.111 and selected for expression on medium lacking methionine. Gaps in the coverage may result from regions that cannot be expressed in yeast or from inadequate sampling of the fragments produced. Lines above the gene represent partially overlapping 300-, 450-, and 600-bp fragments produced by PCR amplification with primers spaced every 150 bp. This approach ensures that fragments covering the entire gene are tested for expression in yeast.

Similar results can be achieved using gene-specific primers spaced evenly along the gene, as outlined in Subheading 3.5.2. This approach ensures that all regions of the gene are tested for expression, but requires many primers for each gene and generates less diverse fragments. In both cases, homologous recombination in R2HMet is used to clone the fragments into pOBD.111, and selection on medium lacking methionine is used to determine if the fragment is expressed.

3.5.1. Random Fragmentation by DNAse I

1. Evaluate the DNAse I activity. Purchase a fresh sample of DNAse I and dedicate it for the use in DNA fragmentation. Before use, test each tube of DNAse I in the partial fragmentation assay. For most consistent results, perform incubations in a thermal cycler.
2. Amplify the gene to be fragmented and purify the PCR product. Measure the DNA concentration. Do not resuspend or elute the DNA in a buffer containing EDTA.
3. Prepare 1:200, 1:1,000, 1:5,000, 1:10,000, and 1:20,000 dilutions of DNAse I in 1× DNAse I buffer.
4. In a thin-wall PCR tube on ice, mix 2 μl of 10× DNAse I buffer with 2 μl 100 mM $MnCl_2$, 1.0 μg DNA, sufficient dH_2O to bring the final volume to 20 μl, and 2 μl of diluted DNAse I.
5. Incubate 20 min at 25°C in a preheated thermal cycler.
6. Place sample on ice and add 2 μl of 0.5 M EDTA to stop the reaction.

7. Load 10 μl on a 1.2% agarose gel to evaluate the extent of DNA fragmentation. Choose the dilution in which a small amount of full-length PCR product remains and an even smear of fragments extending down to ~ 100 bp is produced.
8. Using the optimal concentration of DNAse I, digest 2.5 μg DNA as in step 4. Adjust the amounts of each component proportionally to yield a final volume of 50 μl.
9. Purify DNA on a Qiagen MinElute Reaction Cleanup column. Remove and save 1 μl.
10. In a thin-wall PCR tube on ice, mix the remaining 9.0 μl of DNA with 3.2 μl 10× T4 DNA polymerase buffer, 1.6 μl 1 mg/ml BSA, 3.2 μl 1 mM dNTPs, 14.5 μl dH_2O, and 0.5 μl T4 DNA polymerase.
11. Incubate 15 min at 12°C in a thermal cycler.
12. Place sample on ice and add 1.0 μl 500 mM EDTA.
13. Purify DNA on a Qiagen MinElute Reaction Cleanup column.
14. Ligate 5′ and 3′ adapter oligos to the gene fragments. Mix 10 μl DNA from step 10 with 10.0 μl 5× ligation buffer, 1.0 μl annealed 5′ BD Rec Tail oligos, 1.0 μl annealed 3′ BD Rec Tail oligos, 25.5 μl dH_2O, and 2.5 μl T4 DNA ligase.
15. Incubate 18–24 h at room temperature.
16. Fractionate the fragmented DNA on a Sephacryl S500 column as described in Subheading 3.4.2. The presence of the fragmented DNA can be confirmed by PCR with oligos 5′ BD RecTail Top and 3′ BD RecTail Top. Pool fractions containing fragments ≥300 bp. Ethanol precipitate and resuspend the pellet in 10 μl TE.
17. Clone the *P. falciparum* gene fragments into pOBD.111 by in vivo recombination in R2HMet. Prepare a culture of R2HMet cells for transformation as described in Subheading 3.4.4, steps 2–5. Set up three transformations:

    Transformation #1: 10 ng circular pOBD.111 (positive control and test for transformation efficiency)

    Transformation #2: 250 μg linear pOBD.111 (negative control)

    Transformation #3: 250 μg linearized pOBD.111 plus 5 μl DNA from step 16.

    Process the samples as in steps 7–10 of Subheading 3.4.4
18. Plate 100 μl of transformations 1 and 2 on SD−TRP−MET+ADE and SD−TRP+ADE. Plate 100 μl of transformation 3 on four SD−TRP−MET+ADE plates and 100 μl of 10^{-1} and 10^{-2} diluted transformation mix on SD−TRP+ADE. Incubate 3 days at 30°C.

19. Count the number of colonies. Transformation 1 is expected to yield 30–100 colonies on SD−TRP+ADE and zero on SD−TRP−MET+ADE because *MET2* is out-of-frame with the *GAL4* DBD in this plasmid. Ideally, there will be no colonies from transformation 2 on SD−TRP+ADE or SD−TRP−MET+ADE, although in practice there frequently are. If so, expect to see about three times more colonies on SD−TRP+ADE than on SD−TRP−MET+ADE. Background colony formation from linear pOBD.111 that recircularizes is not a problem as long as there are at least five to ten times as many colonies from transformation 3 on the same medium. About 2% of the yeast colonies from transformation 3 that grow on SD−TRP+ADE are expected to grow on SD−TRP−MET+ADE since 1 out of 4 fragments will have the correct combination of oligos annealed to the 5′ and 3′ ends, 1 out of 9 fragments will be cloned in frame with both the DNA binding domain and the MET2 gene, and not all fragments will be expressed. If no colonies grow on SD−TRP−MET+ADE from transformation #1, but many colonies are present on SD−TRP+ADE (see Note 4).
20. Pick yeast colonies from the SD−TRP−MET+ADE plates and process them as described in Subheading 3.4.4, steps 15–25. Depending on the size of the gene and the amount of coverage you desire, analyze 24–96 colonies.
21. PCR amplify insert using oligos BD111-F1 and BD111-R1 and standard PCR conditions, except that the extension temperature must be reduced to 65°C.
22. Sequence PCR products using standard protocols to determine the size and locations of the fragments.

3.5.2. Defined Fragmentation of P. falciparum Genes

1. Design primers to generate fragments of ~150 bp. It may be necessary to vary the length of the amplified fragment to avoid long stretches of As or Ts.
2. To the 5′ end of each forward oligo, add the following sequence: GAGGATTGAGCGGTAGCGGA.
3. To the 5′ end of each reverse oligo, add the following sequence: TCACGCGATGAGCCTACCCT.
4. PCR amplify all possible 300-, 450-, and 600-bp fragments using the appropriate combinations of forward and reverse oligos and a high-fidelity DNA polymerase.
5. Clone the fragments into plasmid pOBD.111 as described in Subheading 3.5.1, steps 17–19. Substitute 5 μl of PCR product for the DNAse I-digested DNA in transformation 3. Each fragment will require a separate transformation.
6. Fragments that are well-expressed in yeast will yield approximately equal numbers of colonies on SD−TRP−MET+ADE

and SD−TRP+ADE. If many fewer colonies are present on SD−TRP−MET+ADE than on SD−TRP+ADE (see Note 5).

7. Pick ~ 4 yeast colonies from the SD−TRP−MET+ADE plates and process them as described in Subheading 3.5.1, steps 20–22.

3.5.3. Self-Activation Test

Before performing a yeast two-hybrid screen with any DBD construct, it is necessary to test the construct for its potential to activate expression of the reporter genes in the absence of an interacting AD fusion protein (self-activation). Ideally, this should be done under conditions that closely mimic the screening conditions that will be used. Thus, self-activation is evaluated in diploid yeast in the presence of an empty AD plasmid in which *Sp-URA4* has been cloned in frame with the AD (pOAD.102-IF, Fig. 1c).

1. On the same spot on a YPDA plate, streak R2HMet cells containing the DBD construct to be tested and BK100 cells containing pOAD.102-IF. Incubate overnight at 30°C to allow mating.
2. Streak cells from the patch on SD−TRP−LEU+ADE to select for diploid yeast colonies that contain both plasmids.
3. Streak yeast on SD−TRP−MET−LEU−URA−HIS medium supplemented with increasing concentrations of 3-AT (1, 3, 5, 10, 25, 50, and 100 mM). As a control for yeast growth, also streak the diploid yeast on SD−TRP−MET−LEU−URA +ADE.
4. Compare the growth on SD−TRP−MET−LEU−URA+ADE to SD−TRP−MET−LEU−URA−HIS plus 3-AT. Identify the minimum 3-AT concentration that suppresses all yeast growth. In most cases, this will be 1 or 3 mM. DBD fusions that confer growth on high concentrations of 3-AT are strong self-activators and should not be screened in the yeast two-hybrid assay.

3.6. Yeast Two-Hybrid Screens

The *P. falciparum* yeast two-hybrid screens are performed by mating the DBD and AD library strains to produce diploid yeast containing both plasmids. The method described here can be used to screen ~24 fragments at a time. See Note 6 for modifications that enable high-throughput screening.

3.6.1. Mating of DBD Fusion Strains with AD Library Strains

1. Grow DBD strains in 5 ml SD−TRP+ADE to an OD_{600} of 1.0–1.5.
2. Thaw an aliquot of AD library on ice. Begin thawing the sample ~1.5 h before DBD strains will reach the desired density.
3. For each screen to be performed, add 5.0×10^6 cfu of the AD library to 1 ml of YPDA. Incubate 60 min at 30°C with shaking.

4. In a Falcon 2059 tube, add 1.0×10^7 cfu (1 OD_{600} = ~2×10^7 cells/ml; see Note 7) of the DBD strain and 1.0 ml of AD library.
5. Centrifuge 3 min at 1,400 × *g* to pellet cells. Decant supernatant.
6. Resuspend in 3 ml of YPD, pH 3.5.
7. Incubate 90 min at 30°C with rotation.
8. Centrifuge 3 min at 1,400 × *g* to pellet cells. Decant supernatant.
9. Add 3 ml YPDA, pH 4.5. Centrifuge 3 min at 1,400 × *g* to pellet cells. Do not remove supernatant.
10. Incubate overnight at room temperature (16–18 h total).
11. Resuspend cells by gentle vortexing.
12. Remove 10 μl, add to 990 μl dH_2O (10^{-2} dilution), and plate 100 μl on SD−TRP−LEU + ADE to determine the number of diploid yeast present. Incubate at 30°C for 2–3 days and count the number of colonies. Expect to obtain a total of $3–5 \times 10^5$ diploid cells.
13. Centrifuge 3 min at 1,400 × *g* to pellet cells from step 11. Decant supernatant.
14. Wash cells in 5 ml of dH_2O. Centrifuge 3 min at 1,400 × *g* to pellet cells. Decant supernatant.
15. Resuspend in 100 μl dH_2O and plate equal volumes (~100 μl) on two yeast two-hybrid selection plates (SD−TRP−MET−LEU−URA−HIS plus the concentration of 3-AT determined in Subheading 3.5.3). Distribute yeast using glass beads.
16. After plates have dried, remove the glass beads, place the plates in plastic bag, and incubate at 30°C.
17. Monitor the yeast two-hybrid selection plates for the appearance of colonies. Colonies may appear as early as day 3 or 4, or as late as day 14. Pick positive colonies every few days to prevent over growth or loss due to contamination of the plate (see Note 8).
18. Pick yeast colonies from the selection plates and process them as described in Subheading 3.4.4, steps 15–26.
19. Sequence PCR products using standard protocols. An economical approach utilizes exonuclease I and shrimp alkaline phosphatase to remove unincorporated dNTPs and primers and 1/10th Big Dye v3.1 (Applied Biosystems, 0.4 μl BigDye per reaction) reactions for sequencing.

3.7. Confirmation of Yeast Two-Hybrid Positives

Confirmation of yeast two-hybrid positives requires retesting the pairs of DBD and AD fusion proteins in fresh yeast cells. To streamline the retest procedure, the AD insert PCR product from Subheading 3.6 step 19 is cloned into pOAD.102 by homologous recombination in yeast (as opposed to transforming the AD plasmid into *E. coli* and then back into yeast). The method listed below is appropriate for confirmation of a small number of positives. For retesting a larger number of positives (see Note 9).

1. Prepare a culture of BK100 cells for transformation as described in Subheading 3.4.4.
2. Set up three transformations:

   Transformation #1: 10 ng circular pOAD.102

   Transformation #2: 250 μg linear pOAD.102 (negative control)

   Transformation #3: 250 μg linearized pOAD.102 plus 5 μl AD insert PCR product from Subheading 3.6 step 23.

   Complete the transformation as described in Subheading 3.4.4.
3. Plate 50 μl of transformations 1 and 2 on SD−LEU−URA+ADE and SD−LEU+ADE. For transformation 3, plate 50 μl on SD−LEU−URA+ADE and on SD−LEU+ADE plates. Incubate 3 days at 30°C.
4. Count the number of colonies on each plate. The number of colonies from transformation 3 should be equal on SD−LEU−URA+ADE and SD−LEU+ADE plates since this fragment was previously selected for expression in yeast.
5. Pick colonies on the SD−LEU−URA+ADE plate from tranformation 3. Verify the presence of insert by PCR and sequencing.
6. Mate AD strains with appropriate DBD strain and with R2HMet containing pOBD.111-IF (negative control) by patching on YPDA medium.
7. Select diploid yeast on SD−TRP−LEU+ADE plates.
8. Streak diploid cells on SD−TRP−LEU+ADE, SD−TRP−MET−LEU−URA−HIS plus 3-AT, and SD−TRP−MET−LEU−URA−ADE plates.
9. AD fusions that activate expression of both the *HIS3* and *ADE2* reporter genes (i.e., that promote growth on media lacking HIS and ADE) in combination with the DBD fusion, but not with pOBD111-IF, are considered true positives. Constructs that activate expression of the reporter genes in combination with pOBD.111-IF are false-positives and should be discarded. In some cases, you may wish to pursue AD fusions that activate expression of the *HIS3* reporter gene but not the *ADE2* reporter. The *ADE2* reporter gene in BK100 is more stringent than the *HIS3* gene, so some weak positives may not activate both. However, activation of a single reporter gene is associated with a higher false-positive rate.

4. Notes

1. The diversity of the AD cDNA library can be increased by using 5′ or 3′ oligos that have one or two extra bases inserted between the recombination tail and the region homologous to the cDNA. Using all nine combinations of 5′ and 3′ oligos to amplify the cDNA will enable all cDNA fragments to be cloned in-frame with the *GAL4* AD and *URA3*.
2. The volume of PCR product listed assumes that 10 μl of PCR product yielded the greatest number of colonies on SD−LEU−URA+ADE. Scale the amount of PCR product up or down depending on your results.
3. In some cases, promiscuous clones from an AD library can be avoided by amplifying the cDNA with a different set of PCR primers. The pair of PCR primers used in this protocol specify only one out of nine possible reading frames – i.e., only one out of three fragments will be in frame with the AD and one out of three will be in frame with the *URA3* gene. By using 5′ or 3′ oligos that have one or two extra bases inserted between the recombination tail and the region homologous to the cDNA, it is possible to select other reading frames for cloning and to eliminate the promiscuous clone from the library.
4. If no colonies grow on SD−TRP−MET+ADE from transformation 3, but many colonies are present on SD−TRP+ADE, check 12 colonies from SD−TRP+ADE to ensure that the plasmids contain an insert. If no inserts are present, the most likely explanation is that small DNA fragments remained after fractionation on the Sephacryl S500 column and were selected at a higher rate than larger fragments after recombination into pOBD.111. If inserts of >300 bp are observed in >90% of the clones analyzed, first confirm that the plates are indeed SD−TRP−MET+ADE and then replica plate colonies from SD−TRP+ADE to SD−TRP−MET+ADE plates to identify colonies that are able to grow on medium lacking MET and thus contain a plasmid from which *MET2* is expressed. It is possible, though unlikely, that no fragments will be expressed.
5. If many fewer colonies are present on SD−TRP−MET+ADE than on SD−TRP+ADE, the fragment is likely not expressed well. In this case, verify that the colonies on SD−TRP+ADE contain plasmids in which the insert has recombined appropriately. If so, test for growth of these yeast strains on SD−TRP−MET+ADE by streaking or replica plating. In many cases, the level of *MET2* expression is too low to allow colonies to form after transformation, but high enough to enable slow growth when the yeast are plated on SD−TRP−MET+ADE

medium . If no growth is apparent on SD–TRP–MET+ADE, the fragment is not expressed and should not be screened in the yeast two-hybrid assay.

6. The yeast two-hybrid mating assay can be adapted to 96-well format using 2-ml deep-well dishes. Keep the total number of cfus of DBD and AD library strains the same, but reduce the volume to 0.5 ml in steps 4 and 5. Reduce the volumes to 1 ml in steps 7, 10, and 15.
7. Different spectrophotometers give slightly different numbers of cfu per 1 OD_{600} unit. This protocol assumes that 1 $OD_{600} = 2.0 \times 10^7$ cfu. It is worthwhile to check this value for the spectrophotometer being used.
8. In our experience, colonies that appear after day 14 are unlikely to represent reproducible interactions. In some cases, a large number of small colonies will begin to appear around day 14 apparently due to weak self-activation.
9. The volumes of the components in the transformation mix can be scaled-down to enable the transformations to be performed in a sterile 96-well microtiter dish. Use 17.6 μl BK100, 64 μl 50% PEG, 8 μl 1 M LiOAc, 3.2 μl heat denatured carrier DNA (2 mg/ml), 80 μg linear pOAD.102, and 2 μl of PCR product. Incubate 30 min at 30°C, add 7.4 μl DMSO, and heat shock 15 min at 42°C. Centrifuge to pellet cells, remove supernatant, resuspend cells in 5 μl dH_2O, and spot 5 μl on SD–LEU–URA+ADE medium in a Nunc OmniTray. It is critical that the linear vector does not yield any colonies under these conditions since it will not be possible to pick individual colonies from the spot.

Acknowledgments

I thank Hakeenah Brown, Sudip Khadka, Leonie Leduc, and Wesley Penn for critically reading the manuscript. Purdue University provided financial support.

References

1. Fields, S., and Song, O. (1989) A novel genetic system to detect protein-protein interactions. *Nature* **340**, 245–6.
2. Gardner, M. J., Hall, N., Fung, E., et al. (2002) Genome sequence of the human malaria parasite *Plasmodium falciparum*. *Nature* **419**, 498–511.
3. Mandel, C. R., Bai, Y., and Tong, L. (2008) Protein factors in pre-mRNA 3'-end processing. *Cell Mol Life Sci* **65**, 1099–122.
4. LaCount, D. J., Schoenfeld, L. W., and Fields, S. (2009) Selection of yeast strains with enhanced expression of *Plasmodium falciparum* proteins. *Mol Biochem Parasitol* **163**, 119–22.

5. Milek, R. L., Stunnenberg, H. G., and Konings, R. N. (2000) Assembly and expression of a synthetic gene encoding the antigen Pfs48/45 of the human malaria parasite *Plasmodium falciparum* in yeast. *Vaccine* **18**, 1402–11.
6. Sibley, C. H., Brophy, V. H., Cheesman, S., et al. (1997) Yeast as a model system to study drugs effective against apicomplexan proteins. *Methods* **13**, 190–207.
7. Frischmeyer, P. A., van Hoof, A., O'Donnell, K., et al. (2002) An mRNA surveillance mechanism that eliminates transcripts lacking termination codons. *Science* **295**, 2258–61.
8. LaCount, D. J., Vignali, M., Chettier, R., et al. (2005) A protein interaction network of the malaria parasite *Plasmodium falciparum*. *Nature* **438**, 103–7.
9. Vidalain, P. O., Boxem, M., Ge, H., et al. (2004) Increasing specificity in high-throughput yeast two-hybrid experiments. *Methods* **32**, 363–70.
10. Boxem, M., Maliga, Z., Klitgord, N., et al. (2008) A protein domain-based interactome network for *C. elegans* early embryogenesis. *Cell* **134**, 534–45.
11. Flajolet, M., Rotondo, G., Daviet, L., et al. (2000) A genomic approach of the hepatitis C virus generates a protein interaction map. *Gene* **242**, 369–79.
12. Ma, H., Kunes, S., Schatz, P. J., et al. (1987) Plasmid construction by homologous recombination in yeast. *Gene* **58**, 201–16.
13. James, P., Halladay, J., and Craig, E. A. (1996) Genomic libraries and a host strain designed for highly efficient two-hybrid selection in yeast. *Genetics* **144**, 1425–36.
14. Su, X. Z., Wu, Y., Sifri, C. D., et al. (1996) Reduced extension temperatures required for PCR amplification of extremely A+T-rich DNA. *Nucleic Acids Res* **24**, 1574–5.

Chapter 8

Mapping Interactomes with High Coverage and Efficiency Using the Shifted Transversal Design

Xiaofeng Xin, Charles Boone, and Nicolas Thierry-Mieg

Abstract

"Smart-pooling" is a strategy to achieve high efficiency, sensitivity, and specificity in large-scale yeast two-hybrid screening. In smart-pooling, reagents are multiplexed in a highly redundant manner and the positives can be read out on the final selection plates without the requirement of any extra experimental steps. We have shown that the Shifted Transversal Design (STD), a powerful theoretical construction for smart-pooling, can be used in yeast two-hybrid interactome mapping. STD pooling can achieve similar levels of sensitivity and specificity as one-on-one array-based yeast two-hybrid, while the costs and workloads are much lower. This chapter focuses on the construction and usage of STD arrays for large-scale yeast two-hybrid interactome mapping.

Key words: Yeast two-hybrid, Protein–protein interaction, Interactome, Pooling, Smart-pooling, Multiplexing, Redundancy, Shifted transversal design

1. Introduction

"Smart-pooling" (or "group testing") aims to increase efficiency, accuracy, and coverage in high-throughput screening projects, including yeast two-hybrid (Y2H) interactome mapping (1). It involves the building and assaying of defined pools of specific items, including Y2H preys, such that each item is present in several pools and hence tested several times. In contrast to the traditional pooling approaches, positive preys can be read out from the pattern of positive pools on the final selection plates, without the need to perform a secondary step for identification, such as individually testing the preys from positive pools or colony-picking and sequencing the positive colonies in Y2H. The smart-pooling approach is general and has been applied recently to identify protein–protein

Bernhard Suter and Erich E. Wanker (eds.), *Two Hybrid Technologies: Methods and Protocols*, Methods in Molecular Biology, vol. 812, DOI 10.1007/978-1-61779-455-1_8, © Springer Science+Business Media, LLC 2012

interactions (2, 3) and protein–DNA interactions (4), and combined with second-generation DNA sequencing to study rare sequence variants (5, 6).

Central to the smart-pooling procedure is the choice of the pooling design, i.e., which pools contain which preys. The Shifted Transversal Design (STD) is a flexible and powerful algorithm for designing smart-pools (7). It significantly outperforms other published combinatorial smart-pooling designs under a standard theoretical model (7, 8). In our previous study (3), we have shown that STD pooling is a method of choice for obtaining high-coverage protein–protein interactomes, and could prove effective in a wide range of high-throughput experiments.

Choosing an appropriate STD design requires estimating the maximum usable pool size (at the assay densities of interest), the expected rates of false positives and false negatives when using pools of that size, and the number of positives that one wishes to detect (accepting the fact that some positives may be missed for protein–protein interaction hubs that have more positives). Based on these estimates, computer simulations are performed to select the most appropriate STD parameters (redundancy, and batch size which is closely dependent on the pool size) as described (8). "Redundancy" is the number of pools that contain any given item. Part of the redundancy is necessary to identify a single positive item, and the remaining "extra redundancy" allows the system to deal with noise (false-positive and false-negative pools) and multiple positive items (within a particular batch of pools). To choose the most appropriate STD design, one needs to keep in mind that there is always a trade-off: a design with a higher redundancy and/or smaller batches will enable to detect more positives unambiguously and correct for more false positives and false negatives, but this comes at the price of a larger number of pools, hence increasing the screening workloads and costs.

To limit the cost of building the STD pools and to increase their flexibility of usage, in our previous study (3) we took advantage of inherent STD symmetries by designing and building small intermediary micropools. In Fig. 1, an example illustrates the process of designing and building a simple STD: 18 preys are pooled according to a small STD design. Initially, the 18 preys are split into two groups of nine preys (groups A and B), and each group is pooled independently according to its corresponding STD subdesign to obtain two sets of micropools (sets A and B). Each micropool contains three different preys (pool size of 3), and each prey is contained in three different micropools (redundancy of 3), which form this prey's unique signature. In fact, any two micropools are sufficient to uniquely identify a prey, so these micropools have an extra redundancy of 1. Finally, each pair of same-numbered micropools from sets A and B are superposed to obtain one batch of STD pools (p1–p9). These STD pools still possess a redundancy of 3, but their pool size is now 6, and the nine STD pools accommodate

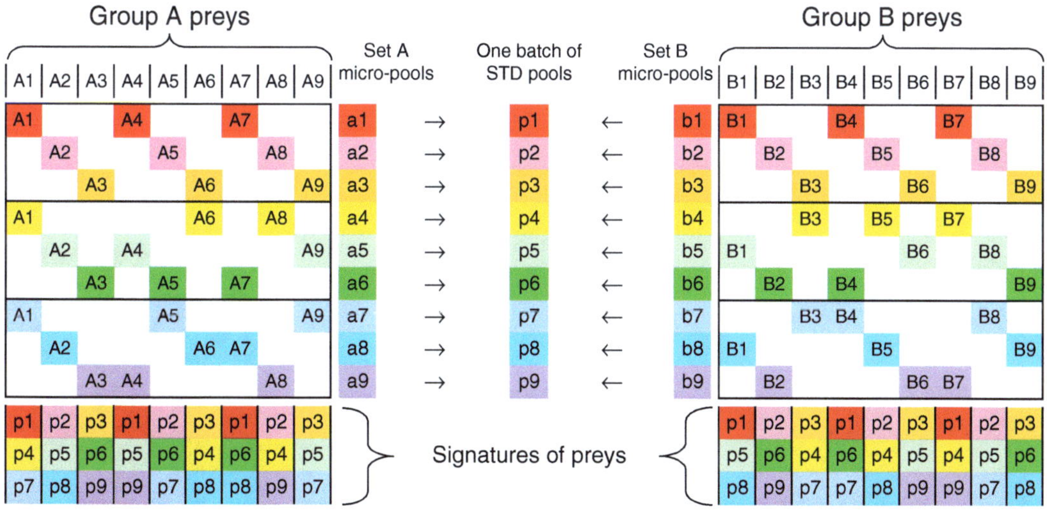

Fig. 1. A simple STD pooling design. Two groups of nine preys (group A: A1–A9 and group B: B1–B9) are separately pooled into nine color-coded micropools (set A: a1–a9 and set B: b1–b9), according to subdesigns of a larger STD design (which can accommodate all 18 preys). Each micropool contains three preys (e.g., a1 contains A1, A4, and A7), and each prey is present in a unique combination of three micropools (a unique signature, e.g., a1|a5|a9 for prey A4). The pairs of same-numbered micropools from sets A and B can then be superposed to generate STD pools (one batch: p1–p9), each containing six preys, and each prey still has a unique signature in the STD pools. Reproduced from ref. 3 with permission from Cold Spring Harbor Laboratory Press.

all 18 preys. Of note, each prey still has a unique signature, although the extra redundancy is now 0 because all three pools are required to identify each prey uniquely.

We built the worm STD pools in a similar manner but on a higher scale (Figs. 2 and 3). The prey library was conceptually split into smaller groups, e.g., 75 groups of 169 preys in our previous study (3). Each group was STD-pooled independently to obtain a set of micropools. Every 12 sets of micro-pools were built according to subdesigns of a larger STD design, so that the micropools are superposable to generate larger STD pools (Fig. 2). In practice, we superposed every two micropools to generate STD-1536, every six micropools to generate STD-384, and every 12 micropools to generate STD-96 (Fig. 3).

Once the STD arrays have been built, STD Y2H screening is performed similar to one-on-one array-based Y2H (1-on-1 Y2H) (i.e., mating, diploid selection, and final growth selection) (9) (see Note 1), with an additional in silico decoding step (Fig. 3).

Building an ORFeome-scale STD array with high redundancy requires a significant initial investment, but it results in a valuable resource that can be distributed and used repeatedly. Subsequently, the workload and cost for STD Y2H screening are much less than those for one-on-one Y2H. For example in our previous study (3), screening costs and workload were approximately threefold higher for one-on-one than for STD Y2H, while sensitivity and specificity were similar.

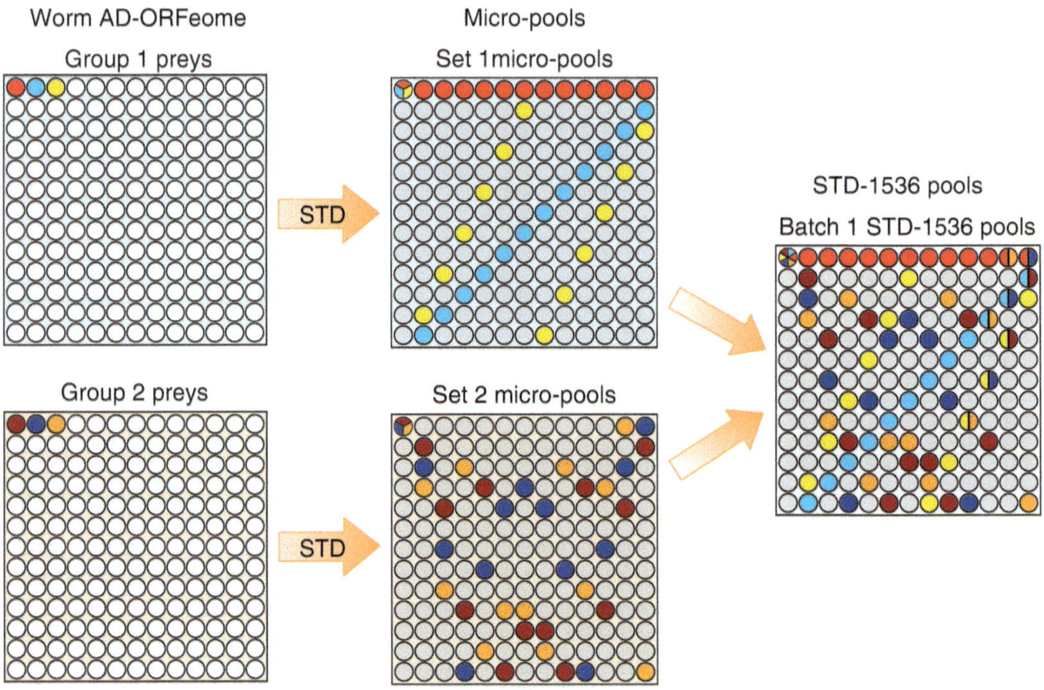

Fig. 2. STD pooling design described in this chapter. 12,675 Worm AD-ORFeome preys were split into 75 groups each containing 169 preys. Each group was STD pooled into a set of 169 micropools. Each micropool contains 13 preys (micropool size: 13), and each prey is contained in a unique combination of 13 micropools (a unique signature; redundancy: 13), as illustrated with three color-coded preys in sets 1 or 2. Two preys co-occur in at most one micropool, so that each prey is uniquely identified by any two of the 13 pools that contain it; therefore, these micropool designs have an extra redundancy of 11. In addition, the micropool signatures of preys with identical AD-ORFeome coordinates from groups 1 and 2 are very different (e.g., light red in set 1 and dark red in set 2): every two sets of micropools can be superposed to obtain one batch of STD-1536 pools, such that two preys from different groups co-occur in at most two common pools. Consequently, in STD-1536 each prey is uniquely identified by any 3 of the 13 pools that contain it: STD-1536 pools possess an extra redundancy of 10. Reproduced from ref. 3 with permission from Cold Spring Harbor Laboratory Press.

For details on STD design, please refer to (7, 8). In this chapter, we assume that an appropriate STD design has been selected, and we focus on the experimental protocols for constructing and screening the chosen STD pools. We illustrate the protocol with the designs that we used in our previous work (3), but all of the procedures would be very similar with other STD designs.

2. Materials

2.1. Media and Stock Solutions

1. 3-AT (3-amino-1,2,4-triazole, Sigma): Prepare 2 M solution in water, filter to sterilize, and store at 4°C (see Note 2).
2. Glycerol: Prepare 40% (v/v) solution in water, filter to sterilize, and store at room temperature.
3. Glucose: Prepare 40% solution in water, autoclave, and store at room temperature.

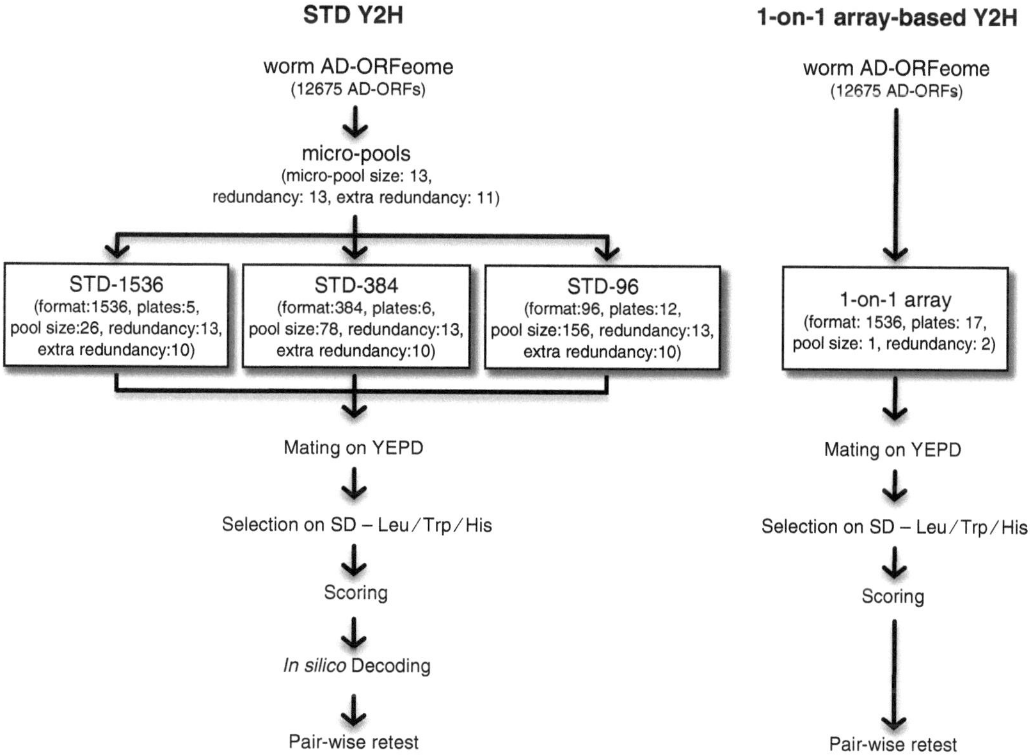

Fig. 3. Flowchart of STD Y2H side-by-side with one-on-one array-based Y2H.

4. Amino acid supplementary powder mixture for synthetic media (complete): 3 g adenine, 2 g uracil, 2 g inositol, 2 g *p*-aminobenzoic acid, 2 g alanine, 2 g arginine, 2 g asparagine, 2 g aspartic acid, 2 g cysteine, 2 g glutamic acid, 2 g glutamine, 2 g glycine, 2 g histidine, 2 g isoleucine, 10 g leucine, 2 g lysine, 2 g methionine, 2 g phenylalanine, 2 g proline, 2 g serine, 2 g theronine, 2 g tryptophan, 2 g tyrosine, and 2 g valine (Fisher). Dropout (DO) mix is a combination of the aforementioned ingredients minus the appropriate supplement (see Note 3). For example, to make DO – Trp, all ingredients except tryptophan need to be mixed; to make DO – Leu/Trp, all ingredients except leucine and tryptophan need to be mixed. The resultant DO mixture can be stored in tinted glass bottles and stored at room temperature. 2 g of the DO powder mixture is used per liter of medium.

5. SD – Leu liquid medium: Add 6.7 g yeast nitrogen base without amino acids (BD Difco), 2 g amino acid supplementary powder mixture lacking leucine (DO – Leu) and 950 mL water in a 2-L flask. Autoclave. Add 50 mL 40% glucose, and mix thoroughly (see Note 4).

6. SD – Trp liquid medium: Add 6.7 g yeast nitrogen base without amino acids, 2 g amino acid supplementary powder mixture lacking tryptophan (DO – Trp), and 950 mL water in a 2-L flask. Autoclave. Add 50 mL 40% glucose, and mix thoroughly.
7. YEPD agar medium: Add 120 mg adenine (Sigma), 10 g yeast extract, 20 g peptone, and 20 g bacto agar (BD Difco) to 950 mL water in a 2-L flask. Autoclave. Add 50 mL 40% glucose, mix thoroughly, cool to approximately 55°C, and pour plates.
8. SD – Leu agar medium: Add 6.7 g yeast nitrogen base without amino acids, 2 g amino acid supplementary powder mixture lacking leucine (DO – Leu), and 150 mL water in a 250 mL flask. Add 20 g bacto agar to 800 mL water in a 2-L flask. Autoclave separately. Combine the autoclaved solutions, add 50 mL 40% glucose, mix thoroughly, cool to approximately 55°C, and pour plates.
9. SD – Trp agar medium: Add 6.7 g yeast nitrogen base without amino acids, 2 g amino acid supplementary powder mixture lacking tryptophan (DO – Trp), and 150 mL water in a 250-mL flask. Add 20 g bacto agar to 800 mL water in a 2-L flask. Autoclave separately. Combine the autoclaved solutions, add 50 mL 40% glucose, mix thoroughly, cool to approximately 55°C, and pour plates.
10. SD – Leu/Trp/His agar medium: Add 6.7 g yeast nitrogen base without amino acids, 2 g amino acid supplementary powder mixture lacking leucine, tryptophan and histidine (DO – Leu/Trp/His), and 150 mL water in a 250 mL flask. Add 20 g bacto agar to 800 mL water in a 2-L flask. Autoclave separately. Combine the autoclaved solutions, add 50 mL 40% glucose, cool to approximately 55°C, add proper volume of 2 M 3-AT solution, mix thoroughly, and pour plates. For example, to make medium with a final concentration of 3-AT at 2 mM, add 1 mL of the 2 M 3-AT stock to the medium (see Note 2).

2.2. Plates and Accessories

1. 96-Format deep well plates (Nunc), Tecan Freedom EVO liquid handling robot, and Tecan Aquarius MultiChannel Pipetting robot (Tecan Group Ltd.) are used for building micropools and STD pools. 96-Well and 384-well plates (ABGene) are used for storing the resultant micropools and STD pools at −80°C.
2. The BioMatrix robot (S&P Robotics Inc., Toronto, ON) and OmniTrays (Nunc) are used for all replica pinning procedures with the 384- and 1536-format screens (see Note 5).
3. 150 mm Petri dishes (Fisher) are used for all procedures with the 96-format screens and pairwise retest. Yeast transferring from liquid culture to the agar plates is performed using the Tecan Aquarius MultiChannel Pipetting robot (Tecan Group Ltd.)

or a multichannel pipette (Fisher). The plate replication procedures are performed using velvet replica plating apparatus Cora Styles Lab Supplies, OR.

2.3. Strains, Plasmids, and AD-ORFeome Libraries

1. Yeast two-hybrid (Y2H) host strains: prey strain Y8800 (*MATa trp1-901 leu2-3,112 his3-200 ura3-52 gal4Δ gal80Δ LYS2::GAL1-HIS3 GAL2-ADE2 met2::GAL7-lacZ cyh*^{R}) and bait strain Y8930 (*MATα trp1-901 leu2-3,112 his3-200 ura3-52 gal4Δ gal80Δ LYS2::GAL1-HIS3 GAL2-ADE2 met2::GAL7-lacZ cyh*^{R}).
2. pPC97 (*GAL4-DBD*, *LEU2*) is the *GAL4* DNA-binding domain (DBD) plasmid for Y2H bait construction.
3. pMV257 (also called pPC86-CYH, *GAL4-AD*, *TRP1*) is the *GAL4* activation domain (AD) plasmid for Y2H prey construction.
4. AD-ORFeome libraries: Many AD-ORFeome collections (and other ORFeome collections) are available through the ORFeome Collaboration (http://www.orfeomecollaboration.org/). We used the *C. elegans* AD-ORFeome v1.1 and v3.1 in our previous study (3), which consists of 12,675 worm AD-ORFs.

3. Methods

3.1. STD Designs

1. Before designing the STD arrays, we performed dilution tests at two different densities, 384 and 1,536 spots per plate, in order to identify the maximum pool sizes that enable detection of positive controls. In conjunction with simulations performed with interpool (8), we choose pool sizes of 78 for the 384-format and 26 for the 1536-format arrays.
2. We did not investigate the maximum pool size as thoroughly for 96-format arrays, because we only screened the *C. elegans* STD pools in 384- and 1536-formats. However, our preliminary experiments in 96-format with human ORFs indicated that pool size of up to 500 may be used (3), employing a screening protocol similar to pairwise retest steps (see Subheading 3.5 below). We still built 96-format worm STD pools, because due to our micropool strategy they were an essentially free side-product. Since our focus was on high-density formats, we did not optimize these pools but chose a pool size of 156 for 96-format (see Note 6).
3. The prey library (worm AD-ORFeome v1.1 and v3.1), which contains 12,675 unique AD-ORFs, was conceptually split into 75 groups of 169 preys (75 × 169 = 12,675). Each group was STD-pooled independently to obtain a set of 169 worm

micropools containing 13 preys each (micro-pool size of 13). Every 12 sets of micropools were built according to sub-designs of a larger STD design, so that the micropools are superposable to generate larger STD pools (Fig. 2).

4. Based on the chosen 1536-, 384- and 96-format pool sizes, the micropools were (1) superposed in pairs (as shown in Fig. 2), to produce the STD-1536 pools containing 26 preys per pool, or (2) superposed in sextuplets, to generate the STD-384 pools with 78 preys per pool, or (3) superposed in duodecaplets, to generate the STD-96 pools with 156 preys per pool. In the resulting STD pooling designs, the 12,675 preys are either split into 40 batches of STD-1536 pools with up to 338 preys per batch, or 13 batches of STD-384 pools with up to 1,014 preys per batch, or seven batches of STD-96 pools with up to 2,028 preys per batch. The STD-1536, STD-384, and STD-96 batches each contain 169 pools. All batches within an STD design are arrayed as spots on a series of plates, but the batches are disjoint and decoded independently. These three designs possess an extra redundancy of 10, which provides high noise-correction capabilities.

In fact, every 12 sets of worm micropools (169 preys per set) is a collection of subdesigns of an STD design with 2,028 preys per batch (whose extra redundancy is 10). They can therefore be superposed to obtain different STD designs. Each individual set of micropools is isomorphic to a smaller STD design and can be used as an STD pooling batch in its own right, with an extra redundancy of 11. When 2, 6, or 12 consecutive micropool sets are superposed to obtain STD-1536, or STD-384, or STD-96, the resulting designs are again isomorphic to an STD design with 228, 1,014, or 2,028 preys per batch, respectively, and therefore they all have an extra redundancy of 10. More specifically, worm micropools are subdesigns of STD(2,028;13;13) isomorphic to STD(169; 13;13), and were superposed to obtain designs isomorphic to STD(338;13;13) for STD-1536, STD(1,014;13;13) for STD-384, and STD(2,028;13;13) for STD-96 (see ref. 8 for details).

3.2. Building STD Pools

1. The sources were all worm ORFeome v1.1 and v3.1 AD plates (11,001–11,114 and 31,001–31,022).
2. The source plates were thawed at room temperature, inoculated in 96-format deep well plates containing SD – Trp liquid media, and incubated at 30°C for 2 days.
3. Before pooling, the yeast cells in the freshly cultured source plates were mixed thoroughly with a plate shaker or by pipetting with a Tecan Aquarius MultiChannel Pipetting robot.
4. Micropools were assembled in 96-format deep well plates by cherry-picking from the source to the target plates using a

Tecan Freedom EVO liquid handling robot. The robot was programmed directly in GWL (rather than using CSV files), which optimized the process. Indeed, each column of eight source wells was aspirated (and dispensed to the appropriate destination plates and wells) 13 times in a row, before moving on to the next column of source wells. With GWL we were able to resuspend and thoroughly mix the yeast in source wells by performing a series of up-and-down operations before the first aspiration (using a specific "liquid class"); while the 12 subsequent aspirations were performed without any up-and-down cycles, which gained a lot of time. All scripts are available (see Note 10).

5. STD-1536, STD-384, and STD-96 pools were generated in 384-format and 96-format by superposing the appropriate micropool plates with a Tecan Aquarius MultiChannel Pipetting robot in 384-well and 96-well storage plates.
6. All pools were frozen and stored at −80°C with 20% glycerol, by adding an equal volume of 40% (v/v) glycerol solution to the yeast culture, and mixing thoroughly.

3.3. Handling STD Arrays

1. Before Y2H screening, STD arrays glycerol stock plates were thawed at room temperature, mixed thoroughly with a plate shaker and transferred to SD – Trp agar plates with a BioMatrix robot, and incubated at 30°C for 2 days.
2. After incubation, this set of "master" agar plates was replicated into multiple copies (up to eight copies), which could be used either for screening or as a source for further replications (see Note 7).

3.4. Y2H Screening with STD Arrays

The Y2H screening with STD arrays was performed following the steps below. For 1536-format and 384-format screens, all pinning steps are performed with a BioMatrix robot, using a 1536-format pinhead with 0.7 mm diameter pins or a 384-format pinhead with 1 mm diameter pins. For 96-format, cell transfer from liquid culture to agar plates is performed using a Tecan Aquarius Multi-Channel Pipetting robot, and plate replication is performed using velvet replication.

1. Bait culture: Bait is inoculated in 5 mL SD – Leu liquid medium and cultured at 30°C, 200 rpm, for 2 days. Bait culture is then transferred to SD – Leu plates. One bait plate is sufficient for up to eight mating plates in step 3 below. The bait plates are bagged and incubated upside-down at 30°C for 2 days.
2. Prey culture: Prey plates are replicated from STD stock array plates from Subheading 3.3 above. The prey plates are bagged and incubated upside-down at 30°C for 2 days.

3. Mating: Baits are first transferred to YEPD plates by robotic pinning (see Note 8). Preys are then transferred to the mating plates on top of the bait spots. The mating plates are bagged and incubated upside-down at 30°C for 16–24 h.
4. Final selection: The incubated mating plates are replicated to SD – Leu/Trp/His + 3-AT final selection plates. The selection plates are bagged and incubated upside-down at 30°C for 7–14 days (see Note 9). Of note, the diploid selection step on SD – Leu/Trp, which is normally used in similar array-based Y2H screens, was skipped (see Note 1).
5. Scoring: STD spots were scored manually using the in-house ColonyImager image-processing program using four discrete levels for each spot: strong (clear positive) or weak (smaller than strong but well above background) for positives, and none (no detectable signal) or faint (barely above background, most likely negative) for negatives. These results were transformed into a suitable XML format with Perl scripts, and decoded with interpool (8). The "distance" parameter δ was chosen to fit the experimental conditions. For example, in our previous work (3), false positives were relatively rare while false negatives were common, leading us to use a very sensitive distance ($\delta_{NONE} = 2$, $\delta_{FAINT} = 1$, $\delta_{WEAK} = 4$, $\delta_{STRONG} = 6$). This choice did not compromise specificity. All relevant scripts, programs, and data files are available (see Note 10). A confidence score was attributed to each STD hit, depending solely on the number of putative false-negative spots for the hit. Specifically, "none" spots carry a cost of 2 and "faint" spots 1, and summing over all false negatives for a hit yields a total cost; if this total cost is at most 4 the confidence score is 5, if it is up to 8 the score is 4, and so on until reaching the lowest confidence score of 1 if the total cost is between 17 and 20.

3.5. Pairwise Retest

1. Pairwise retests were performed in quadruplicate by scoring a single phenotype of the *HIS* reporter in 96-format on agar plates.
2. Put 5 μL bait fresh culture on the YEPD plates using a Tecan Aquarius MultiChannel Pipetting robot or a multichannel pipette, let the plate dry at room temperature. Put 5 μL prey fresh culture on top of the bait spots, dry at room temperature. Incubate at 30°C for 16–24 h.
3. Replicate the YEPD plates to SD – Leu/Trp/His plates using velvet replica plating. Incubate the target plates at 30°C for 2 days.
4. Remove most of the yeast cells on the plates using velvet replica until the yeast cell spots can be barely see on the plates. Incubate at 30°C for 2 days.

5. Each retest was scored as negative, weak, or strong (0, 1, or 2, respectively). Summing over the four replicates, one can obtain a retest score between 0 and 8 for each hit. Positive hits are those whose retest score was at least six, while hits with scores at most two were classified as false positives and the remaining hits with intermediate retest scores were classified as dubious, which can either be retested or be grouped as true positives or false positives depending on one's discrimination and screening stringency.

4. Notes

1. Diploid step skipping: Preliminary experiments showed that, in our hands, including a diploid selection step (on SD – Leu/Trp between mating and final selection) did not result in any improvements, perhaps because the robotic replication step may not be fully effective in transferring all components of a colony spot, so that the additional replication step compensates any gains from the diploid selection.
2. 3-AT can also be added as powder, but it will take longer to dissolve. We recommend to use 2 M 3-AT solution for media with 3-AT concentrations up to 20 mM, and use 3-AT powder for media with >20 mM 3-AT.
3. When making up the amino acid supplementary mixture (complete or dropout), the solid ingredients should be combined and mixed very thoroughly by turning end-over-end for at least 15 min, while adding a couple of clean marbles or glass beads helps the mixing.
4. The yeast medium used in this work can be made using any generic method for making yeast medium, not limited to the one described in Subheading 2 of this chapter.
5. Other robotic pinning systems can also be used. For example, the Singer RoTor DHA benchtop robot uses disposable replicators RePads, and PlusPlates (Singer Instruments, UK). Although manual pinning should also work in theory, to guarantee high quality and high reproducibility of the screening results, robotics pinning is highly recommended for high-density formats such as 384-format, 768-format, 1536-format, or higher density formats.
6. Users focusing on 96-format assays should choose an STD design optimized for this density. In particular, dilution tests should be performed to find the largest reliable pool size in their hands. In our experience, pool size is often the limiting

factor in smart-pooling, and using pools as large as possible should yield the most efficient STD pools without compromising sensitivity – although this also depends on the expected fraction of positives and error-rates, and should be confirmed through simulations as described. The micropools described here can produce pools of size at most 156 (2,028/13), since they are subdesigns of STD(2,028;13;13), but the approach can be easily extended to allow the production of larger pools. For example micropools constructed as subdesigns of STD(12,675;13;13) can be superposed to obtain STD pools of size up to 975, while smaller pools for high-density formats, obtained by superposing two or six sets of these new micropools, would be equivalent to those produced in this study.

7. However, fresh arrays should still be occasionally remade from glycerol stock, because the STD arrays begin losing representativity after more than five sequential replications. The arrays appear fully functional after being stored at 4°C for at least 2 months, though in our previous study we used them within 1 week after replication to avoid confounding factors and guarantee the highest data quality.
8. We notice the wetness of agar plates significantly contributes to the screening variation. Wet plates could lead to too much yeast cells been transferred and result in high background; dry plates could lead to too few and uneven transferring and result in unreadable final plates. It would be a good idea to make and process the plates in a uniform matter, and test plates with different wetness before large-scale screening.
9. The difference of incubation time might vary for different baits. We suggest starting to check the plates after 5 days incubation, and continue the monitoring every 2 days afterward. Multiple sets of pictures might be taken for choosing the best scoring time.
10. The scripts and programs mentioned in this chapter are freely available online: they have been published as a Supplementary Data file associated to our previous study (3), and can also be downloaded from NT-M's website (http://www-timc.imag.fr/Nicolas.Thierry-Mieg).

Acknowledgment

This work was supported by Canadian Cancer Society grant # 015311 awarded to CB.

References

1. Thierry-Mieg, N. (2006) Pooling in systems biology becomes smart, *Nat Methods 3*, 161–162.
2. Jin, F., Avramova, L., Huang, J., and Hazbun, T. (2007) A yeast two-hybrid smart-pool-array system for protein-interaction mapping, *Nat Methods 4*, 405–407.
3. Xin, X., Rual, J. F., Hirozane-Kishikawa, T., Hill, D. E., Vidal, M., Boone, C., and Thierry-Mieg, N. (2009) Shifted Transversal Design smart-pooling for high coverage interactome mapping, *Genome Res 19*, 1262–1269.
4. Vermeirssen, V., Deplancke, B., Barrasa, M. I., Reece-Hoyes, J. S., Arda, H. E., Grove, C. A., Martinez, N. J., Sequerra, R., Doucette-Stamm, L., Brent, M. R., and Walhout, A. J. (2007) Matrix and Steiner-triple-system smart pooling assays for high-performance transcription regulatory network mapping, *Nat Methods 4*, 659–664.
5. Erlich, Y., Chang, K., Gordon, A., Ronen, R., Navon, O., Rooks, M., and Hannon, G. J. (2009) DNA Sudoku--harnessing high-throughput sequencing for multiplexed specimen analysis, *Genome Res 19*, 1243–1253.
6. Prabhu, S., and Pe'er, I. (2009) Overlapping pools for high-throughput targeted resequencing, *Genome Res 19*, 1254–1261.
7. Thierry-Mieg, N. (2006) A new pooling strategy for high-throughput screening: the Shifted Transversal Design, *BMC Bioinformatics 7*, 28.
8. Thierry-Mieg, N., and Bailly, G. (2008) Interpool: interpreting smart-pooling results, *Bioinformatics 24*, 696–703.
9. Uetz, P., Giot, L., Cagney, G., Mansfield, T. A., Judson, R. S., Knight, J. R., Lockshon, D., Narayan, V., Srinivasan, M., Pochart, P., Qureshi-Emili, A., Li, Y., Godwin, B., Conover, D., Kalbfleisch, T., Vijayadamodar, G., Yang, M., Johnston, M., Fields, S., and Rothberg, J. M. (2000) A comprehensive analysis of protein-protein interactions in Saccharomyces cerevisiae, *Nature 403*, 623–627.

Chapter 9

Assigning Confidence Scores to Protein–Protein Interactions

Jingkai Yu, Thilakam Murali, and Russell L. Finley Jr.

Abstract

Screens for protein–protein interactions using assays like the yeast two-hybrid system have generated volumes of useful data. The protein interactions from these screens have been used to develop a better understanding of the functions of individual proteins, regulatory pathways, molecular machines, and entire biological systems. The value of this data, however, is limited by the inherent frequency of false positives that arise in most protein interaction screens. Appreciable numbers of false positives can crop up in both low-throughput and high-throughput screens, and even in screens that employ stringent criteria for defining a positive. A number of classification systems have been used to help distinguish false positives from biologically relevant true positives. This chapter describes a system for assigning a confidence score to each interaction based on the probability that it is a true positive. Such confidence scores can be used to prioritize interactions for validation. The scores are also useful for network analysis methods that take advantage of probabilistic edge weights. The scoring method does not rely on gold standard datasets of reliable true positives and true negatives, and thus circumvents the challenges associated with obtaining such datasets. Moreover, the scoring method uses data features that are largely assay-independent, making it useful for interactions obtained from a variety of different technologies and screening methods.

Key words: Interactome mapping, Protein–protein interaction, Protein networks, Confidence scores

1. Introduction

1.1. The Value of Confidence Scores

The purpose of a confidence scoring system is to help distinguish biologically relevant protein interactions from the false positives that arise in yeast two-hybrid and other assays commonly used to detect protein–protein interactions (PPIs) (1–4). In this chapter, we define a false positive as an interaction between a pair of proteins that do not form both a physical and a functional interaction in any natural biological setting. A true positive, on the other hand, is a physical interaction that occurs *in vivo* and that has a function that may be known or discoverable. The

Bernhard Suter and Erich E. Wanker (eds.), *Two Hybrid Technologies: Methods and Protocols*, Methods in Molecular Biology, vol. 812, DOI 10.1007/978-1-61779-455-1_9, © Springer Science+Business Media, LLC 2012

requirement for function in these definitions accommodates the theoretical possibility that some protein interactions may have no apparent biological function even though they naturally occur *in vivo* (5); these would be considered false positives by the scoring system described here. This definition of false positive also makes clear that an interaction may be a false positive even though it was genuinely detected in a particular assay. Practitioners of the yeast two-hybrid system, for example, have become increasingly adept at ensuring that reporter activity corresponds to an actual PPI occurring in the yeast nucleus. Nevertheless, a PPI that can happen when the two proteins are expressed in a yeast nucleus may not occur in a natural setting (i.e., *in vivo*) or may occur but have no discernable function in the native organism. A similar argument applies to other assay systems; co-affinity purification (co-AP), for example, is frequently performed by ectopic expression of affinity tagged-proteins that may not be at physiological levels or locations. Thus, protein interactions detected even in well-executed assays may be false positives and it is rarely possible to distinguish true positives from false positives based solely on information from a particular assay.

The most common approach to distinguishing a false positive from a true positive PPI has been to conduct a second assay, preferably with a different technique. For example, an interaction detected in a yeast two-hybrid screen might be directly tested by co-AP followed by immunoblotting or mass spectrometry. The problem with this course of action is that often there are too many interactions to test in a second assay. This is especially the case in high-throughput screens that generate thousands of new PPI in need of validation. A second problem is that each assay is subject to false positives and false negatives. An individual interaction detected in one assay but not in another could be a false positive in the first assay or a false negative in the second. Nevertheless, the general intuition that interactions detected by multiple assays are more likely to be true positives than those detected by one assay has been validated through many thousands of observations (e.g., see ref. 3). Thus, detection in one or more additional assays is a valid approach to increasing confidence in an interaction, which is even true of interactions detected multiple times by the same assay technique, in multiple screens, or multiple times in a given screen (e.g., see ref. 6, 7). This fact can be used in crude confidence scoring systems that partition detected interactions into sets with different levels of confidence based on how many times each interaction was detected independently. The main disadvantage of this approach is the lack of resolution; many of the PPI classified as high confidence will be false positives, and many of the low confidence interactions will be true positives. More sophisticated approaches aim to assign each individual interaction a score that reflects the probability that it is a true positive.

1.2. General Principles of Assigning Confidence

A number of computational approaches to assigning probability scores to PPI have been described (e.g., see ref. 8 for an evaluation of several approaches and (9–11) for more recent examples). These scoring systems generally search for features or attributes of the interaction data that correlate with true positives or false positives. For example, the number of times that an interaction was detected is one attribute that correlates with true positives. Another attribute is the extent to which the two proteins are coexpressed; e.g., if two proteins are never or only rarely coexpressed, they are not likely to interact *in vivo*. Studies have also shown that the position of a PPI within an interaction network correlates with likelihood of being a true or false positive (12, 13). Any number of other attributes of PPI can be tested for such a correlation, whether or not there is an *a priori* hypothesis for why there should be a correlation. Machine learning approaches can determine the extent to which each attribute correlates with true positives and false positives, and then use the correlations to develop a statistical model that assigns confidence scores. The influence that each attribute has on the final score is based on its level of correlation with true or false positives.

A drawback to many scoring systems is that they require "gold standards" or sets of interactions that are known to be true positives and false positives. The scoring algorithms use these datasets as training data to discover which attributes are associated with true positives or false positives. Generally, gold standard positives are interactions that have been detected by a number of different approaches and for which there is ample support from published studies. In some cases, structural information has been used to identify highly reliable interactions (1, 14, 15). There are three related problems with the use of gold standards. First, they may not fully represent the full spectrum of true interactions. Use of PPI from literature, for example, will lead to a bias toward highly studied proteins, while the use of structural information leads to a bias toward interactions in stable complexes that can be crystallized. Second, gold standard datasets tend to be small relative to the expected number of true interactions. This can lead to sampling error where the gold standards do not sufficiently represent true interaction space. Finally, despite efforts to identify highly reliable interactions, the gold standard datasets will still have false positives. Moreover, the rate of false positives is difficult to determine without using another gold standard dataset, which itself will have biases and an unknown rate of false positives.

An alternative to using a single gold standard dataset is to use several training sets that are each enriched for biologically relevant interactions (11). If the training sets are derived using independent criteria, which can be demonstrated by showing that they minimally overlap with each other, they are likely to represent a broader spectrum of true interaction space than any individual set of gold standards. The method described here uses multiple sets of positives to train independent models, the results of which are then combined to obtain a final confidence score for each individual interaction.

1.3. Overview of the Method

The scoring procedure is outlined in Fig. 1. An important aspect of this scoring system is that it scores interactions in the context of all available interaction data for a particular organism of interest. Since the position of a PPI in the entire interaction network influences its likelihood of being a true positive, it is useful to start with the most complete network possible. Thus, the first step in the procedure (Subheading 3.1) is to collect PPI data from public resources, many of which are listed in Subheading 2.1. Any number of new PPI can be added to the collected data, including PPI identified in a new two-hybrid screen. All interactions will be scored regardless of source or method used to identify them. Interactions predicted based on experimental data in other organisms (interologs) can also be collected and scored. The next step in the procedure is to annotate all of the collected interactions with values for the attributes that will be used to score them. Subheading 3.2 lists some useful attributes and describes how to compute them. The next step is to derive positive and negative training sets from the collected interactions. Four positive training sets are described in Subheading 3.3. Corresponding negative training sets of equal size are generated by selecting random interactions not found in the positive training set. The final step is to train a logistic regression model using each positive and negative training set (Subheading 3.4). Four independent models are trained on

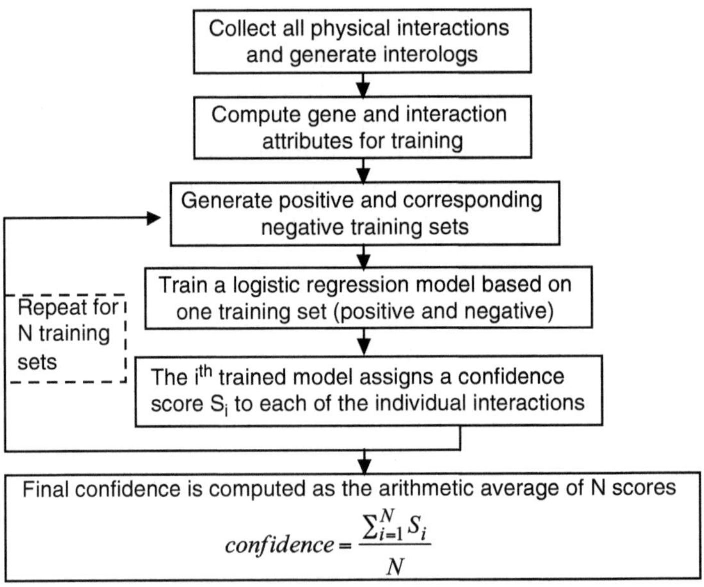

Fig. 1. Protein–protein interaction scoring procedure.

each positive set combined with its corresponding negative set. The four learned models are then used to score all interactions. Finally, a single score is obtained by combining the four confidence scores for each interaction. Final confidence scores have values between 0 and 1, representing the possibility that an interaction is a biological true interaction.

2. Information and Software Resources

2.1. Protein Interaction Databases

PPI data can be downloaded from several sources, including any of the following databases:

BioGRID (http://www.thebiogrid.org): curated PPI and genetic interactions from model organisms.

IntAct (http://www.ebi.ac.uk/intact/main.xhtml): literature curated and directly submitted PPI.

MINT (http://www.mint.bio.uniroma2.it/mint/Welcome.do): literature curated PPI.

DIP (http://www.dip.doe-mbi.ucla.edu/dip/Main.cgi): literature curated PPI.

DroID (http://www.droidb.org): *Drosophila*-specific PPI, genetic and interolog interactions.

MIPS (http://www.mips.helmholtz-muenchen.de/genre/proj/yeast/): Yeast genetic interactions and PPI.

HPRD (http://www.hprd.org): Human-specific PPI.

PDZbase (http://www.icb.med.cornell.edu/services/pdz/start): curated PPI involving PDZ domains.

Reactome (http://www.reactome.org): curated biological pathways.

Other PPI databases (http://www.proteome.wayne.edu/PIDBL.html): lists of PPI and related databases.

2.2. Other Relevant Biological Databases

Ensembl (http://www.ensembl.org): eukaryotic genome data, gene identifier (ID) mappings.

3DID (http://www.3did.irbbarcelona.org): domain–domain interactions (DDI) from three-dimensional structures.

GEO (http://www.ncbi.nlm.nih.gov/geo/): NCBI gene expression database.

Interpro (http://www.ebi.ac.uk/interpro): protein domain and protein ID mapping.

Inparanoid (http://www.inparanoid.sbc.su.se/cgi-bin/index.cgi): orthology mapping.

Homologene (http://www.ncbi.nlm.nih.gov/homologene): eukaryotic genome orthology mapping

UniProt (http://www.uniprot.org): protein data, ID mappings.
Gene Ontology (GO) (http://www.geneontology.org): Gene Ontology (GO) annotations.
KEGG (http://www.genome.jp): pathway database.
Prolinks (http://www. prolinks.mbi.ucla.edu): PPI predicted from genome data.

2.3. Programming Language and Software Used

Data processing can be done in Python (http://www.python.org) or other programming languages that users are comfortable with; other useful languages, for example, include Perl (http://www.perl.org), Ruby (http://www.ruby-lang.org), or Java (http://www.java.sun.com). Bioperl (http://www.bioperl.org) and biopython (http://www.biopython.org) are two programming communities providing tools for biological domains. Model learning and statistical analysis can be done in the R environment (http://www.r-project.org) or MATLAB (http://www.mathworks.com/products/matlab/). In order to train models in R according to the method presented here, the MASS package must be installed (http://www.cran.r-project.org/web/packages/MASS/MASS.pdf).

3. Methods

3.1. Collect Physical Interaction Data and Compute Interologs (see Note 1)

1. Collect physical protein interactions for the organism of interest from all available sources and combine them into a single file. Novel or unpublished interactions not yet available in the major databases may be added to this list (see Note 2).
2. Convert gene or protein identifiers to a common ID (see Note 3).
3. Eliminate redundancy. Duplicate interactions obtained from different data sources should be removed. The final list of interactions should include unique pairs of gene or protein IDs, each annotated as indicated in step 1.
4. Generate interolog interaction data by collecting and combining interaction data available for other organisms, as in steps 1–3 above. To compute interologs, use an orthology mapping tool such as the InParanoid database (see Note 4) or the Homologene database.

3.2. Compute and Tabulate Interaction Attributes

Each interaction has a set of attributes that may help predict whether or not it is a true positive. The following attributes have been shown to have predictive value for some interaction data sets. Additional attributes may be used if available. Compute the following attributes for all of the interactions collected in Subheading 3.1 (see Note 5).

1. Gene expression correlation (corr) (see Note 6). Proteins encoded by genes having similar expression patterns are more

likely to interact than those having very different expression patterns (16, 17). The extent of gene expression similarity can be measured by computing Pearson's correlation of gene expression values across many different samples (e.g., tissues, developmental times, conditions, etc.). Download all gene expression datasets for the species of interest from the NCBI GEO database (GEO data sets – GDS). A GDS contains data for gene expression from a collection of samples processed using the same platform, making them biologically and statistically comparable. To minimize the effect of trivial correlations, remove any GDS with fewer than five samples and eliminate records with null values. Within each GDS, compute Pearson correlation coefficients for interacting gene pairs from Subheading 3.1. If a gene has more than one set of expression values, the values can be averaged. Compute the final correlation coefficient for a gene pair by combining the correlation coefficients from each of the individual datasets according to the following formula

$$\text{corr} = \frac{\sum_{i=1}^{n} x_i \times s_i}{\sum_{i=1}^{n} s_i}$$

where *xi* represents the correlation coefficient computed based on dataset *i*; *si* is the number of samples or conditions in the same dataset; *n* is number of datasets.

2. Number of PubMed Identifiers (PMIDs) associated with an interaction (see Note 6). Several studies have shown a positive correlation between the probability that an interaction is a true positive and the number of publications in which it was reported. Most online interaction databases record the PMIDs of the publications that originally reported each interaction. Count the number of PMIDs associated with each interaction in the lists from Subheading 3.1.
3. DDI. Two proteins are more likely to be involved in a true positive interaction if they contain domains that are known to interact. A pair-wise list of interacting domains can be obtained from 3DID (18), a database of domain interaction data based on high-resolution three-dimensional structures of proteins and protein complexes (see Note 7). Annotate the gene pair list from 3.1 with a value for DDI of "1" if the two proteins have domains known to interact and "0" if they do not.
4. Number of interacting neighbors or degree of each protein in an interaction (d1, d2). The degree of a protein is the number of proteins directly connected to it. Remove self-interactions from the list then compute the degree for each protein (d1 and d2) in each interaction.

5. Clustering coefficients (cc1, cc2) of each protein in an interaction. The clustering coefficient (cc) of a protein in an interaction network is a measure of the connectivity among its direct neighbors. For each protein A, cc is the ratio of the number of connections between the nodes that are directly connected to A to the total number of possible connections. Remove self-interactions from the interaction list and then compute cc for each protein (cc1 and cc2) using the formula:

$$\text{cc} = \frac{\text{Number of connections between protein A's neighbors}}{n \times \frac{n-1}{2}}$$

where protein A has n direct interacting neighbors in the interaction map.

6. Fraction of network neighbors in common between two interacting proteins (jaccard). Two proteins are more likely to interact if they share many interactors in common (19). The extent of shared interactions can be computed using a form of the jaccard similarity index. Remove self-interactions from the interaction list and then calculate jaccard for each interaction using this formula:

$$\text{jaccard} = \frac{\text{Number of proteins connected to both A and B}}{\text{Number of proteins connected to either A or B}}.$$

3.3. Generate Positive and Negative Training Sets

Positive training sets of PPI are significantly enriched for true positives relative to random pairs of proteins. The positive training sets are selected from the master list of interactions collected in Subheading 3.1. Ideally, the scoring system will use several different positive training data sets to represent different regions of true interaction space. The four positive training sets listed below have proven useful for scoring interactions from yeast, *Drosophila* and humans (11). For each positive training set, generate a negative set of equal size by sampling from the set of all interactions, excluding that particular set of training positives (see Note 8).

1. Interactions reported in many publications: From the list of protein interactions, extract those that are annotated with at least *N* PMIDs. Choice of *N* should be based on the interaction data to be scored. A larger *N* may make the training set more reliable, but will also limit the number of interactions in the training set and reduce its power for training a model. Strive to set *N* such that the number of interactions in this positive training set is of the same magnitude as the other training sets. In scoring *Drosophila* and human interactions, $N=10$ has been shown to be effective.
2. Potentially conserved interactions: Many functional PPIs are conserved through evolution. From the list of protein interactions,

extract those that were found in two or more organisms. This can be done by comparing the physical interactions for the organism of interest and the interologs obtained from other organisms.

3. Interactions with expression correlation higher than a certain threshold: From the list of protein interactions, extract those with corr values above a defined threshold. As with PMIDs, the choice of threshold should be based on the interaction data to be scored. Using higher correlation thresholds to generate a positive training set generally gives better performance; however, a threshold that is too high results in a data set that is too small to effectively train a model. In scoring *Drosophila* and human interactions, an expression correlation cutoff of 0.6 was shown to be effective.
4. High confidence high-throughput interactions: High-throughput studies generally divide data into high and low confidence interactions based on screen- or data-specific scoring schemes. This training set can be derived by combining high confidence interactions from high-throughput screens (see Note 9).

3.4. Train Models and Compute Confidence Scores

1. Combine each positive training set with its corresponding negative training set into a single file (e.g., "trainingdata.txt"). Within this file the positive and negative training interactions should be assigned class categories of 1 and 0, respectively.
2. Use each training set file to train a model and score interactions using the MASS package in R. First the training data file is loaded and an initial model is trained using the glm() function. The glm() function is used to fit generalized linear models. This simple model is then used as a starting point for the stepAIC() function to perform step-wise model selection. Finally, the predict() function, a generic function for prediction using the model fitted through the step-wise procedure, assigns confidence values. A simplified version of the training process in R script is shown below.

```
(1) library(MASS)
(2) training <- read.table("trainingdata.txt",header=TRUE)
(3) initialmodel <- glm(formula = category ~ d1+d2, family=binomial, data=training,
    y=TRUE)
(4) finalmodel <- stepAIC(initialmodel,scope=list(upper=~d1+d2+cc1+cc2+jaccarc
    PMID+DDI+corr,lower=~1),direction="both")
(5) confidence <- predict(finalmodel, newdata=alldata, type="response")
```

Statement 1 provides function stepAIC() needed in statement 4. Statement 2 loads training data from the input data file, trainingdata.txt. Statement 3 trains an initial simple model depending only on node degrees. In a step-wise fashion, statement 4 searches through all possible combinations of different attributes. Statement 5 generates confidence scores (between 0 and 1.0) for all interactions based on the model learned in statement 4.

3. Repeat the training and scoring with each of the four training set files. Note that the R script shown above is applicable to positive training sets 2 (potentially conserved PPI) and 4 (high throughput high confidence). For training sets 1 (many publications or PMIDs) and 3 (expression correlation), the PMID and corr attributes should be removed from their respective statements.
4. Combine the four scores into a single confidence score for each interaction by taking the arithmetic mean.

3.5. Evaluate Computed Interaction Scores

The approach described above has been shown to effectively assign scores that reflect the probability that an interaction is a true positive for particular PPI data from yeast, *Drosophila*, and human (11). The effectiveness of the approach on any new data set, however, must be evaluated. One way to evaluate the scores is to divide the PPI into bins based on their scores and then determine which bins have interactions that look more like true positives based on information that is not used in the training or scoring process. For example, the data may be divided into two bins, a high confidence set (HCS) with PPIs scoring over 0.5, and a low confidence set (LCS), consisting of the remaining PPI. Gene Ontology (GO) annotations can be used to evaluate the PPI. The high confidence PPI should have a larger fraction of interactions involving two proteins that share GO annotations than the low confidence PPI. Similarly, the different subsets of PPI can be evaluated by determining the fraction of PPI involving two proteins known to be in the same biological pathway. Such pathway information is available from the KEGG or Reactome databases. Another way to evaluate subsets of PPI with different confidence scores is to determine their rate of overlap with independent data sets that should be enriched for true positives. For example, several studies have shown that genetic interactions are enriched for genes that encode interacting proteins (20, 21). Thus, higher confidence interactions should have a larger overlap with genetic interactions than the lower confidence PPI. Scored PPI can also be compared with computationally predicted PPI that are predicted using genomic information not used in the training and scoring process, such as the predicted PPI in the Prolinks database (22). Readers are referred to (11) for further details on evaluating confidence scores.

4. Notes

1. The result of Subheading 3.1, steps 1–4 will be a file containing a list of physical PPI records for the species of interest from published experimental data, and a list of predicted interologs based on experimental data in other organisms. Each record will include a pair of gene IDs or protein IDs, the PubMed IDs of the papers from which the interaction was derived, a pair of taxonomy IDs for the organism the proteins are derived from, and the physical detection methods; care should be taken to filter out interactions that are not experimentally derived physical PPI. For high-throughput datasets, experiment-specific confidence scores should also be collected where available. Combining data from multiple large databases, such as BioGrid, IntAct, and MINT, and species-specific databases such as DroID, MIPS, and HPRD, will maximize the number of interactions collected. Databases such as BioGrid, IntAct, and MINT, apart from providing downloadable files in various formats, also have interfaces for programmable access using parameters and filters to download a user-defined subset of the data.
2. When obtaining data from online databases, be sure to also obtain PubMed identifiers for each interaction and any available confidence scores, particularly experiment-specific confidence values.
3. Interactions obtained from different databases may use different gene or protein identifiers. Conversion to one common ID type is necessary so that all interactions can be treated similarly in subsequent steps. Some useful universal gene or protein IDs to consider include Flybase Gene IDs (FBgn) for *D. melanogaster*, Ensembl Gene IDs for *H. sapiens* (ENSG), WormBase Gene IDs for *C. elegans* (WBGene), and Systematic names (protein coding ORF) for *S. cerevisiae* (e.g., YHR014W). Because PPI data frequently does not include information on specific protein isoforms, use of gene IDs rather than protein IDs may allow more of the data to be mapped to the common ID. Several gene or protein ID mapping tools or look-up tables are available (e.g., mapping files are available at the UniProt database; ID mapping is also provided by various portals in the BioMart project: http://www.biomart.org). It will be useful for later steps to maintain the UniProt IDs, even if some other common ID is chosen to track the genes and proteins.
4. In many cases, a gene in one organism maps to multiple parologous genes in another organism because the specific functional ortholog is not obvious from sequence comparisons alone. Thus, an interaction in one organism may be represented by multiple potential interologs in another organism. In

the InParanoid database, orthologous genes are organized into clusters based on an InParanoid scoring system; each cluster may contain multiple paralogous genes from a given organism. Thus, the number of interologs to generate for each interaction in the original organism will depend on the number of genes from the target organism in each cluster containing the original two proteins.

5. Some of the attributes are assigned to interactions while others are assigned individually to the two proteins involved in an interaction. One way to tabulate the attributes is to use the file of interacting protein pairs generated in Subheading 3.1 and add columns or data fields for each attribute. For calculating network attributes (4–6) in Subheading 3.2 it may be useful to convert gene identifiers to integer values to save memory. Also self-interactions should be removed before calculating network attributes.
6. Numbers of publications (PMIDs) and expression correlation attributes are also used as criteria for the generation of two different positive training sets. Thus, these two attributes need to be excluded when training with the two respective positive training sets.
7. The 3DID database provides the domain identifiers of interacting domains. To determine which protein pairs in the interaction list have interacting domains, the proteins need to be annotated with the domains that they contain. Domain information can be obtained from the InterPro database. Download the data and parse out UniProt IDs and the domain identifiers for each protein.
8. Negative training data are lists of interactions that are unlikely to be biologically relevant. Since the number of biologically relevant interactions for any species is expected to be only a small fraction of the total number of possible interactions, any two random proteins are likely to be noninteractors. Thus, it has become common practice to generate negative training data by randomly selecting pairs of proteins. For the scoring system outlined here, the negative training data must have all of the interaction attribute information found in the positive training data. Thus, the negative training set is drawn randomly from the list of interactions rather than from all possible protein pairs.
9. Higher confidence subsets of high-throughput data are often determined from screen-specific or method-specific information. For two-hybrid screens, for example, interactions have been classified as high confidence based on reporter scores or the number of times the interaction was detected. High-throughput co-AP assays also use screen-specific or method-specific classification schemes. In several cases, confidence values are given for each

interaction using a continuous scale or several discrete values. As with other positive training sets, a choice must be made about the cut offs used to select the high-throughput HCSs. In scoring *Drosophila* interactions, for example, we chose to combine interactions from three different large-scale two-hybrid screens, each with a different confidence scoring scheme: for data from Giot (23) we used interactions that they had scores >0.5; for data from Stanyon (24) we used interactions with reporter scores >3; for data from Formstetcher (25) we used interactions that the authors had classified into one of the top three out of five different confidence levels.

References

1. Edwards, A. M., Kus, B., Jansen, R., Greenbaum, D., Greenblatt, J., and Gerstein, M. (2002) Bridging structural biology and genomics: assessing protein interaction data with known complexes, *Trends Genet* **18**, 529–536.
2. Huang, H., Jedynak, B. M., and Bader, J. S. (2007) Where have all the interactions gone? Estimating the coverage of two-hybrid protein interaction maps, *PLoS Comput Biol* **3**, e214.
3. von Mering, C., Krause, R., Snel, B., Cornell, M., Oliver, S. G., Fields, S., and Bork, P. (2002) Comparative assessment of large-scale data sets of protein-protein interactions, *Nature* **417**, 399–403.
4. Uetz, P., and Finley, R. L., Jr. (2005) From protein networks to biological systems, *FEBS Lett* **579**, 1821–1827.
5. Levy, E. D., Landry, C. R., and Michnick, S. W. (2009) How perfect can protein interactomes be?, *Sci Signal* **2**, pe11.
6. Yu, H., Braun, P., Yildirim, M. A., Lemmens, I., Venkatesan, K., Sahalie, J., Hirozane-Kishikawa, T., Gebreab, F., Li, N., Simonis, N., Hao, T., Rual, J. F., Dricot, A., Vazquez, A., Murray, R. R., Simon, C., Tardivo, L., Tam, S., Svrzikapa, N., Fan, C., de Smet, A. S., Motyl, A., Hudson, M. E., Park, J., Xin, X., Cusick, M. E., Moore, T., Boone, C., Snyder, M., Roth, F. P., Barabasi, A. L., Tavernier, J., Hill, D. E., and Vidal, M. (2008) High-quality binary protein interaction map of the yeast interactome network, *Science* **322**, 104–110.
7. Schwartz, A. S., Yu, J., Gardenour, K. R., Finley, R. L., Jr., and Ideker, T. (2009) Cost-effective strategies for completing the interactome, *Nature methods* **6**, 55–61.
8. Suthram, S., Shlomi, T., Ruppin, E., Sharan, R., and Ideker, T. (2006) A direct comparison of protein interaction confidence assignment schemes, *BMC Bioinformatics* **7**, 360.
9. Braun, P., Tasan, M., Dreze, M., Barrios-Rodiles, M., Lemmens, I., Yu, H., Sahalie, J. M., Murray, R. R., Roncari, L., de Smet, A. S., Venkatesan, K., Rual, J. F., Vandenhaute, J., Cusick, M. E., Pawson, T., Hill, D. E., Tavernier, J., Wrana, J. L., Roth, F. P., and Vidal, M. (2009) An experimentally derived confidence score for binary protein-protein interactions, *Nature methods* **6**, 91–97.
10. Lin, X., Liu, M., and Chen, X. W. (2009) Assessing reliability of protein-protein interactions by integrative analysis of data in model organisms, *BMC Bioinformatics* **10 Suppl 4**, S5.
11. Yu, J., and Finley, R. L., Jr. (2009) Combining multiple positive training sets to generate confidence scores for protein-protein interactions, *Bioinformatics* **25**, 105–111.
12. Bader, G. D., and Hogue, C. W. (2003) An automated method for finding molecular complexes in large protein interaction networks, *BMC Bioinformatics* **4**, 2.
13. Goldberg, D. S., and Roth, F. P. (2003) Assessing experimentally derived interactions in a small world, *Proc Natl Acad Sci USA* **100**, 4372–4376.
14. Mosca, R., Pons, C., Fernandez-Recio, J., and Aloy, P. (2009) Pushing structural information into the yeast interactome by high-throughput protein docking experiments, *PLoS Comput Biol* **5**, e1000490.
15. Prieto, C., and De Las Rivas, J. (2010) Structural domain-domain interactions: assessment and comparison with protein-protein interaction data to improve the interactome, *Proteins* **78**, 109–117.
16. Ge, H., Liu, Z., Church, G. M., and Vidal, M. (2001) Correlation between transcriptome and interactome mapping data from Saccharomyces cerevisiae, *Nat Genet* **29**, 482–486.

17. Rhodes, D. R., Tomlins, S. A., Varambally, S., Mahavisno, V., Barrette, T., Kalyana-Sundaram, S., Ghosh, D., Pandey, A., and Chinnaiyan, A. M. (2005) Probabilistic model of the human protein-protein interaction network, *Nat Biotechnol* **23**, 951–959.
18. Stein, A., Russell, R. B., and Aloy, P. (2005) 3did: interacting protein domains of known three-dimensional structure, *Nucleic Acids Res* **33**, D413–417.
19. Bader, J. S., Chaudhuri, A., Rothberg, J. M., and Chant, J. (2004) Gaining confidence in high-throughput protein interaction networks, *Nat Biotechnol* **22**, 78–85.
20. Beyer, A., Bandyopadhyay, S., and Ideker, T. (2007) Integrating physical and genetic maps: from genomes to interaction networks, *Nat Rev Genet* **8**, 699–710.
21. Tong, A. H., Lesage, G., Bader, G. D., Ding, H., Xu, H., Xin, X., Young, J., Berriz, G. F., Brost, R. L., Chang, M., Chen, Y., Cheng, X., Chua, G., Friesen, H., Goldberg, D. S., Haynes, J., Humphries, C., He, G., Hussein, S., Ke, L., Krogan, N., Li, Z., Levinson, J. N., Lu, H., Menard, P., Munyana, C., Parsons, A. B., Ryan, O., Tonikian, R., Roberts, T., Sdicu, A. M., Shapiro, J., Sheikh, B., Suter, B., Wong, S. L., Zhang, L. V., Zhu, H., Burd, C. G., Munro, S., Sander, C., Rine, J., Greenblatt, J., Peter, M., Bretscher, A., Bell, G., Roth, F. P., Brown, G. W., Andrews, B., Bussey, H., and Boone, C. (2004) Global mapping of the yeast genetic interaction network, *Science* **303**, 808–813.
22. Bowers, P. M., Pellegrini, M., Thompson, M. J., Fierro, J., Yeates, T. O., and Eisenberg, D. (2004) Prolinks: a database of protein functional linkages derived from coevolution, *Genome Biol* **5**, R35.
23. Giot, L., Bader, J. S., Brouwer, C., Chaudhuri, A., Kuang, B., Li, Y., Hao, Y. L., Ooi, C. E., Godwin, B., Vitols, E., Vijayadamodar, G., Pochart, P., Machineni, H., Welsh, M., Kong, Y., Zerhusen, B., Malcolm, R., Varrone, Z., Collis, A., Minto, M., Burgess, S., McDaniel, L., Stimpson, E., Spriggs, F., Williams, J., Neurath, K., Ioime, N., Agee, M., Voss, E., Furtak, K., Renzulli, R., Aanensen, N., Carrolla, S., Bickelhaupt, E., Lazovatsky, Y., DaSilva, A., Zhong, J., Stanyon, C. A., Finley, R. L., Jr., White, K. P., Braverman, M., Jarvie, T., Gold, S., Leach, M., Knight, J., Shimkets, R. A., McKenna, M. P., Chant, J., and Rothberg, J. M. (2003) A protein interaction map of Drosophila melanogaster, *Science* **302**, 1727–1736.
24. Stanyon, C. A., Liu, G., Mangiola, B. A., Patel, N., Giot, L., Kuang, B., Zhang, H., Zhong, J., and Finley, R. L., Jr. (2004) A Drosophila protein-interaction map centered on cell-cycle regulators, *Genome Biol* **5**, R96.
25. Formstecher, E., Aresta, S., Collura, V., Hamburger, A., Meil, A., Trehin, A., Reverdy, C., Betin, V., Maire, S., Brun, C., Jacq, B., Arpin, M., Bellaiche, Y., Bellusci, S., Benaroch, P., Bornens, M., Chanet, R., Chavrier, P., Delattre, O., Doye, V., Fehon, R., Faye, G., Galli, T., Girault, J. A., Goud, B., de Gunzburg, J., Johannes, L., Junier, M. P., Mirouse, V., Mukherjee, A., Papadopoulo, D., Perez, F., Plessis, A., Rosse, C., Saule, S., Stoppa-Lyonnet, D., Vincent, A., White, M., Legrain, P., Wojcik, J., Camonis, J., and Daviet, L. (2005) Protein interaction mapping: a Drosophila case study, *Genome Res* **15**, 376–384.

Chapter 10

The Integration and Annotation of the Human Interactome in the UniHI Database

Gautam Chaurasia and Matthias Futschik

Abstract

In recent years, remarkable progress has been made toward the systematic charting of human protein interactions. The utilization of the generated interaction data remained however challenging for biomedical researchers due to lack of integration of currently available resources. To facilitate the direct access and analysis of the human interactome, we have developed the Unified Human Interactome (UniHI) database. It provides researchers with a user-friendly Web-interface and integrates interaction data from 12 major resources in its latest version, establishing one of the largest catalogs for human PPIs worldwide. At present, UniHI houses over 250,000 distinct interactions between 22,300 unique proteins and is publically available at http://www.unihi.org.

Key words: Protein–protein interactions, Database, Data integration, Web applications, Bioinformatics

1. Introduction

In a living organism, proteins interact with other proteins to carry out vital cellular functions, such as signal transduction, DNA replication, transcription, protein transport, or metabolic catalysis. Also, many major diseases such as neurological disorders or cancer are characterized by complex interactions of multiple proteins (1–9). The study of human protein interactions may therefore help (i) to improve our general understanding of biological processes, (ii) to decipher the molecular basis of complex diseases, and (iii) to provide new potential therapeutic targets.

For many years, interactions between proteins have been studied in small-scale experiments. This situation has, however, dramatically changed in the last decade. The availability of fully sequenced

Bernhard Suter and Erich E. Wanker (eds.), *Two Hybrid Technologies: Methods and Protocols*, Methods in Molecular Biology, vol. 812, DOI 10.1007/978-1-61779-455-1_10,

genomes (10) and advances in high-throughput approaches (11–14) have led to large-scale studies of protein–protein interactions on a genome-wide scale and to efforts to map the complete protein–protein interaction (PPI) network for an organism, termed also as interactome. Indeed, we have recently witnessed many large-scale protein interaction mapping projects in several model organisms such as *Saccharomyces cerevisiae* (15–18), *Drosophila melanogaster* (19), and *Caenorhabditis elegans* (20). Now, the focus has moved toward a systematic mapping of human protein–protein interactions (21–33). The constructed human PPI maps have been derived from both experimental and computational approaches (11–14, 34, 35) and not only offer a wealth of information but are also expected to a valuable tool for the biomedical research community (6, 7, 36, 37). However, utilization of these interaction maps is impeded by manifold limitations:

A major limitation of current approaches to map the interactome is their limited capability to capture interactions in a comprehensive manner (38, 39). Several studies have shown that the current human PPI maps are incomplete and highly unsaturated (40–42). A further shortcoming of human PPI maps has been a lack of integration. Although many large human PPI maps are publicly accessible, they are distributed at multiple locations. To find comprehensive information on potential interaction partners, researchers have to perform multiple searches at different databases. This, however, can be very time-consuming due to many factors such as various query formats and identifiers used in different interaction databases. Another drawback of many PPI databases is their restriction to the search of interaction partners only for a single query protein. By contrast, comprehensive molecular studies frequently require complex network-oriented searches for interactions of multiple proteins.

Finally, determining the biological function of networks has remained a bottleneck in the analysis of PPI data. In order to understand the complexity of interaction networks, it is necessary to assess not only the function of individual proteins but also their molecular context. Thus, PPI data have to be integrated with other functional data and information to facilitate their interpretation.

UniHI (43–45) was developed to overcome these limitations and to provide a large community of researchers in biomedicine direct access to comprehensive information on human proteins and their potential interaction partners. UniHI currently stores more than a quarter of a million human protein interactions that have been experimentally detected or computationally predicted (see Note 1). Several Web applications are included in UniHI offering a convenient interface for the network analysis, especially for researchers less acquainted with computational investigations. For functional interpretation of interaction networks, PPI data were integrated with biological pathway data from Encyclopedia of Genes and Genomes (KEGG) (46, 47) and gene expression data from Human Gene Atlas (48). The integrated data can be utilized through advanced tools (i.e., UniHI Express and UniHI Scanner) that enable users to construct tissue-specific interaction networks or to annotate edges with the specific type of interaction.

2. Materials

2.1. PPI Data

Currently, protein interactions in UniHI are derived from 12 large-scale human protein–protein interaction maps (21–33). These maps were generated using yeast-two-hybrid (Y2H) assays, literature review or orthology-based approaches. UniHI includes interaction maps that are based on curated literature (i.e., IntAct (21), BIND (22), BioGrid (23), and REACTOME (27), HPRD (29), DIP (32), on computational text-mining (i.e., COCIT (30)), on large Y2H-based screens (i.e., CCSB-H1 (31) and MDC-Y2H (33)) and on computational prediction (i.e., OPHID (24) and ORTHO (26), HOMOMINT (28)). Further details on these maps are given in the Table 1.

2.2. Gene Expression Data

Gene expression data were collected from Gene Atlas database (48). The data set comprises of expression profiles for 79 human tissues measured in replicates using Affymetrix HG-U133A and custom-designed GNF1H arrays. Expression summaries for the 44,775 transcripts were derived utilizing the MAS5 algorithm by Affymetrix.

2.3. Gene Annotation

Additional information on proteins (e.g., functional annotations and description of proteins, chromosomal location, and potential disease association of corresponding genes) were imported from National Center for Biotechnology Information (NCBI) (49), Online Mendelian Inheritance in Man (OMIM) (50), and Gene Ontology (GO) (51, 52) databases.

2.4. Pathway Information

Association of genes with pathways were derived from the KEGG pathway database (46, 47), which constitutes a collection of manually drawn pathway maps representing accumulated knowledge of molecular interaction and reaction networks for metabolism, cellular processes, and human diseases, as well as for genetic and environmental information processing. In addition, information about the relations between pathway elements such as regulatory or physical interactions (e.g., phosphorylation, dephosphorylation, activation, inhibition, and binding association) was collected from the KEGG pathways database.

Table 1
Current list of PPI datasets, integrated in UniHI

Dataset	Proteins	Interactions	Method	Database location
MDC-Y2H	1,703	3,186	Y2H SCREEN (33)	http://www.mdc-berlin.de/neuroprot
CCSB-Y2H	1,549	2,754	Y2H SCREEN (31)	http://vidal.dfci.harvard.edu (flat file only)
HPRD-BIN	8,788	32,776	LITERATURE (29)	http://www.hprd.org
DIP	1,085	1,397	LITERATURE (32)	http://dip.doe-mbi.ucla.edu
BioGrid	7,953	24,624	LITERATURE (23)	http://www.thebiogrid.org
IntAct	7,273	19,404	LITERATURE (21)	http://www.ebi.ac.uk/intact
BIND	5,286	7,394	LITERATURE (22)	http://www.bind.ca
COCIT	3,737	6,580	TEXT MINING (30)	http://www.reactome.org
REACTOME	1,554	37,332	LITERATURE (27)	http://Bioinformatics.icmb.utexas.edu/idserve/
ORTHO	6,225	71,466	ORTHOLOGY (26)	http://www.sanger.ac.uk/PostGenomics/signaltransduction/interactionmap
HOMOMINT	4,127	10,174	ORTHOLOGY (28)	http://mint.bio.uniroma2.it
OPHID	4,785	24,991	ORTHOLOGY (24)	http://ophid.utoronto.ca
UniHI	22,307	200,473	INTEGRATION (45)	http://www.unihi.org

Number of proteins and interactions in each dataset as well as construction approach are given

2.5. Gene and Protein Identifiers

Since PPI data were collected from different sources, one of the major problems was to establish common identifiers for their aggregation. For this purpose, lists of different protein identifiers were downloaded from the Web sites of NCBI, HUGO Gene Nomenclature Committee (HGNC) (53), and Ensembl EnsMart (54, 55).

3. Methods

3.1. Architecture of UniHI

The UniHI system should offer users a convenient entry gate to a comprehensive collection of PPI data. With this aim in mind, we have designed its architecture to comply with following requirements: (i) The structure of the platform should be extensible to new data without changing its data structure as the amount of data of protein interactions is growing rapidly; (ii) it should be easily accessible; (iii) queries should be processed in minimal time;

(iv) data should be accurately integrated from different sources; and (v) it should also be consistently updated.

To meet these demands, UniHI system is based on a multitier architecture with different layers, each assigned to a certain task. A data access layer is responsible for writing and retrieving data to and from a locally installed database. A business layer processes the applications and queries by the user. Finally, a presentation layer provides a Web interface and visualization tools for accessing and viewing interactions. The advantage of the implemented UniHI architecture is its modularity and portability. The technical details of the current UniHI system were described in a recent publication (43).

3.2. Mapping of Proteins

As UniHI integrates a large number of different PPI resources, aggregation of heterogeneous interaction maps and building a unique identifier indexing system is a foremost task. For unification of primary data, we first computed complete lists of proteins for each interaction map separately. Subsequently, these lists were compared employing information from NCBI, HGNC (53, 54), and EnsMart (55) to map corresponding identifiers between interaction datasets. After mapping, identical protein identifiers were merged together in a horizontal manner where each protein is a unique entry in a table. A unique identifier was assigned to each protein entry of this table. These unique identifiers were further used for grouping of the redundant interactions from all interaction datasets.

3.3. Search Interface and Filtering Methods

The UniHI system has been designed to facilitate network-oriented functional analyses. It provides several tools to perform different types of network searches and to map the expression and pathway data onto retrieved networks. Several gene and protein identifiers, e.g., gene symbol, Entrez Gene, Uniprot, NCBI Geneinfo, Ensembl, Biogrid, and HPRD IDs, can be used for searching interaction partners. Notably, these identifiers can now also be employed for direct hyperlinks to UniHI. Users can also choose the PPI resource, from which interactions data should be retrieved. This permits, for instance, the exclusion of less validated mapping approaches such as computation prediction (see Note 1). As primary output, a list of interaction partners for the query proteins is presented (Fig. 1). This list can subsequently be filtered based on additional criteria. For instance, interactions can be excluded if they are not associated with a minimum number of PubMed references in which they have been reported (Fig. 2a).

3.4. Visualization of Protein Interactions

Visualization of the retrieved interaction networks remains crucial for the evaluation of query results. Thus, we have implemented a visualization tool for interaction data that offer many attractive features for rapid analysis and adjustment of the extracted information (see Note 2). For filtering of the network and manual adjustment

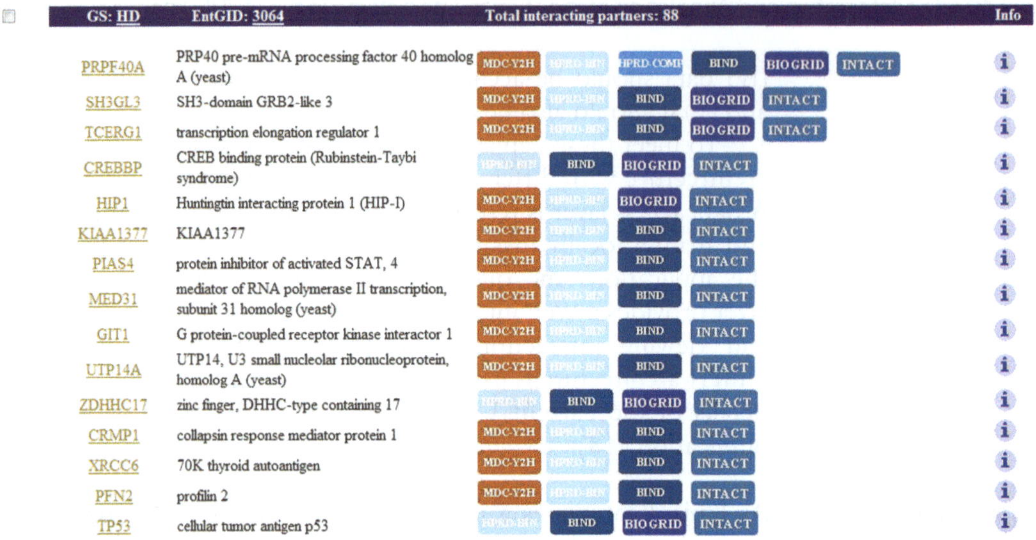

GS: HD	EntGID: 3064	Total interacting partners: 88	Info
PRPF40A	PRP40 pre-mRNA processing factor 40 homolog A (yeast)	MDC-Y2H, HPRD-BIN, HPRD-COMP, BIND, BIOGRID, INTACT	i
SH3GL3	SH3-domain GRB2-like 3	MDC-Y2H, HPRD-BIN, BIND, BIOGRID, INTACT	i
TCERG1	transcription elongation regulator 1	MDC-Y2H, HPRD-BIN, BIND, BIOGRID, INTACT	i
CREBBP	CREB binding protein (Rubinstein-Taybi syndrome)	HPRD-BIN, BIND, BIOGRID, INTACT	i
HIP1	Huntingtin interacting protein 1 (HIP-I)	MDC-Y2H, HPRD-BIN, BIOGRID, INTACT	i
KIAA1377	KIAA1377	MDC-Y2H, HPRD-BIN, BIND, INTACT	i
PIAS4	protein inhibitor of activated STAT, 4	MDC-Y2H, HPRD-BIN, BIND, INTACT	i
MED31	mediator of RNA polymerase II transcription, subunit 31 homolog (yeast)	MDC-Y2H, HPRD-BIN, BIND, INTACT	i
GIT1	G protein-coupled receptor kinase interactor 1	MDC-Y2H, HPRD-BIN, BIND, INTACT	i
UTP14A	UTP14, U3 small nucleolar ribonucleoprotein, homolog A (yeast)	MDC-Y2H, HPRD-BIN, BIND, INTACT	i
ZDHHC17	zinc finger, DHHC-type containing 17	HPRD-BIN, BIND, BIOGRID, INTACT	i
CRMP1	collapsin response mediator protein 1	MDC-Y2H, HPRD-BIN, BIND, INTACT	i
XRCC6	70K thyroid autoantigen	MDC-Y2H, HPRD-BIN, BIND, INTACT	i
PFN2	profilin 2	MDC-Y2H, HPRD-BIN, BIND, INTACT	i
TP53	cellular tumor antigen p53	HPRD-BIN, BIND, BIOGRID, INTACT	i

Fig. 1. Textual representation of a query result for protein interactions in UniHI. Multiple links indicate identification of the interaction by different methods. For easy discrimination between maps, specific colors have been assigned. *Shades of blue* have been used for datasets derived by literature search, *shades of green* for orthology-based maps, *shades of red* for maps derived from Y2H screens.

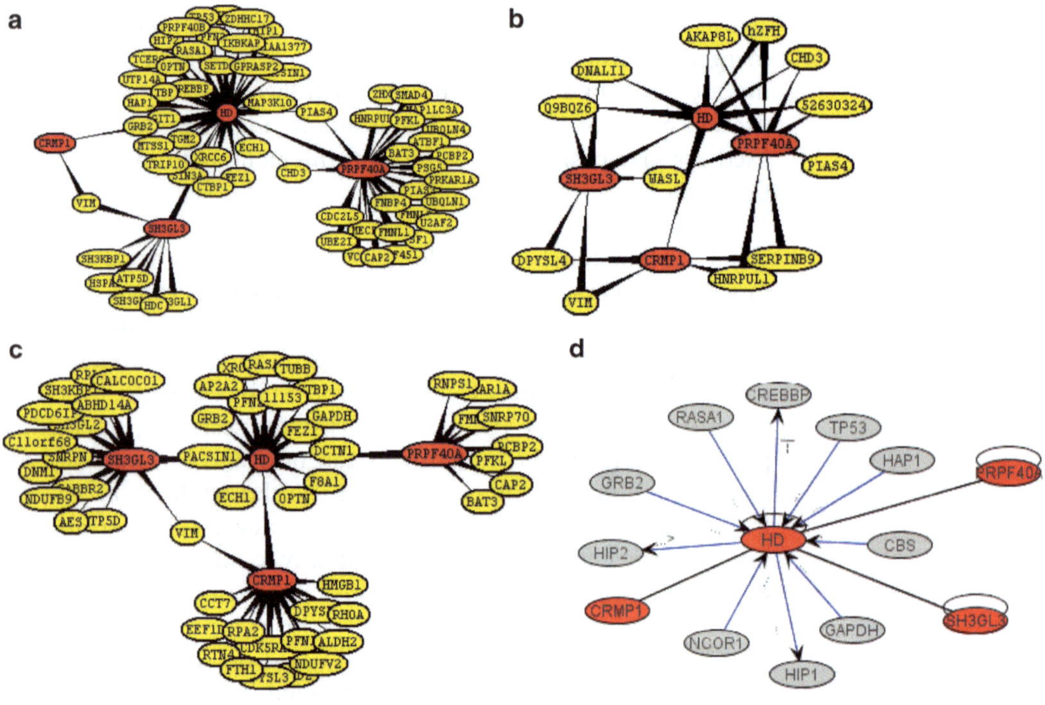

Fig. 2. Graphical representation and analysis of PPI networks using UniHI Search, UniHI Express and UniHI Scanner tools. Display of the interaction partners (*yellow or gray*) of the query proteins (*red*) HD, CRMP1, PRPF40A, and SH3GL3. (**a**) UniHI Search: Using the number of PubMed as a criterion, only those interaction partners are shown, which have been reported in more than two publications. (**b**) Only common or direct interactions between query proteins are displayed. (**c**) UniHI Express: Creation of the brain-specific networks with a medium expression level of 1000 units. (**d**) UniHI Scanner: Annotation of found network using KEGG pathway. *Gray nodes* represent proteins included in KEGG "Huntington Disease" pathway. UniHI Scanner allows a display of the intersection between the retrieved PPI network and the pathway. Additional information is given regarding the type of interaction (e.g., phosphorylation (+P), dephosphorylation (–P), activation (–>), or inhibition (–l)) facilitating the assessment of the retrieved interactions.

of the layout, nodes (representing proteins) can be anchored or hidden. Also, information about proteins and interactions can be accessed directly via the visualization interface, thereby avoiding cumbersome review of textual output. Moreover, the display can be restricted to direct interactions between query proteins or extended to include bridging proteins (Fig. 2b).

3.5. Advanced Tools for Network Analysis

Addressing the need of a more dynamic interactome and its functional interpretation, we developed UniHI Express and UniHI Pathway Scanner as two new tools in our database (Fig. 2c, d). UniHI Express allows the filtering of PPI based on the expression in a selected tissue and thus enables the construction of tissue-specific networks. First preliminary studies show that the usage of UniHI express can be highly efficient to prioritize interactions. Furthermore, UniHI Pathway Scanner provides the possibility to examine the intersection of canonical pathways from KEGG with the extracted networks. Thus, it enables researchers to detect possible modifiers of pathways as well as proteins involved in the cross talk between different pathways.

3.5.1. UniHI Express

Biomedical researchers frequently study processes that occur in specific tissues or cell types, e.g., degeneration of neuronal tissue. By contrast, current collections of protein interactions are derived from experiments using various cell and tissue types. Thus, PPI networks retrieved from these resources represent rather a gross summary of possible interactions, thereby neglecting the actual conditions in a specific tissue of interest. In fact, since only a small percentage of proteins correspond to ubiquitously expressed housekeeping genes, the probability might be high, that many proteins included in retrieved networks are not expressed in a chosen tissue. Hence, researchers are required to examine carefully the presence of proteins in their model system of interest. This is a considerable task considering the large number of interaction partners that queries with even a small number of proteins can produce. We, therefore, integrated gene expression data to allow researchers the construction of tissue-specific networks (see Note 3). As tissue expression data set, the Human Gene Expression Atlas was utilized (18).

To enable the integration with PPI data, microarray probes were mapped to their corresponding Entrez Gene IDs using the annotation generated by the curators of the Gene Expression atlas. Expression values were averaged over probes which correspond to the same nonredundant Entrez Gene IDs. To facilitate the use of UniHI Express, the samples were assigned to 19 main tissue classes. These main classes comprise adrenal gland, brain, heart, kidney, liver, lung, prostate, pancreas, placenta, muscle, salivary gland, thymus, thyroid, tonsil, lymph node, testis, trachea, uterus, and uterine corpus. To obtain a unique tissue expression profile, we averaged expression values of tissues samples belonging to the same class.

Employing UniHI Express, users can filter the interacting proteins by requiring a minimum expression in the selected tissue class. Note that the cutoff value is not applied to the query protein. By adjusting the expression threshold, the PPI network can be reduced to include only highly expressed or can be extended to include also lowly expressed proteins (Fig. 2c). Additionally, the PPI resources to be queried can be specified.

3.5.2. UniHI Scanner

PPI maps represent the chain of molecular events connected with each other. Functional analyses of these events can help us to understand the roles of individual proteins and the interplay between them (56–58). Indeed, PPI networks need to be integrated with other functional data to gain new insights into molecular mechanisms. Therefore, we have integrated the UniHI protein interaction data with pathway data from KEGG (46, 47).

As platform for integration of PPI with pathway data, we have implemented a new tool termed UniHI Pathway Scanner, which offers a search Web interface. Here, users can provide a list of query proteins and choose a list of pathways to be scanned against identified network. Additionally, the source of interaction data can be selected. Subsequently, UniHI Scanner performs three functions. First, interaction partners of search proteins are identified and a PPI network is created. Second, a pathway network is created from the KEGG pathway IDs provided by user. Edges of this pathway network contain the explicit information about the mode of interactions (such as phosphorylation, activation, or inhibition). Finally, the PPI and pathway networks are intersected. Nodes and edges in the PPI network are annotated if they are found in the pathway network. Thus, directed edges in an annotated network represent the molecular relations found in KEGG pathways between proteins (Fig. 2d). For viewing both types of networks, an interactive visualization tool is provided. Users can choose between the display of the full PPI network or the annotated intersection. Additionally, researchers can rearrange the layout of PPI network. For instance, the whole network or individuals nodes can be moved and their location adjusted. Details of networks can be magnified by the zoom function. With these features, UniHI Pathway Scanner is anticipated to be a valuable tool to identify modifiers of pathways or to detect proteins involved in cross talk between pathways.

4. Notes

1. In the postgenomic era, one of the daunting tasks of proteomics is to chart the complete protein–protein interaction networks that occur within organisms (59, 60). Although, the availability of many large genome sequences and advances in

high-throughput methods provided us a platform to construct PPI maps, the hitherto constructed interactome are far from complete (61, 62).

Not surprisingly, this is also the case for the human interactome. Additionally, currently available human PPI data had been stored in various locations. In a recent survey, we observed that interaction maps derived from different resources are remarkably distinct (45). In fact, less than 19% of all interactions occurred in multiple maps indicating a low degree of saturation of single maps. The small number of shared interactions is striking considering that more than 50% of all proteins were included in two or more maps. Thus, current PPI datasets are highly complementary sharing few interactions between many common proteins. For researchers, it is clearly tempting to use as many resources as possible for the in silico construction of their networks of interest. However, it is important to keep in mind that the single human PPI maps are generated using different approaches and that they might contain characteristic biases, i.e., over- or underrepresentation of proteins from certain categories, due to the sensitivity and specificity of applied methods toward specific classes of proteins. In a comparative analysis, we indeed detected a strong sampling and detection biases linked to the method used to generate a PPI map (41). For example, RNA binding proteins were overrepresented in orthology-based maps (e.g., ORTHO), whereas signal transducers were overproportionally sampled in literature-based maps. A significant depletion of membrane proteins was observed in all networks and not only in Y2H-based maps as previously expected. Moreover, different maps of different origin showed considerable divergence with respect to their internal structure (40). These findings are necessary to be considered to avoid pitfalls in the application of PPI maps. We therefore advise researchers to carefully check whether the results of their network analyses might be affected by the choice of resource. Note that this kind of examination is facilitated by the option implemented in UniHI to select or deselect particular primary PPI resources.

2. The quality of PPI data remains a crucial issue. It has been frequently stated that especially data produced by high-throughput methods contain a high rate of false positive or negative interactions (40, 41, 61, 62). However, a recent study questioned this view and rather suggested that PPI data sets based on small scale studies include a considerable rate of unreliable interactions (63). A major reason for this controversy is the lack of an unbiased high confidence human interaction data set which would allow a stringent assessment of the quality of different mapping approaches. Although the existence of such "gold standard" is obviously desirable, the possibility of

its generation in the nearby future remains doubtful. Moreover, as protein interactions frequently require specific protein modifications, the structure of interaction networks depends critically on the cellular conditions and is highly dynamic.

Since there is no unique quality index for interaction data, we have provided in UniHI additional information that enables researchers to assess the reliability of the retrieved interactions. In particular, we have computed two different measures: coannotation and coexpression. Coannotation assumes that two proteins are likely to interact if they share same functions, involved in same biological process or are allocated to the same cellular compartment. For computing coannotation, we assessed the similarity of GO categories assigned to interacting proteins. The similarity of GO categories was approximated by calculating the length of the shared path from the root category. Large shared path lengths indicate that the GO categories are in proximity to each other within the GO graph and thus can be considered similar. In case of multiple GO assignments for proteins, the largest shared path length is counted.

Similarly, coexpression provides an indication whether two proteins are coregulated on transcriptional level, and has been shown to correlate with the probability of interaction (64, 65). To measure coexpression for interacting proteins, Spearman rank correlation of expression levels was calculated. Additionally, the corresponding quantiles for correlation coefficients were derived allowing users to assess the significance of coexpression. For instance, a quantile of 0.05 signifies that the corresponding correlation coefficient is within the top 5% of the total distribution of observed coefficients.

3. Evidently, the usage of gene expression as proxy for protein abundance has its limitation. However, it can give researchers valuable first indications to prioritize interactions in possible follow-up studies. For future releases, we expect that the inclusion of quantitative measurement of protein abundance in tissues – once such data become available on a proteome-wide scale – will lead to considerably more accurate tissue-specific PPI networks. Nevertheless, the current release of UniHI constitutes a first important step toward a dynamic representation of the human interactome.

Acknowledgments

The development and implementation of UniHI was funded by the SFB 618 grant of the *Deutsche Forschungsgesellschaft* (DFG). We would also like to acknowledge the support of M.F by the Portuguese *Fundação para a Ciência e Tecnologia* (IBB/CBME, LA.

References

1. Baranzini, S. E., Galwey, N. W., Wang, J., Khankhanian, P., Lindberg, R., Pelletier, D., Wu, W., Uitdehaag, B. M., Kappos, L., Polman, C. H., Matthews, P. M., Hauser, S. L., Gibson, R. A., Oksenberg, J. R., and Barnes, M. R. (2009) Pathway and network-based analysis of genome-wide association studies in multiple sclerosis, *Hum Mol Genet 18*, 2078–2090.
2. Calvano, S. E., Xiao, W., Richards, D. R., Felciano, R. M., Baker, H. V., Cho, R. J., Chen, R. O., Brownstein, B. H., Cobb, J. P., Tschoeke, S. K., Miller-Graziano, C., Moldawer, L. L., Mindrinos, M. N., Davis, R. W., Tompkins, R. G., and Lowry, S. F. (2005) A network-based analysis of systemic inflammation in humans, *Nature 437*, 1032–1037.
3. Chuang, H. Y., Lee, E., Liu, Y. T., Lee, D., and Ideker, T. (2007) Network-based classification of breast cancer metastasis, *Mol Syst Biol 3*, 140.
4. Ergun, A., Lawrence, C. A., Kohanski, M. A., Brennan, T. A., and Collins, J. J. (2007) A network biology approach to prostate cancer, *Mol Syst Biol 3*, 82.
5. Goehler, H., Lalowski, M., Stelzl, U., Waelter, S., Stroedicke, M., Worm, U., Droege, A., Lindenberg, K. S., Knoblich, M., Haenig, C., Herbst, M., Suopanki, J., Scherzinger, E., Abraham, C., Bauer, B., Hasenbank, R., Fritzsche, A., Ludewig, A. H., Bussow, K., Coleman, S. H., Gutekunst, C. A., Landwehrmeyer, B. G., Lehrach, H., and Wanker, E. E. (2004) A protein interaction network links GIT1, an enhancer of huntingtin aggregation, to Huntington's disease, *Mol Cell 15*, 853–865.
6. Goh, K. I., Cusick, M. E., Valle, D., Childs, B., Vidal, M., and Barabasi, A. L. (2007) The human disease network, *Proc Natl Acad Sci USA 104*, 8685–8690.
7. Ideker, T., and Sharan, R. (2008) Protein networks in disease, *Genome Res 18*, 644–652.
8. Lage, K., Karlberg EO., Størling, Z., Ólason, P., Pedersen, A., Rigina, O., Hinsby, A., Tümer, Z., Pociot, F., Tommerup, N., Moreau, Y., and Brunak, S. (2007) A human phenome-interactome network of protein complexes implicated in genetic disorders, *Nature Biotechnology 25*(3), 309–316.
9. Lim, J., Hao, T., Shaw, C., Patel, A. J., Szabo, G., Rual, J. F., Fisk, C. J., Li, N., Smolyar, A., Hill, D. E., Barabasi, A. L., Vidal, M., and Zoghbi, H. Y. (2006) A protein-protein interaction network for human inherited ataxias and disorders of Purkinje cell degeneration, *Cell 125*, 801–814.
10. Ruder, K., and Winstead, E. A Quick Guide to Sequenced Genomes, http://www.genomenewsnetwork.org/resources/sequenced_genomes/.
11. Fields, S., and Song, O. (1989) A novel genetic system to detect protein-protein interactions, *Nature 340*, 245–246.
12. Figeys, D., McBroom, L. D., and Moran, M. F. (2001) Mass spectrometry for the study of protein-protein interactions, *Methods 24*, 230–239.
13. Puig, O., Caspary, F., Rigaut, G., Rutz, B., Bouveret, E., Bragado-Nilsson, E., Wilm, M., and Seraphin, B. (2001) The tandem affinity purification (TAP) method: a general procedure of protein complex purification, *Methods 24*, 218–229.
14. Suter, B., Kittanakom, S., and Stagljar, I. (2008) Two-hybrid technologies in proteomics research, *Curr Opin Biotechnol 19*, 316–323.
15. Ho, Y., Gruhler, A., Heilbut, A., Bader, G. D., Moore, L., Adams, S. L., Millar, A., Taylor, P., Bennett, K., Boutilier, K., Yang, L., Wolting, C., Donaldson, I., Schandorff, S., Shewnarane, J., Vo, M., Taggart, J., Goudreault, M., Muskat, B., Alfarano, C., Dewar, D., Lin, Z., Michalickova, K., Willems, A. R., Sassi, H., Nielsen, P. A., Rasmussen, K. J., Andersen, J. R., Johansen, L. E., Hansen, L. H., Jespersen, H., Podtelejnikov, A., Nielsen, E., Crawford, J., Poulsen, V., Sorensen, B. D., Matthiesen, J., Hendrickson, R. C., Gleeson, F., Pawson, T., Moran, M. F., Durocher, D., Mann, M., Hogue, C. W., Figeys, D., and Tyers, M. (2002) Systematic identification of protein complexes in Saccharomyces cerevisiae by mass spectrometry, *Nature 415*, 180–183.
16. Ito, T., Chiba, T., Ozawa, R., Yoshida, M., Hattori, M., and Sakaki, Y. (2001) A comprehensive two-hybrid analysis to explore the yeast protein interactome, *Proc Natl Acad Sci USA 98*, 4569–4574.
17. Uetz, P., Giot, L., Cagney, G., Mansfield, T. A., Judson, R. S., Knight, J. R., Lockshon, D., Narayan, V., Srinivasan, M., Pochart, P., Qureshi-Emili, A., Li, Y., Godwin, B., Conover, D., Kalbfleisch, T., Vijayadamodar, G., Yang, M., Johnston, M., Fields, S., and Rothberg, J. M. (2000) A comprehensive analysis of protein-protein interactions in Saccharomyces cerevisiae, *Nature 403*, 623–627.
18. Yu, H., Braun, P., Yildirim, M. A., Lemmens, I., Venkatesan, K., Sahalie, J., Hirozane-Kishikawa, T., Gebreab, F., Li, N., Simonis, N., Hao, T., Rual, J. F., Dricot, A., Vazquez, A., Murray, R. R., Simon, C., Tardivo, L., Tam, S., Svrzikapa, N., Fan, C., de Smet, A. S., Motyl,

A., Hudson, M. E., Park, J., Xin, X., Cusick, M. E., Moore, T., Boone, C., Snyder, M., Roth, F. P., Barabasi, A. L., Tavernier, J., Hill, D. E., and Vidal, M. (2008) High-quality binary protein interaction map of the yeast interactome network, *Science 322*, 104–110.

19. Giot, L., Bader, J. S., Brouwer, C., Chaudhuri, A., Kuang, B., Li, Y., Hao, Y. L., Ooi, C. E., Godwin, B., Vitols, E., Vijayadamodar, G., Pochart, P., Machineni, H., Welsh, M., Kong, Y., Zerhusen, B., Malcolm, R., Varrone, Z., Collis, A., Minto, M., Burgess, S., McDaniel, L., Stimpson, E., Spriggs, F., Williams, J., Neurath, K., Ioime, N., Agee, M., Voss, E., Furtak, K., Renzulli, R., Aanensen, N., Carrolla, S., Bickelhaupt, E., Lazovatsky, Y., DaSilva, A., Zhong, J., Stanyon, C. A., Finley, R. L., Jr., White, K. P., Braverman, M., Jarvie, T., Gold, S., Leach, M., Knight, J., Shimkets, R. A., McKenna, M. P., Chant, J., and Rothberg, J. M. (2003) A protein interaction map of Drosophila melanogaster, *Science 302*, 1727–1736.

20. Li, S., Armstrong, C. M., Bertin, N., Ge, H., Milstein, S., Boxem, M., Vidalain, P. O., Han, J. D., Chesneau, A., Hao, T., Goldberg, D. S., Li, N., Martinez, M., Rual, J. F., Lamesch, P., Xu, L., Tewari, M., Wong, S. L., Zhang, L. V., Berriz, G. F., Jacotot, L., Vaglio, P., Reboul, J., Hirozane-Kishikawa, T., Li, Q., Gabel, H. W., Elewa, A., Baumgartner, B., Rose, D. J., Yu, H., Bosak, S., Sequerra, R., Fraser, A., Mango, S. E., Saxton, W. M., Strome, S., Van Den Heuvel, S., Piano, F., Vandenhaute, J., Sardet, C., Gerstein, M., Doucette-Stamm, L., Gunsalus, K. C., Harper, J. W., Cusick, M. E., Roth, F. P., Hill, D. E., and Vidal, M. (2004) A map of the interactome network of the metazoan C. elegans, *Science 303*, 540–543.

21. Aranda, B., Achuthan, P., Alam-Faruque, Y., Armean, I., Bridge, A., Derow, C., Feuermann, M., Ghanbarian, A. T., Kerrien, S., Khadake, J., Kerssemakers, J., Leroy, C., Menden, M., Michaut, M., Montecchi-Palazzi, L., Neuhauser, S. N., Orchard, S., Perreau, V., Roechert, B., van Eijk, K., and Hermjakob, H. The IntAct molecular interaction database in 2010, *Nucleic Acids Res 38*, D525–531.

22. Bader, G. D., Betel, D., and Hogue, C. W. (2003) BIND: the Biomolecular Interaction Network Database, *Nucleic Acids Res 31*, 248–250.

23. Breitkreutz, B. J., Stark, C., Reguly, T., Boucher, L., Breitkreutz, A., Livstone, M., Oughtred, R., Lackner, D. H., Bahler, J., Wood, V., Dolinski, K., and Tyers, M. (2008) The BioGRID Interaction Database: 2008 update, *Nucleic Acids Res 36*, D637–640.

24. Brown, K. R., and Jurisica, I. (2005) Online predicted human interaction database, *Bioinformatics 21*, 2076–2082.

25. Ewing, R. M., Chu, P., Elisma, F., Li, H., Taylor, P., Climie, S., McBroom-Cerajewski, L., Robinson, M. D., O'Connor, L., Li, M., Taylor, R., Dharsee, M., Ho, Y., Heilbut, A., Moore, L., Zhang, S., Ornatsky, O., Bukhman, Y. V., Ethier, M., Sheng, Y., Vasilescu, J., Abu-Farha, M., Lambert, J. P., Duewel, H. S., Stewart, II, Kuehl, B., Hogue, K., Colwill, K., Gladwish, K., Muskat, B., Kinach, R., Adams, S. L., Moran, M. F., Morin, G. B., Topaloglou, T., and Figeys, D. (2007) Large-scale mapping of human protein-protein interactions by mass spectrometry, *Mol Syst Biol 3*, 89.

26. Lehner, B., and Fraser, A. G. (2004) A first-draft human protein-interaction map, *Genome Biol 5*, R63.

27. Matthews, L., Gopinath, G., Gillespie, M., Caudy, M., Croft, D., de Bono, B., Garapati, P., Hemish, J., Hermjakob, H., Jassal, B., Kanapin, A., Lewis, S., Mahajan, S., May, B., Schmidt, E., Vastrik, I., Wu, G., Birney, E., Stein, L., and D'Eustachio, P. (2009) Reactome knowledgebase of human biological pathways and processes, *Nucleic Acids Res 37*, D619–622.

28. Persico, M., Ceol, A., Gavrila, C., Hoffmann, R., Florio, A., and Cesareni, G. (2005) HomoMINT: an inferred human network based on orthology mapping of protein interactions discovered in model organisms, *BMC Bioinformatics 6 Suppl 4*, S21.

29. Prasad, T. S., Kandasamy, K., and Pandey, A. (2009) Human Protein Reference Database and Human Proteinpedia as discovery tools for systems biology, *Methods Mol Biol 577*, 67–79.

30. Ramani, A. K., Bunescu, R. C., Mooney, R. J., and Marcotte, E. M. (2005) Consolidating the set of known human protein-protein interactions in preparation for large-scale mapping of the human interactome, *Genome Biol 6*, R40.

31. Rual, J. F., Venkatesan, K., Hao, T., Hirozane-Kishikawa, T., Dricot, A., Li, N., Berriz, G. F., Gibbons, F. D., Dreze, M., Ayivi-Guedehoussou, N., Klitgord, N., Simon, C., Boxem, M., Milstein, S., Rosenberg, J., Goldberg, D. S., Zhang, L. V., Wong, S. L., Franklin, G., Li, S., Albala, J. S., Lim, J., Fraughton, C., Llamosas, E., Cevik, S., Bex, C., Lamesch, P., Sikorski, R. S., Vandenhaute, J., Zoghbi, H. Y., Smolyar, A., Bosak, S., Sequerra, R., Doucette-Stamm, L., Cusick, M. E., Hill, D. E., Roth, F. P., and Vidal, M. (2005) Towards a proteome-scale map of the human protein-protein interaction network, *Nature 437*, 1173–1178.

32. Salwinski, L., Miller, C. S., Smith, A. J., Pettit, F. K., Bowie, J. U., and Eisenberg, D. (2004) The Database of Interacting Proteins: 2004 update, *Nucleic Acids Res 32*, D449–451.
33. Stelzl, U., Worm, U., Lalowski, M., Haenig, C., Brembeck, F. H., Goehler, H., Stroedicke, M., Zenkner, M., Schoenherr, A., Koeppen, S., Timm, J., Mintzlaff, S., Abraham, C., Bock, N., Kietzmann, S., Goedde, A., Toksoz, E., Droege, A., Krobitsch, S., Korn, B., Birchmeier, W., Lehrach, H., and Wanker, E. E. (2005) A human protein-protein interaction network: a resource for annotating the proteome, *Cell 122*, 957–968.
34. Jansen, R., Yu, H., Greenbaum, D., Kluger, Y., Krogan, N. J., Chung, S., Emili, A., Snyder, M., Greenblatt, J. F., and Gerstein, M. (2003) A Bayesian networks approach for predicting protein-protein interactions from genomic data, *Science 302*, 449–453.
35. Matthews, L. R., Vaglio, P., Reboul, J., Ge, H., Davis, B. P., Garrels, J., Vincent, S., and Vidal, M. (2001) Identification of potential interaction networks using sequence-based searches for conserved protein-protein interactions or "interologs", *Genome Res 11*, 2120–2126.
36. Braun, P., Rietman, E., and Vidal, M. (2008) Networking metabolites and diseases, *Proc Natl Acad Sci USA 105*, 9849–9850.
37. Yildirim, M. A., Goh, K. I., Cusick, M. E., Barabasi, A. L., and Vidal, M. (2007) Drug-target network, *Nat Biotechnol 25*, 1119–1126.
38. Koegl, M., and Uetz, P. (2007) Improving yeast two-hybrid screening systems, *Brief Funct Genomic Proteomic 6*, 302–312.
39. Venkatesan, K., Rual, J. F., Vazquez, A., Stelzl, U., Lemmens, I., Hirozane-Kishikawa, T., Hao, T., Zenkner, M., Xin, X., Goh, K. I., Yildirim, M. A., Simonis, N., Heinzmann, K., Gebreab, F., Sahalie, J. M., Cevik, S., Simon, C., de Smet, A. S., Dann, E., Smolyar, A., Vinayagam, A., Yu, H., Szeto, D., Borick, H., Dricot, A., Klitgord, N., Murray, R. R., Lin, C., Lalowski, M., Timm, J., Rau, K., Boone, C., Braun, P., Cusick, M. E., Roth, F. P., Hill, D. E., Tavernier, J., Wanker, E. E., Barabasi, A. L., and Vidal, M. (2009) An empirical framework for binary interactome mapping, *Nat Methods 6*, 83–90.
40. Chaurasia, G., Herzel, H., Wanker, E. E., and Futschik, M. E. (2006) Systematic functional assessment of human protein-protein interaction maps, *Genome Inform 17*, 36–45.
41. Futschik, M. E., Chaurasia, G., and Herzel, H. (2007) Comparison of human protein-protein interaction maps, *Bioinformatics 23*, 605–611.
42. Futschik, M. E., Tschaut, A., Chaurasia, G., and Herzel, H. (2007) Graph-theoretical comparison reveals structural divergence of human protein interaction networks, *Genome Inform 18*, 141–151.
43. Chaurasia, G., Iqbal, Y., Hänig, C., Herzel, H., Wanker, E., and Futschik, M. Flexible web-based integration of distributed large-scale human protein interaction maps, *Journal of Integrative Bioinformatics 4*, 51.
44. Chaurasia, G., Iqbal, Y., Hanig, C., Herzel, H., Wanker, E. E., and Futschik, M. E. (2007) UniHI: an entry gate to the human protein interactome, *Nucleic Acids Res 35*, D590–594.
45. Chaurasia, G., Malhotra, S., Russ, J., Schnoegl, S., Hanig, C., Wanker, E. E., and Futschik, M. E. (2009) UniHI 4: new tools for query, analysis and visualization of the human protein-protein interactome, *Nucleic Acids Res 37*, D657–660.
46. Kanehisa, M., and Goto, S. (2000) KEGG: kyoto encyclopedia of genes and genomes, *Nucleic Acids Res 28*, 27–30.
47. Kanehisa, M., Goto, S., Furumichi, M., Tanabe, M., and Hirakawa, M. KEGG for representation and analysis of molecular networks involving diseases and drugs, *Nucleic Acids Res 38*, D355–360.
48. Su, A. I., Wiltshire, T., Batalov, S., Lapp, H., Ching, K. A., Block, D., Zhang, J., Soden, R., Hayakawa, M., Kreiman, G., Cooke, M. P., Walker, J. R., and Hogenesch, J. B. (2004) A gene atlas of the mouse and human protein-encoding transcriptomes, *Proc Natl Acad Sci USA 101*, 6062–6067.
49. Sayers, E. W., Barrett, T., Benson, D. A., Bolton, E., Bryant, S. H., Canese, K., Chetvernin, V., Church, D. M., Dicuccio, M., Federhen, S., Feolo, M., Geer, L. Y., Helmberg, W., Kapustin, Y., Landsman, D., Lipman, D. J., Lu, Z., Madden, T. L., Madej, T., Maglott, D. R., Marchler-Bauer, A., Miller, V., Mizrachi, I., Ostell, J., Panchenko, A., Pruitt, K. D., Schuler, G. D., Sequeira, E., Sherry, S. T., Shumway, M., Sirotkin, K., Slotta, D., Souvorov, A., Starchenko, G., Tatusova, T. A., Wagner, L., Wang, Y., John Wilbur, W., Yaschenko, E., and Ye, J. Database resources of the National Center for Biotechnology Information, *Nucleic Acids Res 38*, D5–16
50. Lenffer, J., Nicholas, F. W., Castle, K., Rao, A., Gregory, S., Poidinger, M., Mailman, M. D., and Ranganathan, S. (2006) OMIA (Online Mendelian Inheritance in Animals): an enhanced platform and integration into the Entrez search interface at NCBI, *Nucleic Acids Res 34*, D599–601.

51. Ashburner, M., Ball, C. A., Blake, J. A., Botstein, D., Butler, H., Cherry, J. M., Davis, A. P., Dolinski, K., Dwight, S. S., Eppig, J. T., Harris, M. A., Hill, D. P., Issel-Tarver, L., Kasarskis, A., Lewis, S., Matese, J. C., Richardson, J. E., Ringwald, M., Rubin, G. M., and Sherlock, G. (2000) Gene ontology: tool for the unification of biology. The Gene Ontology Consortium, *Nat Genet 25*, 25–29.
52. Harris, M. A., Clark, J., Ireland, A., Lomax, J., Ashburner, M., Foulger, R., Eilbeck, K., Lewis, S., Marshall, B., Mungall, C., Richter, J., Rubin, G. M., Blake, J. A., Bult, C., Dolan, M., Drabkin, H., Eppig, J. T., Hill, D. P., Ni, L., Ringwald, M., Balakrishnan, R., Cherry, J. M., Christie, K. R., Costanzo, M. C., Dwight, S. S., Engel, S., Fisk, D. G., Hirschman, J. E., Hong, E. L., Nash, R. S., Sethuraman, A., Theesfeld, C. L., Botstein, D., Dolinski, K., Feierbach, B., Berardini, T., Mundodi, S., Rhee, S. Y., Apweiler, R., Barrell, D., Camon, E., Dimmer, E., Lee, V., Chisholm, R., Gaudet, P., Kibbe, W., Kishore, R., Schwarz, E. M., Sternberg, P., Gwinn, M., Hannick, L., Wortman, J., Berriman, M., Wood, V., de la Cruz, N., Tonellato, P., Jaiswal, P., Seigfried, T., and White, R. (2004) The Gene Ontology (GO) database and informatics resource, *Nucleic Acids Res 32*, D258–261.
53. Bruford, E. A., Lush, M. J., Wright, M. W., Sneddon, T. P., Povey, S., and Birney, E. (2008) The HGNC Database in 2008: a resource for the human genome, *Nucleic Acids Res 36*, D445–448.
54. Birney, E., Andrews, D., Bevan, P., Caccamo, M., Cameron, G., Chen, Y., Clarke, L., Coates, G., Cox, T., Cuff, J., Curwen, V., Cutts, T., Down, T., Durbin, R., Eyras, E., Fernandez-Suarez, X. M., Gane, P., Gibbins, B., Gilbert, J., Hammond, M., Hotz, H., Iyer, V., Kahari, A., Jekosch, K., Kasprzyk, A., Keefe, D., Keenan, S., Lehvaslaiho, H., McVicker, G., Melsopp, C., Meidl, P., Mongin, E., Pettett, R., Potter, S., Proctor, G., Rae, M., Searle, S., Slater, G., Smedley, D., Smith, J., Spooner, W., Stabenau, A., Stalker, J., Storey, R., Ureta-Vidal, A., Woodwark, C., Clamp, M., and Hubbard, T. (2004) Ensembl 2004, *Nucleic Acids Res 32*, D468–470.
55. Kasprzyk, A., Keefe, D., Smedley, D., London, D., Spooner, W., Melsopp, C., Hammond, M., Rocca-Serra, P., Cox, T., and Birney, E. (2004) EnsMart: a generic system for fast and flexible access to biological data, *Genome Res 14*, 160–169.
56. Barabasi, A. L., and Oltvai, Z. N. (2004) Network biology: understanding the cell's functional organization, *Nat Rev Genet 5*, 101–113.
57. Spirin, V., and Mirny, L. A. (2003) Protein complexes and functional modules in molecular networks, *Proc Natl Acad Sci USA 100*, 12123–12128.
58. Yook, S. H., Oltvai, Z. N., and Barabasi, A. L. (2004) Functional and topological characterization of protein interaction networks, *Proteomics 4*, 928–942.
59. Bork, P., Jensen, L. J., von Mering, C., Ramani, A. K., Lee, I., and Marcotte, E. M. (2004) Protein interaction networks from yeast to human, *Curr Opin Struct Biol 14*, 292–299.
60. Hart, G. T., Ramani, A. K., and Marcotte, E. M. (2006) How complete are current yeast and human protein-interaction networks?, *Genome Biol 7*, 120.
61. Bader, G. D., and Hogue, C. W. (2002) Analyzing yeast protein-protein interaction data obtained from different sources, *Nat Biotechnol 20*, 991–997.
62. von Mering, C., Krause, R., Snel, B., Cornell, M., Oliver, S. G., Fields, S., and Bork, P. (2002) Comparative assessment of large-scale data sets of protein-protein interactions, *Nature 417*, 399–403.
63. Cusick, M. E., Yu, H., Smolyar, A., Venkatesan, K., Carvunis, A. R., Simonis, N., Rual, J. F., Borick, H., Braun, P., Dreze, M., Vandenhaute, J., Galli, M., Yazaki, J., Hill, D. E., Ecker, J. R., Roth, F. P., and Vidal, M. (2009) Literature-curated protein interaction datasets, *Nat Methods 6*, 39–46.
64. Ge, H., Liu, Z., Church, G. M., and Vidal, M. (2001) Correlation between transcriptome and interactome mapping data from Saccharomyces cerevisiae, *Nat Genet 29*, 482–486.
65. Mrowka, R., Patzak, A., and Herzel, H. (2001) Is there a bias in proteome research?, *Genome Res 11*, 1971–1973.

Chapter 11

Gene-Centered Yeast One-Hybrid Assays

John S. Reece-Hoyes and Albertha J.M. Walhout

Abstract

Transcription is regulated by sequence-specific transcription factors (TFs) that bind to short genomic DNA elements that can be located in promoters, enhancers and other *cis*-regulatory modules. Determining which TFs bind where requires techniques that enable the ab initio identification of TF–DNA interactions. These techniques can either be "TF-centered" (protein-to-DNA), where regions of DNA bound by a TF of interest are identified, or "gene-centered" (DNA-to-protein), where TFs that bind a DNA sequence of interest are identified. Here, we describe gene-centered yeast one-hybrid (Y1H) assays. Briefly, in Y1H assays, a DNA fragment is cloned upstream of two different reporters, and these reporter constructs are integrated into the genome of a yeast strain. Next, plasmids expressing TFs as hybrid proteins (hence the name of the assay) fused with the strong transcriptional activation domain (AD) of the yeast TF Gal4 are introduced into the yeast strain. When a TF interacts with the DNA fragment of interest, the AD moiety activates reporter expression in yeast regardless of whether the TF is an activator or repressor in vivo. Sequencing the plasmid in the colonies that exhibit reporter activation reveals the identity of the TFs that can bind the DNA fragment. We have shown Y1H to be a robust method for detecting interactions between a variety of DNA elements and multiple families of TFs.

Key words: Transcription factor, Yeast one-hybrid, Gene expression, Transcription

1. Introduction

The yeast one-hybrid (Y1H) assay is a genetic method for identifying proteins that can interact with a DNA sequence of interest (1, 2) (Fig. 1). The assay is similar to the widely used yeast two-hybrid assay that identifies protein–protein interactions, either in small or large-scale settings (3–5). An outline of the Y1H protocol is provided in Fig. 2. Briefly, a DNA fragment is cloned upstream of two different reporters, and these reporter constructs are integrated into the genome of a yeast strain. Next, plasmids expressing TFs as hybrid proteins fused with the strong transcriptional activation

Bernhard Suter and Erich E. Wanker (eds.), *Two Hybrid Technologies: Methods and Protocols*, Methods in Molecular Biology, vol. 812, DOI 10.1007/978-1-61779-455-1_11, © Springer Science+Business Media, LLC 2012

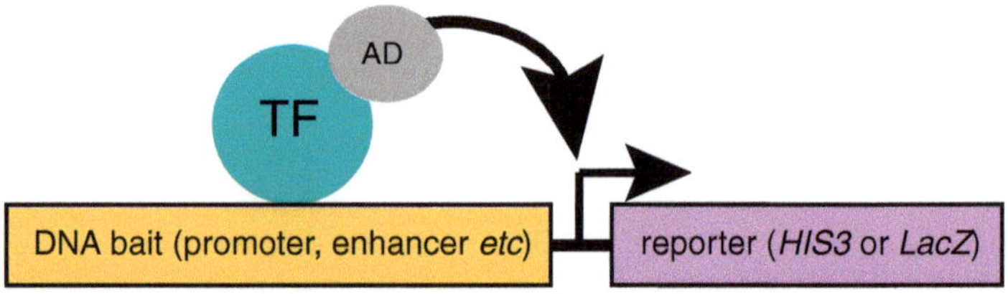

Fig. 1. Cartoon of Y1H assay: *TF* transcription factor; *AD* Gal4 transcription activation domain.

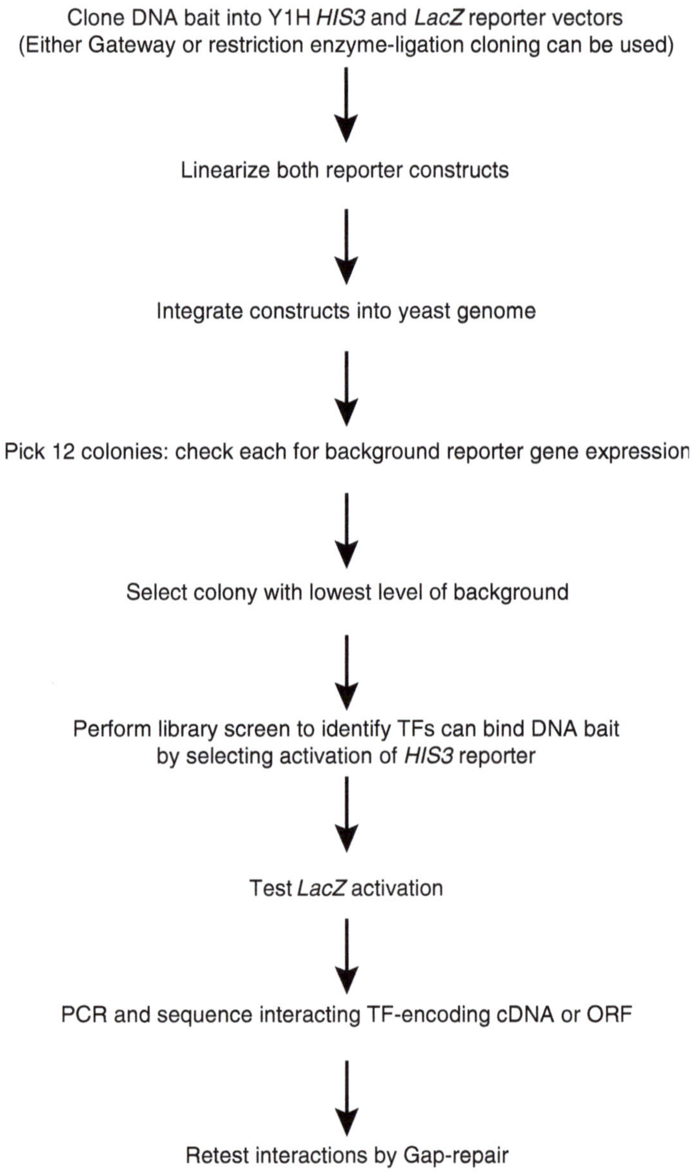

Fig. 2. Outline of Y1H cDNA library screen procedure.

domain (AD) of the yeast Gal4 transcription factor are introduced into the yeast strain. When a TF interacts with the DNA fragment of interest, the AD moiety activates reporter expression in yeast regardless of whether the TF is an activator or repressor in vivo. Sequencing the plasmid in each of these colonies reveals the identity of the TFs that can bind the DNA fragment.

The Y1H assay provides a "gene-centered" (DNA-to-protein) method capable of identifying a collection of TFs that can bind to a DNA fragment of interest. The Y1H assay is therefore complementary to "TF-centered" (protein-to-DNA) methods such as chromatin immunoprecipitation (ChIP) (6), DamID (7) and protein binding microarrays (8) that determine the DNA sequences bound by a protein of interest. While powerful, there are several important conceptual and technical limitations to TF-centered methods (we focus on ChIP, as it is the most widely used). First, and most importantly, ChIP can only be used for one TF at a time and thus the comprehensive identification of a TF binding profile (i.e., the collection of TFs that bind a region of interest) requires performing a ChIP assay for every TF in the system of interest. With hundreds of TFs encoded by eukaryotic genomes, this is as of yet not feasible. Second, the success of ChIP depends on the expression level of the TF (i.e., the more highly or broadly expressed, the more likely it will work in ChIP) as well as the availability of suitable anti-TF antibodies. Currently, only few TFs are amenable to ChIP in most complex eukaryotic organisms and cell systems, making the delineation of TF binding profiles using this technique a daunting task. The Y1H assay does not require antibodies, and every TF is expressed from a plasmid transformed into a yeast strain, so generating TF binding profiles is more feasible using this method. The Y1H assay does, however, have its own set of limitations, including the need to generate TF clone libraries or cDNA libraries, and the possibility that a TF protein may have reduced functionality (e.g., that depend on posttranslational modifications) in the yeast system.

TF-DNA interactions can be studied at different scales. For instance, many researchers are interested in how individual genes affect a specific biological process and often use a variety of assays to study the physical interactions in which these genes and their products engage. On the contrary, more and more biologists aim to understand biological processes at a systems level, by connecting large numbers of genes and their products into gene regulatory networks (9). We have developed Y1H assays for the large-scale identification of TF-DNA interactions by combining conventional Y1H assays with recombinational Gateway cloning (10), enabling the processing of large numbers of target DNA sequences ("DNA baits") simultaneously. We have demonstrated that this system can be efficiently used to generate regulatory networks involving multiple DNA baits and their interactions with many different types of TFs (11–14). However, it is important to note that our Gateway-compatible Y1H system is equally applicable to smaller sets or single DNA fragments of interest.

The protocols in this chapter describe the identification of protein–DNA interactions by screening a cDNA library transformed into a haploid DNA bait strain. While this type of screen is the most popular option because numerous cDNA libraries are commercially available, it should be noted that other technical variations of Y1H are available. For instance, TF clones can be obtained from ORF collections (15, 16) rather than cDNA libraries, and assayed as individual clones (17), multiple small pools of clones (17), or as a TF-only mini-library (10). Further, Y1H assays using individual TF clones can be performed by mating a TF-containing strain with DNA-bait containing strain and detecting the interaction in the resulting diploid yeast (17).

2. Materials

2.1. Yeast Medium

1. Medium additives: 3-amino-triazol (3AT, Sigma) 2 M, filter-sterilized, wrapped in aluminum foil and stored at 4°C. Tryptophan 40 mM, filter-sterilized, wrapped in aluminum foil and stored at 4°C. Leucine 100 mM, stored at room temperature. Histidine 100 mM, stored at room temperature. Uracil 20 mM, stored at room temperature.
2. Synthetic complete (Sc) solid media: In one 2-L flask, 2.6 g synthetic complete amino acid mix lacking Ura, His, Trp, Leu (US Biologicals), 3.4 g yeast nitrogen base (US Biologicals), and 10 g ammonium sulfate are dissolved in 950 mL water, pH to 5.9 by adding (~3 mL) 10 M NaOH, retain stir bar in flask. In a second 2-L flask, 35 g agar is added to 950 mL water. Both flasks are autoclaved for 40 min on liquid cycle, the contents of the first flask (including stir bar) is poured into the agar flask, 100 mL autoclaved 40% (w/v) glucose is then added. The mix is cooled to 55°C before adding media additives as required (16 mL of leucine, histidine, tryptophan, uracil stock solutions will provide the desired final concentrations). 150-mm sterile Petri dishes require about 80 mL per dish, are dried for 3–5 days at room temperature, and stored in air-tight packaging at room temperature for up to 6 months.
3. YAPD liquid: 10 g yeast extract, 0.16 g adenine hemisulphate dihydrate and 20 g peptone are dissolved in 950 mL water. Autoclave and then add 50 mL sterile 40% (w/v) glucose.
4. YAPD plates (solid media): In one 2-L flask, 20 g yeast extract, 0.32 g adenine hemisulphate dihydrate and 40 g peptone are dissolved in 950 mL water; retain stir bar in the flask. In a second

2-L flask 35 g agar is added to 950 mL water. Both flasks are autoclaved for 40 min on liquid cycle, the contents of the first flask (including stir bar) is poured into the agar flask, 100 mL autoclaved 40% (w/v) glucose is added. 150-mm sterile Petri dishes require about 80 mL per dish, are dried for 3–5 days at room temperature, and stored in air-tight packaging at room temperature for up to 6 months.

2.2. Generating a DNA Bait Yeast Strain

1. YM4271 yeast strain (*MATa, ura3-52, his3-Δ200, ade2-101, ade5, lys2-801, leu2-3,112, trp1-901, tyr1-501, gal4Δ, gal80Δ, ade5::hisG*).
2. DNA bait::reporter constructs cloned in pMW#2(*HIS3*) and pMW#3(*LacZ*).

2.3. High-Efficiency Yeast Transformation

1. TE buffer (10×): 100 mM Tris–HCl pH 8.0, 10 mM EDTA.
2. TE/LiAc solution: 1× TE buffer, 100 mM lithium acetate (Sigma).
3. TE/LiAc/PEG solution: 1× TE buffer, 100 mM lithium acetate (Sigma), 40% (w/v) PEG 3350 (Fisher Scientific).
4. Disposable 250-mL conical flasks (Corning).
5. Salmon sperm DNA, 10 mg/mL (Invitrogen).

2.4. Yeast Replica-Plating

1. Velvets: 22 × 22 cm pieces of velveteen (100% cotton velveteen without rayon). When first cut from a larger sheet of velveteen, velvets need to be washed and dried at least ten times (without soap) to remove excess lint. After use in replica plating, velvets need to be first autoclaved, then washed without soap and dried, packed in stacks, and wrapped in aluminum foil and again autoclaved using a dry cycle.
2. Replica-plating apparatus (Cora Styles for 150 mm plates).

2.5. βGal Assay

1. Z-buffer: 60 mM Na_2HPO_4, 60 mM $NaH_2PO_4.H_2O$, 10 mM KCl, 1 mM $MgSO_4$, pH to 7.0 with 10 M NaOH, autoclaved, stored at room temperature.
2. X-gal solution: X-gal (Gold Biotechnology) dissolved at 4% (w/v) in di methyl formamide (DMF) (in fume hood), wrapped in aluminum foil, stored at −20°C.
3. Nitrocellulose (NC) filters (45 μm, 137 mm) (Fisher Scientific).
4. Whatman filters, 125 mm, (VWR).

2.6. Yeast PCR

1. Zymolyase suspension: 200 mg Zymolase-100T (Associates of Cape Cod) is mixed in 100 mL sterile sodium phosphate buffer (0.1 M, pH 7.5) for 30 min (some precipitate will be visible). 1 mL aliquots are stored at −20°C.
2. M13F primer (5′-GTAAAACGACGGCCAGT-3′).
3. 1H1F primer (5′-GTTCGGAGATTACCGAATCAA-3′).

4. HIS293R primer (5′-GGGACCACCCTTTAAAGAGA-3′).
5. LACZ592R primer (5′-ATGCGCTCAGGTCAAATTCAGA-3′).
6. AD primer (5′-CGCGTTTGGAATCACTACAGGG-3′).
7. TERM primer (5′-GGAGACTTGACCAAACCTCTGGCG-3′).

2.7. Y1H Screening an AD-cDNA Library

1. DNA-bait yeast strain.
2. AD-cDNA library DNA at a concentration of 1 μg/μL.

3. Methods

3.1. Generating a DNA Bait Yeast Strain

3.1.1. High-Efficiency Yeast Transformation for Integrating Reporter Constructs

1. Linearize at least 1 μg of the DNA bait pMW#2 reporter construct using one of *Afl*II, *Xho*I, *Nsi*I or *Bse*R1 in a 25 μL reaction. Similarly linearize at least 1 μg of the DNA bait pMW#3 reporter construct using one of *Nco*I, *Apa*I, *Stu*I in a 25 μL reaction (see Note 1).
2. Using a sterile toothpick or disposable plastic loop, thickly spread YM4271 yeast strain onto 150 mm YAPD plates. Incubate overnight at 30°C.
3. Using a sterile toothpick or disposable plastic loop resuspend YM4271 into 40 mL liquid YAPD to an OD600 of 0.15–0.20. Remember to also perform Subheading 3.1.1, steps 3–19 for a "no DNA" control.
4. Incubate the culture in a shaking incubator (30°C, 200 rpm) until the culture reaches an OD600 of 0.4–0.6. This usually takes about 5 h. However, the culture should be checked regularly (every hour).
5. Pellet the cells by centrifugation (700 × *g*) in conical tubes at room temperature for 5 min.
6. Prepare at least 20 μL denatured salmon sperm DNA (ssDNA, 10 mg/mL) by boiling for 5 min, then keeping on ice.
7. Discard the liquid from the pellet formed in Subheading 3.1.1, step 5 (the yeast form a firm pellet), and wash the cells with 20 mL sterile water. The cells can be resuspended by pipetting up and down or by shaking the tube. Do not vortex.
8. Centrifuge the yeast suspension as in Subheading 3.1.1, step 5, discard the supernatant, and resuspend the cells in 5 mL freshly made TE/LiAc solution. Do not vortex.
9. Centrifuge as in Subheading 3.1.1, step 5, remove the supernatant carefully by aspiration, and then resuspend the cells in 200 μL TE/LiAc solution by pipetting up and down.
10. Add 20 μL denatured ssDNA from Subheading 3.1.1, step 6 to the yeast suspension and mix by pipetting up and down.

11. Transfer the TE/LiAc/ssDNA/YM4271 suspension into a sterile 1.5-mL tube.
12. Add 20 μL of both linearized constructs from Subheading 3.1.1, step 1 into the same 200 μL of TE/LiAc/ssDNA/YM4271 suspension.
13. Add 1,250 μL TE/LiAc/PEG solution to the tube and mix by multiple inversions or by pipetting up and down (do not vortex).
14. Incubate all transformation reactions for 30 min at 30°C (see Note 2).
15. Heat-shock the cells for exactly 20 min at 42°C (in a water bath).
16. Centrifuge the tubes for 1 min at room temperature at 700 ×*g* and remove the supernatant by aspiration.
17. Resuspend the pellet in 500 μL sterile water by pipetting up and down (do not vortex).
18. Using sterile glass beads, plate the resuspension onto a 150 mm Sc-Ura,-His plate. Do this also for the "no DNA" control.
19. Incubate all plates at 30°C.
20. After 3 days, yeast colonies should appear on the integration plate and no colonies should grow on the "no DNA" control. If colonies do appear on the latter plate, there has been a contamination in one of the reagents and the integration should be repeated with new reagents (see Note 3).

3.1.2. Testing Autoactivation of DNA Bait Yeast Strains

1. Using sterile toothpicks, transfer 12 integrant colonies (see Note 4) from the plate in Subheading 3.1.1, step 20 to a 150 mm Sc-Ura,-His plate in "96-spot format" (Fig. 3). If positive control yeast strains that express known levels of reporters are available, they should also be transferred to these plates (Fig. 3). Incubate the plates at 30°C for 1–2 days.
2. Replica-plate (see Note 5) the yeast from Subheading 3.1.2, step 1 onto several 150 mm Sc-Ura,-His plates, each containing a different concentration of 3AT (0, 10, 20, 40, 60, 80 mM) (see Note 6), and a 150 mm YAPD plate onto which a nitrocellulose (NC) filter has been placed (so that the yeast will grow on the filter). Replica-clean (see Note 5) the plates containing 3AT and incubate them for 5 days at 30°C. Incubate the other two plates overnight at 30°C. The Sc-Ura,-His plate without 3AT is used to maintain the integrants and can be stored at room temperature for 14 days. The NC filter/YAPD plate is used for the βGal assay (steps 3–7 below).
3. For each NC filter/YAPD plate to be analyzed, place two Whatman filters in an empty 150-mm Petri dish. Move to a fumehood for the next three steps.

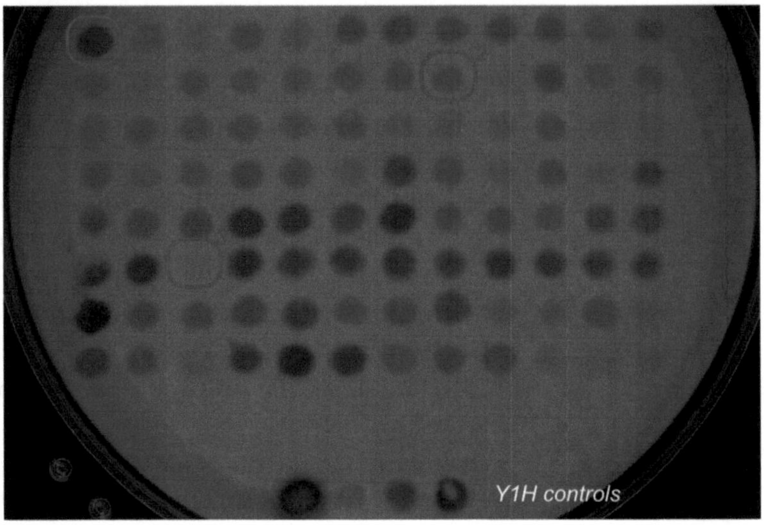

Fig. 3. Example of 96-spot *LacZ* assay. 1 – strong *LacZ* expression; 2 – weak *LacZ* expression; 3 – no *LacZ* expression.

4. For each Petri dish in Subheading 3.1.2, step 3, set up a reaction mix containing 6 mL Z-buffer, 11 μL β-mercapto-ethanol, and 100 μL X-gal solution (see Note 7). Use this entire mix to soak the Whatman filters, and remove any air bubbles using forceps to lift the filters and squeeze the bubbles to the sides, then remove excess liquid into a waste bottle by tipping the plate.
5. Lift the NC filter from the YAPD plate using forceps and place the filter yeast side up in a liquid nitrogen bath for 10 s. Discard the YAPD plate.
6. Use the forceps to place the frozen NC filter with the yeast facing up onto the wet Whatman filters from Subheading 3.1.2, step 4, and use forceps (or a needle) to remove air bubbles under the NC filter quickly as the NC filter (and yeast lysate) thaws.
7. Incubate each βGal assay plate at 37°C overnight. Record the amount of blue compound generated by each yeast lysate.
8. Observe the amount of growth each integrant displays after 5 days on the Sc-Ura,-His plates containing 3AT (Subheading 3.1.2, step 2), recording how much 3AT was required to inhibit the growth of each integrant strain.
9. Choose 1–4 integrant strains showing the lowest autoactivity for both reporters (see Note 8).

3.1.3. Yeast PCR to Confirm DNA Bait Identity

1. Using a sterile toothpick, transfer the chosen yeast from Subheading 3.1.2, step 9 to a fresh YAPD plate (see Note 9). Incubate overnight at 30°C.

2. For each integrant to be analyzed, aliquot 15 μL zymolyase suspension into a sterile PCR tube or the wells of a 96-well PCR plate (see Note 10).
3. Using a sterile toothpick or pipette tip, transfer a small amount (approximately one-fourth the size of a match head) of each yeast colony to its own zymolyase aliquot. Too much yeast will inhibit the PCR reaction.
4. Transfer the PCR tubes/plate to a thermal cycler and incubate the yeast-enzyme mix for 30 min at 37°C, then heat-inactivate the enzyme for 10 min at 95°C.
5. Remove the PCR tubes/plate from the thermal cycler and dilute each lysate by adding 85 μL sterile PCR-grade water. The lysates can be stored at –20°C for up to 1 year for subsequent PCR reactions.
6. For each lysate prepare two PCR reactions, one for the *HIS3* construct and one for the *LacZ* construct (see Note 11). *HIS3*: 5 μL diluted lysate, 1 μL M13F primer (20 μM), 1 μL HIS293R primer (20 μM), 5 μL dNTPs (1 mM), 5 μL PCR buffer (10×), 2 U Thermostable DNA polymerase, PCR-grade water to total reaction volume of 50 μL. *LacZ*: 5 μL diluted lysate, 1 μL 1H1F primer (20 μM), 1 μL LAC592R primer (20 μM), 5 μL dNTPs (1 mM), 5 μL PCR buffer (10×), 2 U Thermostable DNA polymerase, PCR-grade water to total reaction volume of 50 μL.
7. Place the tubes in a thermal cycler and run the following PCR progam: (a) 94°C for 2 min, (b) 94°C for 1 min, (c) 56°C for 1 min, (d) 72°C for 1 min/kb of DNA bait, (e) Repeat steps b–d 34 times, (f) 72°C for 7 min. The conditions of the PCR reaction may need to be optimized.
8. Run 5–10 μL of the PCR reaction on a 1% (w/v) agarose gel in 1×TBE alongside DNA molecular weight markers (see Note 12).
9. Use the appropriate forward primer to sequence each PCR product that is the expected size. Maintain the integrant strains on either YAPD or Sc-His,-Ura plates while waiting for sequence results (the yeast will need to be transferred to fresh media every 2 weeks).
10. Using a sterile toothpick, transfer the integrants that have the correct DNA bait in both reporter constructs to a fresh Sc Ura, His plate and incubate the plate at 30°C overnight. Only one integrant strain per DNA bait is needed for library screens but it is useful to have at least one backup strain.
11. Using a sterile toothpick, transfer a match-head-sized amount of this freshly grown yeast into 200 μL of sterile 15% (v/v) glycerol solution. Vortex the yeast/glycerol solution for 5 s

and store the resulting yeast suspension at –80°C. These frozen yeast stocks can be revived by transferring some (~5 μL) of the frozen stock to a YAPD plate and allowing growth at 30°C for at least 2 days.

3.2. Y1H Screening an AD-cDNA Library

3.2.1. High-Efficiency Yeast Transformation with an AD-cDNA Library

1. Using a sterile toothpick or disposable plastic loop, thickly spread the DNA bait yeast strain onto YAPD plates (see Note 9). Incubate overnight at 30°C.
2. Using a sterile toothpick or disposable plastic loop resuspend the bait strain into 400 mL liquid YAPD to an OD600 of 0.15–0.20.
3. Incubate the culture in a shaking incubator (30°C, 200 rpm) until the culture reaches an OD600 of 0.4–0.6. This usually takes about 5 h. However, the culture should be checked regularly (every hour).
4. Pellet the cells by centrifugation (700 × *g*) in conical 250 mL tubes at room temperature for 5 min.
5. Prepare at least 200 μL denatured salmon sperm DNA (ssDNA, 10 mg/mL) by boiling for 5 min, then keeping on ice.
6. Discard the liquid from the pellet formed in Subheading 3.2.1, step 4 (the yeast form a firm pellet), and wash the cells with 40 mL sterile water. The cells can be resuspended by and pipetting or by shaking the tube. Do not vortex.
7. Centrifuge the yeast suspension as in Subheading 3.2.1, step 4, discard the supernatant, and resuspend the cells in 8 mL freshly made TE/LiAc solution. Do not vortex.
8. Centrifuge as in Subheading 3.2.1, step 4, remove the supernatant carefully by aspiration, and then resuspend the cells in 2 mL TE/LiAc solution by pipetting up and down.
9. Add 200 μL denatured ssDNA to the yeast suspension and mix by pipetting up and down.
10. Aliquot 2 mL of TE/LiAc/ssDNA bait yeast suspension into a sterile 15-mL tube, 30 μL into a sterile 1.5 mL tube, and 100 μL into another sterile 1.5-mL tube.
11. Add 30 μg of AD-cDNA library (e.g., 30 μL of a 1 μg/μL solution) to the 2 mL of TE/LiAc/ssDNA bait yeast suspension, then add 10.5 mL freshly made TE/LiAc/PEG solution, and mix by multiple inversions or pipetting up and down (do not vortex). Aliquot this yeast/library/PEG mix into 12 sterile 1.5-mL tubes (see Note 13).
12. Add 50 ng empty library plasmid to the 30 μL TE/LiAc/ssDNA bait yeast suspension, then add 175 μL TE/LiAc/PEG solution, and mix by pipetting up and down (do not vortex).

13. Add 500 μL TE/LiAc/PEG solution to the tube containing 100 μL of TE/LiAc/ssDNA bait yeast suspension, and mix by multiple inversions or by pipetting up and down (do not vortex).
14. Incubate all transformation reactions for 30 min at 30°C (see Note 2).
15. Heat-shock the cells for exactly 20 min at 42°C (in a water bath).
16. Centrifuge the tubes for 1 min at room temperature at 700 × *g* and remove the supernatant by aspiration.
17. Resuspend the pellet in each of the cDNA library transformation tubes (Subheading 3.2.1, step 11) in 500 μL sterile water by pipetting up and down (do not vortex). Take 5 μL of the library transformation mix from one of these tubes and mix with 495 μL of sterile water to create a 1:100 dilution. Take 50 μL from the 1:100 dilution and add 450 μL sterile water to create a 1:1,000 dilution. Use sterile glass beads to spread the dilutions onto 150 mm Sc-Ura,-His,-Trp plates (see Note 14). These plates will be used to determine the transformation efficiency of the reactions.
18. Using sterile glass beads, plate the remainder of the library transformation across twelve 150 mm Sc-Ura,-His,-Trp + 3AT plates. Multiple large media plates are needed to ensure the growth of well-separated colonies (see Note 15).
19. Resuspend the "empty plasmid" control yeast transformation reaction (Subheading 3.2.1, step 12) in 30 μL sterile water by pipetting up and down (do not vortex). Pipette 5 μL of this suspension as a single spot onto a Sc-Ura,-His,-Trp plate. This yeast transformed with empty library vector will be used later as the "negative control" for the assessment of interactions retrieved from the cDNA library.
20. Resuspend the "no DNA" control yeast transformation (Subheading 3.2.1, step 13) in 500 μL sterile water by pipetting up and down (do not vortex). Use sterile glass beads to spread this suspension onto a single 150 mm Sc-Ura,-His,-Trp plate.
21. Incubate all plates at 30°C.
22. After 3 days, yeast colonies should appear on the empty plasmid (Subheading 3.2.1, step 19) and transformation efficiency determination (Subheading 3.2.1, step 17) plates and no colonies should grow on the no DNA control (Subheading 3.2.1, step 20). If colonies do appear on the no DNA control plate, the screen cannot be used as there has been a contamination in one of the reagents.

23. Calculate the transformation efficiency (number of colonies per microgram of DNA) using the plated dilutions from Subheading 3.2.1, step 17 (see Note 16).
24. If the transformation was successful, the plates from Subheading 3.2.1, step 18 can be incubated at 30°C for up to 14 days. Check the plates regularly for appearing colonies and pick "HIS-positive" colonies using a sterile toothpick to transfer to a 150 mm Sc-Ura-His-Trp plate (see Note 17). This plate of selected colonies can be stored at room temperature for up to 14 days.

3.2.2. Identifying "Double-Positive" Transformants

1. Using sterile toothpicks (or by replica-plating), transfer "*HIS*-positive" colonies from the plate(s) from Subheading 3.2.1, step 24 to a new 150 mm Sc-Ura,-His,-Trp plate in "96-spot format" (Fig. 3). Also transfer some of the empty plasmid control yeast from Subheading 3.2.1, step 19 to these plates – these will be used later to judge reporter gene induction. If positive control yeast strains that express known levels of reporters are available, they should also be transferred to these plates. Incubate the plates at 30°C for 1–2 days.
2. Replica-plate (see Note 5) the yeast from Subheading 3.2.2, step 1 onto a fresh 150 mm Sc-Ura,-His,-Trp plate, a 150 mm Sc-Ura,-His,-Trp plate containing the 3AT concentration used for screening, and a 150 mm YAPD plate onto which a nitrocellulose (NC) filter has been placed (so that the yeast will grow on the filter). Replica-clean (see Note 5) the plate containing 3AT and incubate this plate for 5–10 days at 30°C (to confirm the "*HIS*-positive" result). Incubate the other two plates overnight at 30°C. The Sc-Ura,-His,-Trp plate is used to maintain the transformants and can be stored at room temperature for 14 days. The NC filter/YAPD plate is used for the βGal assay (Subheading 3.1.2, steps 3–6).
3. Check the βGal assays for blue coloring regularly during the 37°C incubation (every hour if necessary) over a maximum 24 h period. Pictures should be taken that demonstrate the amount of blue compound generated by each yeast lysate. For baits that display moderate to high *LacZ* expression in the absence of a TF, it may be necessary to compare potential positive yeast to the empty plasmid controls often to see increased reporter expression.
4. Double-positive yeast will exhibit both growth on 3AT-containing media (Subheading 3.2.2, step 2) and blue coloring in the βGal assay (Subheading 3.2.2, step 3) that is greater than the yeast transformed with empty plasmid (see Note 18).

3.2.3. Gap-Repair Retesting of Interactions

1. Transfer "double-positive" yeast from Subheading 3.2.2, step 4 to a fresh Sc-Ura,-His,-Trp plate(s) with a sterile toothpick (or by replica-plating), and incubate overnight at 30°C (see Note 19).
2. For each positive yeast generate a lysate for yeast PCR (Subheading 3.1.3, steps 2–5).
3. For each amplification, in a sterile PCR tube (or well of a PCR plate) prepare the following PCR reaction mix: 5 μL diluted lysate, 1 μL AD primer (20 μM), 1 μL TERM primer (20 μM), 5 μL dNTPs (1 mM), 5 μL PCR buffer (10×), 2 U Thermostable DNA polymerase, PCR-grade water to total reaction volume of 50 μL (see Note 11).
4. Place the tubes in a thermal cycler and run the following PCR progam: (a) 94°C for 2 min, (b) 94°C for 1 min, (c) 56°C for 1 min, (d) 72°C for 1 min/kb of average library insert size, (e) Repeat steps b–d 34 times, (f) 72°C for 7 min. The conditions of the PCR reaction may need to be optimized.
5. Run 5–10 μL of the PCR reaction on a 1% (w/v) agarose gel in 1×TBE alongside DNA molecular weight markers. The PCR products can be stored at −20°C. Only PCR reactions that amplify a single band can be used in later steps (see Note 20).
6. In a sterile 1.5 mL microcentrifuge tube set up the following restriction endonuclease digest: 10 μg pPC86 plasmid, 4 μL Bovine Serum Albumin (1 mg/mL), 4 μL Restriction Buffer (10×), 2 μL (20 U) *Sal*I, 2 μL (20 U) *Bgl*II, PCR-grade water to a final volume of 40 μL. Mix the digest and incubate overnight at 37°C (see Note 21).
7. Using Subheading 3.2.1, steps 1–9, prepare 2.2 mL of TE/LiAc/ssDNA bait yeast suspension. This will be enough yeast suspension for 96 transformations performed in a 96-well PCR plate.
8. To each well, aliquot 20 μL TE/LiAc/ssDNA yeast suspension, and add 5 μL prey PCR product (from Subheading 3.2.3, step 5) and 40 ng linear pPC86 (from Subheading 3.2.3, step 6). Three negative controls *without* PCR product should be included: no DNA, 40 ng of linear pPC86 alone, and 40 ng of circular (i.e., undigested) pPC86.
9. Add 150 μL freshly made TE/LiAc/PEG solution to each well and mix by pipetting up and down.
10. Incubate the PCR plate for 30 min at 30°C using a PCR machine or water bath (see Note 2).
11. Heat-shock the cells for exactly 20 min at 42°C using a PCR machine or water bath.
12. Centrifuge the tubes for 1 min at room temperature (700×*g*) and remove the supernatant by careful aspiration (using an 8-channel aspirator connected to a vacuum source is optimal).

13. Resuspend each well of yeast in 20 μL sterile water by pipetting up and down. Pipette 3 μL of each suspension as a single spot onto a Sc-Ura,-His,-Trp plate in "96-spot format" (Fig. 3). If positive control yeast strains that express known levels of reporters are available, they should also be transferred to these plates (Fig. 3). Incubate the plates at 30°C.
14. After 3 days, yeast colonies should have appeared. The number of transformants should be an order of magnitude higher in gap-repair samples and uncut pPC86 controls compared to the linear pPC86 alone control. No transformants should be present in the no DNA control (see Note 22).
15. Use Subheading 3.2.2, step 2 with the yeast media plate from Subheading 3.2.3, step 14 to identify double-positive transformants.
16. Determine the identity of these double-positive preys that retest by gap-repair by sequencing the PCR products generated in Subheading 3.2.3, step 5 using the AD primer (see Note 23).

4. Notes

1. This protocol assumes that the DNA bait::reporter constructs have been generated by Gateway cloning in pMW#2 and pMW#3 (10). Non-Gateway versions of these vectors called pHISi-1 and pLACZi are available from Clontech. The restriction enzyme of choice must not cut within the DNA bait sequence. These digests linearize the Y1H reporter constructs such that regions of homology to the yeast genome occur at both ends. When these linear constructs are transformed into the yeast strain, the pMW#2 construct is integrated into the mutant *HIS3* locus (*his3-Δ200*) of the host strain YM4271, while the pMW#3 construct is integrated into the mutant *URA3* locus (*ura3-52*). If alternate reporter constructs are used, restriction digests that result in a conceptually similar DNA molecule must be performed.
2. This incubation step can continue for up to 4 h, but the transformation efficiency is greatly reduced if this incubation step is less than 30 min or more than 4 h.
3. If the entire transformation is spread onto a single 100 mm plate, the integrants will have difficulty growing due to the presence of untransformed yeast – if 100 mm plates are used, spread each transformation over at least three plates. Both Y1H reporter constructs are integrative (YIp) vectors that cannot be replicated within yeast, so any colonies that grow at this stage must be integrants. The number of colonies obtained usually varies between 10 and 100. If no integrants arise, a first option

to try is to repeat the transformation with more linear DNA, a second option is to integrate the reporters sequentially instead of simultaneously [i.e., integrate the *LacZ* reporter first, pick a low autoactive integrant (see below) then integrate the *HIS3* reporter into this strain].

4. Autoactivity is the level of expression of integrated DNA bait::reporters in the absence of an AD-prey clone. The cause of autoactivation is likely due to an endogenous yeast activator binding the DNA bait. It has also been observed that different integrant strains arising from the same transformation can show varying levels of autoactivity, which may be because differing numbers of DNA bait::reporter cassettes have been integrated into each strain (and even at each locus within each strain) (10). It is important to select the integrant with lowest autoactivity for both reporters for subsequent Y1H assay so that TF-DNA interactions are easily retrieved and retested. To achieve this, multiple integrants need to be tested for autoactivation per DNA bait.
5. Replica-plating: A sterile velvet is placed (velvet side up) onto the replica-plating block and locked into place using the collar. The yeast plate is placed yeast side down onto the velvet and evenly pressed to transfer the yeast from the plate to the velvet. Plates should be marked to maintain orientation. The yeast plate is then removed, and a new plate is placed media side down onto the velvet and pressed evenly to transfer the yeast from the velvet to the new plate. Multiple (up to five) transfers can be performed in this way from a single velvet. Sometimes "replica-cleaning" of the new plate is required. Replica-cleaning involves pressing the plate onto a series of sterile velvets (used only once) to remove excess yeast that can interfere with the observation of actively growing yeast – usually two or three velvets are required.
6. 3AT is a competitive inhibitor of the His3 enzyme, and is used in Y1H assays to quench *HIS3* autoactivity.
7. The *LacZ* reporter expresses the enzyme βGalactosidase (βGal) that converts clear X-gal into a blue compound that is observed in this colorimetric assay (Fig. 3).
8. Both reporter constructs contain a minimal promoter so all integrant strains should both generate a little blue compound in the colorimetric assay (Fig. 3), and confer growth on media containing a little 3AT. If no blue compound at all is observed (like spot 3 in Fig. 3), or no growth even on 10 mM 3AT plates is observed, this may indicate a problem with either reporter construct and such integrants should not be selected if possible. The lowest concentration of 3AT that totally prevents growth should be used for library screens (i.e., integrants that grow on 20 mM 3AT but not on 40 mM 3AT

should be screened at 40 mM). Integrants that grow strongly on 80 mM 3AT cannot be used in Y1H library screens because few (if any) protein–DNA interactions can activate the *HIS3* reporter enough to overcome this high background, and because higher concentrations of 3AT are toxic to yeast. However, integrants with low *HIS3* autoactivity and high *LacZ* autoactivity can be screened because interactions are detected by growth assay using the *HIS3* reporter first and the *LacZ* activity in these "HIS-positive" yeast may be higher than background when observed closely. However, results obtained with highly autoactive baits should be judged with caution (see Subheading 3.2.2, step 4). Note that for some DNA baits (10–20%) the autoactivity levels for all integrants will be too high, making Y1H screens impossible. For these highly autoactive DNA baits it may be desirable to use smaller fragments of the DNA bait that confer lower autoactivity.

9. DNA bait strains can be grown on YAPD rich media because the reporter constructs are integrated in the genome, and so the selection of the reporter constructs on media lacking histidine and uracil is not required once a DNA bait strain has been generated, verified and examined for autoactivation.

10. Because zymolyase has low solubility in water, it is important to mix the suspension thoroughly before, and periodically (every 30 s) during distribution into the tubes/wells.

11. Remember to include a negative control PCR reaction that lacks template. For PCR reactions from yeast lysates we recommend Invitrogen Taq DNA Polymerase (Cat. No. 10342-053).

12. The *HIS3* construct primers add about 400 bp to the amplicon, the *LacZ* construct primers add about 800 bp to the amplicon.

13. The amount of library listed here is for a *Caenorhabditis elegans* cDNA library of 4×10^7 clones, representing ~16,000 of the ~20,000 genes, in the non-Gateway pPC86 plasmid (18). This amount should be adjusted if using a higher- or lower-complexity library.

14. Tryptophan is omitted from the media because the TF-expressing plasmids contain a *TRP1* selection marker, and thus, only transformed DNA bait yeast will grow on this media.

15. The concentration of 3AT in these plates should be based on the autoactivation results in Subheading 3.1.2, step 9 such that growth of DNA bait yeast not demonstrating reporter activation is minimal.

16. For relatively simple eukaryotes, an effective library screen involves testing (i.e., transforming) at least one million clones, with higher clone numbers increasing the chance of finding interactions involving less abundant TFs. For more complex organisms such as humans that have more splice variants and

tissues, it may be desired to screen more colonies. The number of colonies to be screened also depends on the complexity of the cDNA library. If fewer than one million colonies were obtained, the screen may need to be repeated until the required number of colonies is attained. The transformation efficiency depends on the condition of the yeast strain; use only freshly grown bait yeast strains, do not vortex any yeast suspensions, and use solutions at room temperature only. The amount and quality of plasmid DNA being transformed is also important, so preparation of library DNA should involve an ethanol precipitation to remove contaminants and the concentration should be determined by spectrophotometry. If no colonies grew at all, the media may have been incorrectly prepared (e.g., no glucose was added, or wrong media).

17. The number and size of colonies obtained varies greatly and depends on the DNA bait used. Different TFs can confer different Y1H interaction phenotypes, so it is important to select larger and smaller colonies where feasible. Some bait strain transformations may not yield any colonies on the library screen plates (Subheading 3.2.1, step 18) despite good transformation efficiency (Subheading 3.2.1, step 17) – such baits may be screened again, perhaps using media with lower concentrations of 3AT or using a different cDNA library (e.g., obtained with mRNA from another tissue/source). Alternatively, it is possible that some bait strains may yield many hundreds of colonies. Such DNA baits may be interacting with many partners, or they interact with a highly abundant protein that is overrepresented in the library (the latter issue is particularly relevant for nonnormalized cDNA libraries). These baits with many colonies may need to be screened again using media with higher concentrations of 3AT.

18. If the bait is moderately autoactive, only clones that confer reporter gene expression that is clearly much higher than background should be considered. If the bait is highly autoactive, it may be extremely difficult to identify clear interactions and the bait may need to be redesigned (see Note 8).

19. It is necessary to retest every interaction identified from a cDNA library screen because some of the "double-positive" yeast phenotypes may not be due to the AD-prey interacting with the bait, but arise when bait yeast turn into (i.e., de novo) autoactivators (19). This is when a small number of individual yeast from a population with generally low autoactivity levels exhibit high reporter activation due to mutations induced either by PCR (when initially cloning the bait) or during propagation of the yeast. When de novo autoactivators occur in screens, they will appear as positive colonies. The easiest method to confirm protein–DNA interactions is by gap-repair. Gap-repair involves using PCR to amplify the ORF from the

pPC86 clone in each positive yeast colony, then cotransforming each amplicon with linear empty pPC86 vector into fresh bait strain yeast. Homologous recombination within the yeast reconstitutes the vector and insert into a new construct, and reporter activation within the resulting transformed yeast is reassayed. Without gap-repair, every interaction would need to be confirmed by transforming the AD-prey construct into fresh yeast, and while extracting the plasmid from the double-positive yeast for such a transformation is technically possible, it is challenging to do this for many interactions simultaneously.

20. At either end of these PCR products is about 100 bp of sequence matching the vector sequence on either side of the pPC86 multicloning site. These sequences will facilitate homologous recombination. Some of each amplicon will be used for the gap-repair retest, and the remainder will be used for DNA sequencing. Therefore, yeast lysates that generate multiple PCR bands likely contain multiple AD-prey clones and do not give clean sequence data to identify the interacting protein.

21. This digest excises a small (~20 bp) fragment from the multicloning site of the vector resulting in a linear fragment with noncompatible ends. If the library of choice is not pPC86-based then restriction enzymes should be chosen for the appropriate vector that results in the same type of linear fragment. There is no need to purify the linear plasmid backbone from the digestion mix.

22. If transformants occur in the "no DNA" control then a contaminant is present and the transformation must be repeated using new solutions (the yeast lysis and PCR may also need to be repeated). If the "uncut pPC86" control does not give an order of magnitude more transformants than the "linear pPC86" control, this suggests that the original vector was not fully digested, and new linear vector needs to be generated. If the gap-repair transformations failed to generate more colonies than the linear vector control, it is likely that there was not enough PCR product in the transformation. For these samples, extra (up to 20 μL) PCR reaction can be transformed or some (0.5 μL) can be used as template in an additional PCR reaction to generate a higher amount of PCR product. If no colonies appear for any transformation, it is likely that the media is incorrect.

23. Verify that the preys are in-frame with the Gal4 AD; if not, a spurious peptide may be responsible for the interaction phenotype. Proteins that are retrieved multiple times likely represent real interactions, at least in the context of the Y1H assay. However, proteins that were identified only once should not be excluded because some TFs may be present at low copy numbers in the cDNA library and are therefore more difficult to retrieve than abundant TFs, particularly since Y1H

screens are rarely saturating. In Y1H screens between zero and 40 interacting proteins can be observed with a single DNA bait (11–14). The majority (>95%) of retrieved proteins will be predicted regulatory TFs. However, proteins without a predicted DNA binding domain can constitute a small proportion (~5%) of interactors. These proteins may possess a novel type of DNA binding domain or may interact with DNA baits indirectly (11, 12).

Acknowledgments

The authors thank members of the Walhout lab for comments and advice. This work was supported by NIH grants R01DK068429 and R01GM082971.

References

1. Wang, M. M., and Reed, R. R. (1993) Molecular cloning of the olfactory neuronal transcription factor Olf-1 by genetic selection in yeast *Nature* **364**, 121–126.
2. Li, J. J., and Herskowitz, I. (1993) Isolation of the ORC6, a component of the yeast origin recognition complex by a one-hybrid system *Science* **262**, 1870–1874.
3. Fields, S., and Song, O. (1989) A novel genetic system to detect protein-protein interactions *Nature* **340**, 245–246.
4. Walhout, A. J. M., Sordella, R., Lu, X., Hartley, J. L., Temple, G. F., Brasch, M. A., Thierry-Mieg, N., and Vidal, M. (2000) Protein interaction mapping in *C. elegans* using proteins involved in vulval development *Science* **287**, 116–122.
5. Rual, J. F., Venkatesan, K., Hao, T., Hirozane-Kishikawa, T., Dricot, A., Li, N., Berriz, G. F., Gibbons, F. D., Dreze, M., Ayivi-Guedehoussou, N., Klitgord, N., Simon, C., Boxem, M., Milstein, S., Rosenberg, J., Goldberg, D. S., Zhang, L. V., Wong, S. L., Franklin, G., Li, S., Albala, J. S., Lim, J., Fraughton, C., Llamosas, E., Cevik, S., Bex, C., Lamesch, P., Sikorski, R. S., Vandenhaute, J., Zoghbi, H. Y., Smolyar, A., Bosak, S., Sequerra, R., Doucette-Stamm, L., Cusick, M. E., Hill, D. E., Roth, F. P., and Vidal, M. (2005) Towards a proteome-scale map of the human protein-protein interaction network *Nature* **437**, 1173–1178.
6. Ren, B., Robert, F., Wyrick, J. J., Aparicio, O., Jennings, E. G., Simon, I., Zeitlinger, J., Schreiber, J., Hannett, N., Kanin, E., Volkert, T. L., Wilson, C. J., Bell, S. P., and Young, R. A. (2000) Genome-wide location and function of DNA binding proteins *Science* **290**, 2306–2309.
7. van Steensel, B., Delrow, J., and Henikoff, S. (2001) Chromatin profiling using targeted DNA adenine methyltransferase *Nat Genet* **27**, 304–8.
8. Berger, M. F., Philippakis, A. A., Qureshi, A. M., He, F. S., Estep, P. W., 3rd, and Bulyk, M. L. (2006) Compact, universal DNA microarrays to comprehensively determine transcription-factor binding site specificities *Nat Biotechnol* **24**, 1429–35.
9. Walhout, A. J. M. (2006) Unraveling transcription regulatory networks by protein-DNA and protein-protein interaction mapping *Genome Res* **16**, 1445–1454.
10. Deplancke, B., Dupuy, D., Vidal, M., and Walhout, A. J. M. (2004) A Gateway-compatible yeast one-hybrid system *Genome Res* **14**, 2093–2101.
11. Deplancke, B., Mukhopadhyay, A., Ao, W., Elewa, A. M., Grove, C. A., Martinez, N. J., Sequerra, R., Doucette-Stam, L., Reece-Hoyes, J. S., Hope, I. A., Tissenbaum, H. A., Mango, S. E., and Walhout, A. J. M. (2006) A gene-centered *C. elegans* protein-DNA interaction network *Cell* **125**, 1193–1205.
12. Vermeirssen, V., Barrasa, M. I., Hidalgo, C., Babon, J. A. B., Sequerra, R., Doucette-Stam, L., Barabasi, A. L., and Walhout, A. J. M. (2007) Transcription factor modularity in a gene-centered *C. elegans* core neuronal protein-DNA interaction network *Genome Res* **17**, 1061–1071.

13. Martinez, N. J., Ow, M. C., Barrasa, M. I., Hammell, M., Sequerra, R., Doucette-Stamm, L., Roth, F. P., Ambros, V., and Walhout, A. J. M. (2008) A *C. elegans* genome-scale microRNA network contains composite feedback motifs with high flux capacity *Genes Dev* **22**, 2535–2549.
14. Arda, H. E., Taubert, S., Conine, C., Tsuda, B., Van Gilst, M. R., Sequerra, R., Doucette-Stam, L., Yamamoto, K. R., and Walhout, A. J. M. (2010) Functional modularity of nuclear hormone receptors in a *C. elegans* gene regulatory network *Molecular Systems Biology in press.*
15. Reboul, J., Vaglio, P., Rual, J. F., Lamesch, P., Martinez, M., Armstrong, C. M., Li, S., Jacotot, L., Bertin, N., Janky, R., Moore, T., Hudson, J. R., Jr., Hartley, J. L., Brasch, M. A., Vandenhaute, J., Boulton, S., Endress, G. A., Jenna, S., Chevet, E., Papasotiropoulos, V., Tolias, P. P., Ptacek, J., Snyder, M., Huang, R., Chance, M. R., Lee, H., Doucette-Stamm, L., Hill, D. E., and Vidal, M. (2003) *C. elegans* ORFeome version 1.1: experimental verification of the genome annotation and resource for proteome-scale protein expression *Nat. Genet.* **34**, 35–41.
16. Rual, J.-F., Hirozane-Kishikawa, T., Hao, T., Bertin, N., Li, S., Dricot, A., Li, N., Rosenberg, J., Lamesch, P., Vidalain, P.-O., Clingingsmith, T. R., Hartley, J. L., Esposito, D., Cheo, D., Moore, T., Simmons, B., Sequerra, R., Bosak, S., Doucette-Stam, L., Le Peuch, C., Vandenhaute, J., Cusick, M. E., Albala, J. S., Hill, D. E., and Vidal, M. Human ORFeome version 1.1: a platform for reverse proteomics (2004) *Genome Res* **14**, 2128–2135.
17. Vermeirssen, V., Deplancke, B., Barrasa, M. I., Reece-Hoyes, J. S., Arda, H. E., Grove, C. A., Martinez, N. J., Sequerra, R., Doucette-Stamm, L., Brent, M., and Walhout, A. J. M. (2007) Matrix and Steiner-triple-system smart pooling assays for high-performance transcription regulatory network mapping *Nat Methods* **4**, 659–664.
18. Walhout, A. J. M., Temple, G. F., Brasch, M. A., Hartley, J. L., Lorson, M. A., van den Heuvel, S., and Vidal, M. (2000) GATEWAY recombinational cloning: application to the cloning of large numbers of open reading frames or ORFeomes *Methods in enzymology: "Chimeric genes and proteins"* **328**, 575–592.
19. Walhout, A. J. M., and Vidal, M. (1999) A genetic strategy to eliminate self-activator baits prior to high-throughput yeast two-hybrid screens *Genome Res.* **9**, 1128–1134.

Chapter 12

One- Plus Two-Hybrid System for the Efficient Selection of Missense Mutant Alleles Defective in Protein–Protein Interactions

Ji Young Kim, Ok Gu Park, and Young Chul Lee

Abstract

In an effort to develop a method for the high-throughput analysis of protein interaction interfaces, we devised a novel yeast genetic screening method, termed the "one- plus two-hybrid system," which efficiently selects specific missense mutations that disrupt known protein–protein interactions. This system modifies the standard yeast two-hybrid system to allow the operation of dual reporter systems within the same cell. The one-hybrid screening system is used first to positively select intact prey proteins, harboring informative missense mutations, from a large library of randomly generated mutant alleles. Next, among the isolated missense mutants of the prey proteins, interaction-defective mutants for a given protein (bait) are selected using the two-hybrid screening system. As a validation of the feasibility of this method, we utilized this technique to rapidly characterize the molecular determinants of the interactions between vitamin D receptor and its transcriptional coactivator protein, thyroid hormone receptor-associated protein 220. This efficient and rapid method should prove useful in the systematic analysis of large numbers of interaction interfaces.

Key words: Protein–protein interaction, One- plus two-hybrid system, Interaction-defective mutant allele, Vitamin D receptor, Thyroid hormone receptor-associated protein 220

1. Introduction

The identification of potential protein–protein interactions is believed to be an essential step in understanding the molecular mechanism underlying a given cellular event. Although yeast two-hybrid methods have been extensively utilized for the detection of protein–protein interactions in vivo, these interactions must first be tested in the relevant biological systems (1, 2). For such functional confirmations, the identification of missense mutations that

Bernhard Suter and Erich E. Wanker (eds.), *Two Hybrid Technologies: Methods and Protocols*, Methods in Molecular Biology, vol. 812, DOI 10.1007/978-1-61779-455-1_12, © Springer Science+Business Media, LLC 2012

specifically disrupt the interaction with a given partner (interaction-defective mutants) is invaluable in determining the functional significance and molecular basis of the interaction. Modified yeast two-hybrid systems, which are also known as "reverse two-hybrid" or "split-hybrid" systems, have been developed for the rapid isolation of mutant proteins that are specifically defective in their interaction with a potential partner (3–6). The "reverse two-hybrid system" employs a *URA3* reporter gene as a counterselective marker, which is activated by protein–protein interactions. In this system, expressed Ura3p inhibits growth of the cells on media containing 5-fluoroorotic acid, which is converted by Ura3p into a toxic compound (4). In the "split-hybrid system," the two-hybrid interaction results in the expression of the TetR repressor, which subsequently blocks expression of the *HIS3* reporter gene and prevents yeast growth in histidine-deficient media (6). Thus, both systems are designed specifically for the positive selection of interaction-defective mutants using specific counterselection markers, and they can both be used to define mutations influencing protein–protein interactions (7).

Despite the advantages of these methods for the selection of noninteractors, there remain certain technical obstacles that limit the widespread adoption of these methods. First, the methods were not designed to positively select for informative missense mutations among interaction-defective alleles generated via random mutagenesis. In the "split-hybrid system," for example, the prey protein is expressed as a triple fusion between the VP16 activation domain and β-galactosidase to allow for the identification of uninformative mutations (truncation) as white colonies on X-gal plates, which corresponds to a negative color selection for the missense mutants (6). As more than 97% of counterselected colonies are expected to harbor uninformative mutations (5, 6, 8), the isolation of a small portion of full-length alleles from a large library of mutant alleles remains a primary technical huddle. Second, in many cases, a high background of false positives (more than 65%) was generated during the first counterselection step due to the loss of reporter gene or bait plasmid (8, 9). All of these factors strongly indicate that the negative selection of full-length allele after the counterselection of noninteractor does not constitute an effective strategy for the isolation of specific missense alleles from a randomly generated mutant library.

Here, we describe a novel yeast genetic method, referred to as the "one- plus two-hybrid system (OPTHiS)," which can dramatically improve the analyses of protein interaction interfaces. OPTHiS combines a modified yeast two-hybrid system with random mutagenesis, effectively eliminating truncation mutations and rapidly identifying missense mutations that specifically disrupt known protein–protein interactions. For this screening system, we designed novel cloning vectors and a yeast strain to operate the dual reporter systems for one-hybrid and two-hybrid assays. As with the conventional yeast two-hybrid system, the bait protein of interest

(X) is expressed as a LexA fusion (LexA-X), whereas the prey protein of interest (Y) is expressed as a triple fusion protein (B42-Y-GBD) between the N-terminal B42 transactivation domain and the C-terminal Gal4 DNA-binding domain (GBD). The host strain YOK400 harbors dual-reporter genes: the yeast *HIS3* gene driven by the UAS_{GAL} promoter (chromosomal reporter for one-hybrid screening) and the *LacZ* gene under the control of *lexA* operators (episomal reporter for two-hybrid screening).

The basic strategy of OPTHiS is as follows (Fig. 1). If X and Y form a protein–protein complex, B42-Y-GBD is recruited to the

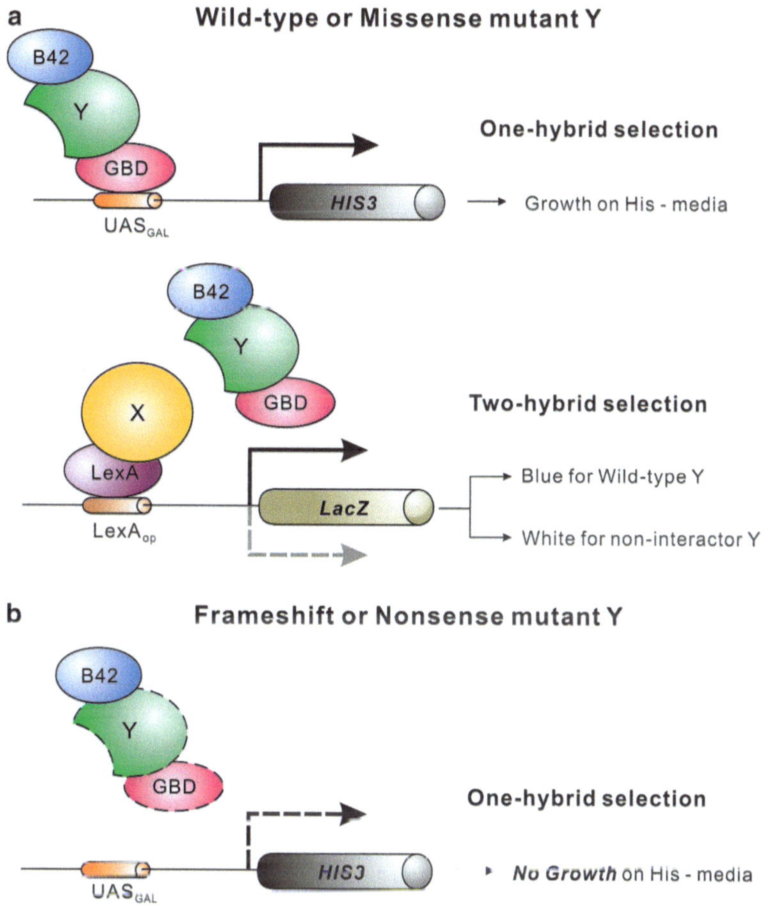

Fig. 1. "One- plus two-hybrid system" strategy. The bait protein (X) is expressed as a LexA fusion, whereas the prey protein (Y) is expressed as a triple fusion between the B42 activation domain and the Gal4-DNA-binding domain (GBD). (**a**) The protein–protein interaction between X and wild-type Y turns on the episomal LexA-driven *LacZ* reporter and generates blue yeast colonies on X-gal plates. In contrast, specific missense mutations in Y that disrupt the association with X generate white colonies on X-gal plates (negative in the two-hybrid selection). In both cases, intact B42-Y-GBD fusions are recruited to the second chromosomal reporter to activate *HIS3* expression via upstream Gal4-binding sites (UAS_{GAL}), resulting in cell survival on media lacking histidine (His^- media) (His^+ phenotype; positive in the one-hybrid selection). (**b**) Any nonsense or frame-shift mutations in Y result in no growth on histidine-deficient media (His^- phenotype; negative in the one-hybrid selection).

episomal *lexA*$_{op}$-*LacZ* reporter and generates blue yeast colonies on X-gal plates. The expression of chromosomal *HIS3* reporter gene requires the intact B42-Y-GBD fusion protein for cell survival in the absence of histidine (His$^+$ phenotype). Cells harboring constructs with specific missense mutations in Y generate the functional B42-Y-GBD fusion and exhibit a strong His$^+$ phenotype, serving as the first positive selection for missense mutation. Among these His$^+$ cells, noninteracting Y mutants can be selected simply by isolating white yeast colonies again growing on X-gal media, which serves as the second color selection for loss of interaction. Conversely, any nonsense or frame-shift mutations in Y result in a lack of functional GBD and concurrent cell death at the first-step selection for the His$^+$ phenotype.

We have exploited this efficient and rapid method to identify the specific amino acid residues within nuclear receptor-interaction domain 2 (NID2) of the thyroid hormone receptor-associated protein 220 (TRAP220), which are specifically required for ligand-dependent interaction with the vitamin D receptor (VDR).

2. Materials

2.1. Yeast Strains and Media

Yeast strains are grown in yeast extract, peptone, dextrose (YPD), or selective medium under standard conditions. Selective medium is a dropout medium in which all nutrients are included, except those required to select for specific auxotrophies. Strain YOK400 (*MAT*α, *leu2*, *trp3*, *ura3*, *lexA*$_{op}$-*LEU2*, UAS$_{GAL}$-*HIS3*) is constructed via the genetic manipulation of strain EGY48 (10). The EGY-LG strain is a derivative of EGY48 and harbors the pLGSD5 plasmid (UAS$_{GAL}$-*LacZ* reporter) rather than pSH18-34.

2.2. Plasmids

The PCR fragments corresponding to TRAP220-NID2 (TN2) were subcloned into the EcoRI and BamHI sites of pRS324UB42-GBD and EcoRI/XhoI sites of pGEX4T-1 (Amersham-Pharmacia Biotech, NJ). The partial human VDR (amino acids 4–428) was subcloned into EcoRI/XhoI sites of pcDNA3HA (Invitrogen Life Technologies, CA), and EcoRI/NotI sites of the pEG202 and pRS325LexA vectors.

2.3. Primers for Mutagenic PCR

Forward primer (oligo-SF), 5′-CC AGC CTC TTG CTG AGT GGA GAT G-3′.

Reverse primer (oligo-SR), 5′-CGG TTT TTC TTT GGA GCA C-3′.

3. Methods

3.1. Vectors for One- Plus Two-Hybrid System

For this screening system, we designed new cloning vectors, pRS324UB42-GBD and pRS325LexA, to operate the dual-reporter systems for one-hybrid and two-hybrid selections (Fig. 2).

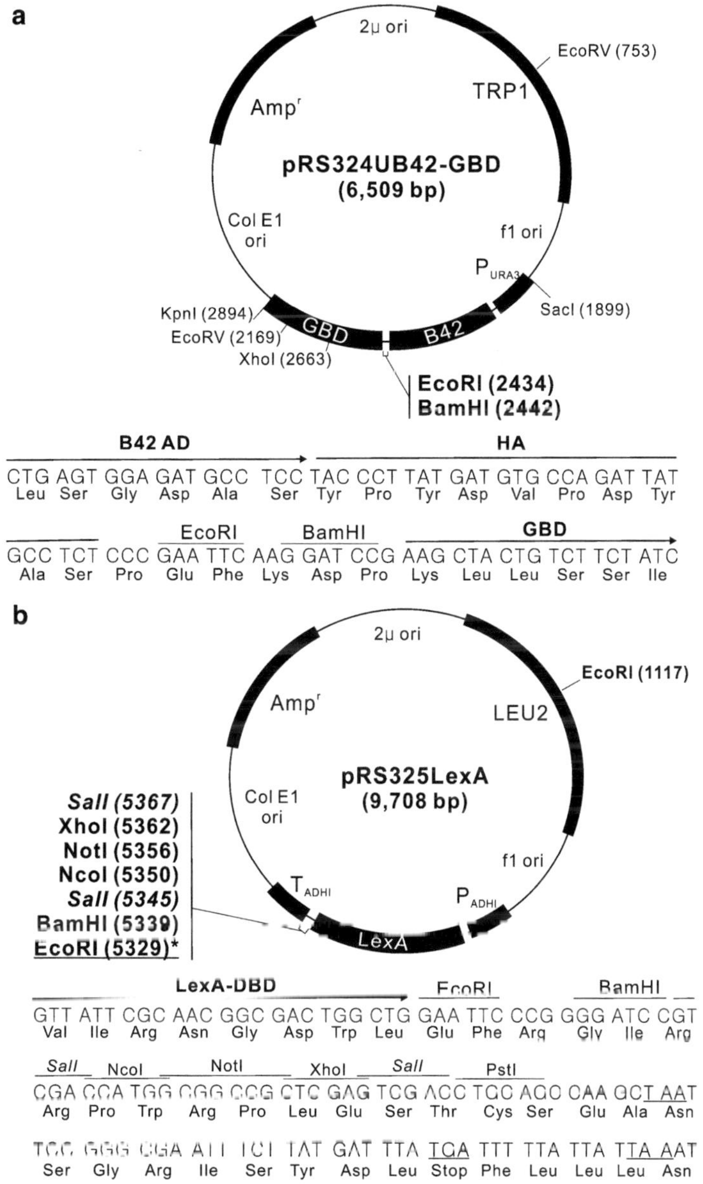

Fig. 2. Schematic diagram of prey (**a**) and bait (**b**) vectors designed for OPTHiS. Each plasmid map is drawn to scale with the indicated restriction sites. Restriction sites with *asterisk* are not unique and are unavailable for cloning. The SalI sites of pRS325LexA vector may be used as a unique site for cloning.

3.1.1. Prey (B42-GBD Fusion) Vector, pRS324UB42-GBD

For the construction of the prey vector pRS324UB42-GBD, the *URA3* promoter region was first inserted between the SacI and BamHI sites of pRS324, thus yielding the pRS324U vector. The DNA fragments corresponding to the B42 activation domain [from pJG4-5 (11)] and the GBD region [from pAS2-1 (MATCHMAKER, Clontech Laboratories Inc., CA)] were generated via PCR and inserted into BglII/KpnI sites of pRS324U. A unique EcoRI/BamHI cloning site (GAA TTC AAG GAT CCG AAG) exists in-frame between the B42 and GBD regions of pRS324UB42-GBD harboring the *TRP1* selection marker (Fig. 2a). In the case of pRS324UB42-FS-GBD plasmid, the B42 and the GBD regions were connected out-of-frame – thus, the GBD portion is not properly expressed. This plasmid can be utilized for the negative control of one-hybrid screening (see below) (see Notes 1 and 2).

3.1.2. Bait (LexA Fusion) Vector, pRS325LexA

The bait plasmid, pRS325LexA, is the yeast shuttle vector derived from the pRS325 vector. This vector contains the *ADH* promoter, the LexA-DNA-binding domain, multicloning sites, and the *ADH* terminator regions obtained from the pEG202 vector (11). The NdeI fragment from pEG202 was blunt ended via Klenow fill-in and inserted into the PvuII site of pRS325, yielding the pRS325LexA plasmid harboring the *LEU2* marker gene (Fig. 2b).

3.2. Construction of YOK400 Having Dual-Reporter Systems: A Host Strain for OPTHiS

1. The chromosomal reporter gene UAS$_{GAL}$-*HIS3* was generated via AflII digestion of pUAS$_{GAL}$-HISi-1 plasmid (Clontech Laboratories Inc., CA).
2. Strain EGY48 (*MAT*α, *leu2*, *trp3*, *ura3*, *lexAop-LEU2*) harboring the pSH18-34 reporter plasmid (8×*lexAop-LacZ* reporter) (11) was transformed with AflII-linearlized pUAS$_{GAL}$-HISi-1 plasmid (see Note 3). In order to ensure the correct integration of the UAS$_{GAL}$ region upstream of the chromosomal *HIS3* gene, one-half of the competent cell–DNA mix was plated on galactose synthetic defined (SD) medium lacking uracil and histidine (SD Gal-Ura-His), whereas the other half of the competent cell–DNA mix was plated on SD glucose-Ura-His. Then, it was confirmed that the transformants did not grow on SD Glu-Ura-His media, but grew on the SD Gal-Ura-His plate. If colonies are formed on the glucose media, step 2 should be repeated.
3. In order to determine whether the chromosomal UAS$_{GAL}$-*HIS3* reporter gene is properly induced by galactose and inhibited by glucose, five to six transformants that grew on the galactose plate were suspended in 100 μL of sterile water and serially diluted (10^0, 10^{-1}, and 10^{-2} dilutions). 10 μL of diluents were serially spotted on both SD Gal-Ura-His and SD Glu-Ura-His plates, which contain 0, 10, or 50 mM 3-amino-1,2,4-triazol (3AT, a competitive inhibitor of His3p). After 3 days of incubation,

the transformant exhibiting the most profound resistance to 3AT on galactose media and no growth on glucose media was selected as the YOK400 strain (see Notes 4 and 5).

3.3. Control Test for One- and Two-Hybrid of the Bait and Prey Vectors

The entire screening process of OPTHiS is outlined in Fig. 3.

3.3.1. One-Hybrid Test of Prey Fusion Using Plasmid Reporter

After the cloning of prey protein (TRAP220-NID2, TN2) into the pRS324UB42-GBD vector, the one-hybrid test must be conducted to confirm whether the prey protein has been properly expressed as the B42-GBD triple fusion (see Note 6).

1. pRS324UB42-TN2-GBD was transformed into the EGY-LG strain harboring the pLGSD5 plasmid (UAS_{GAL}-*LacZ* reporter) and plated on an SD Glu-Ura-Trp plate. The pRS324UB42-FS-GBD vector was also transformed as a negative control.
2. Five to six transformants were patched onto SD Glu-Ura-Trp plate media containing 5-bromo-4-chloro-3-indolyl-β-D-galactopyranoside (X-gal). After 12 h of incubation at 30°C, the transformants harboring pRS324UB42-TN2-GBD exhibited a blue color, thereby indicating that the B42-TN2-GBD fusion was intact. Conversely, colonies harboring pRS324UB42-FS-GBD were white in color due to the lack of expression of the GBD portion.

3.3.2. One-Hybrid Test of Prey Fusion Using the YOK400 Strain

1. Strain YOK400 was transformed with identical amounts of pRS324UB42-TN2-GBD or pRS324UB42-FS-GBD. The transformation mixtures were divided into three parts and plated individually on SD Glu-Ura-Trp media (+His), SD Glu-Ura-His-Trp media (−His), or SD Glu-Ura-His-Trp media containing 10 mM 3AT (−His + 3AT).
2. After 4 days of incubation, the numbers of each of the transformants were counted. The numbers of pRS324UB42-TN2-GBD colonies formed on the +His plate were similar to those of the pRS324UB42-FS-GBD transformants (100%). The number of colonies expressing B42-TN2-GBD was slightly reduced on the −His plate (70%), but this number was maintained on −His + 3AT plate (65–70%). By contrast, the number of colonies harboring pRS324UB42-FS-GBD was found to be profoundly reduced on the −His plate (less than 20–30%) and eliminated completely on the −His + 3AT plate (0%).

3.3.3. Two-Hybrid Test of Bait–Prey Fusion

1. Strain EGY48 harboring the pSH18-34 reporter plasmid was cotransformed with pRS325LexA-VDR and pRS324UB42-TN2-GBD

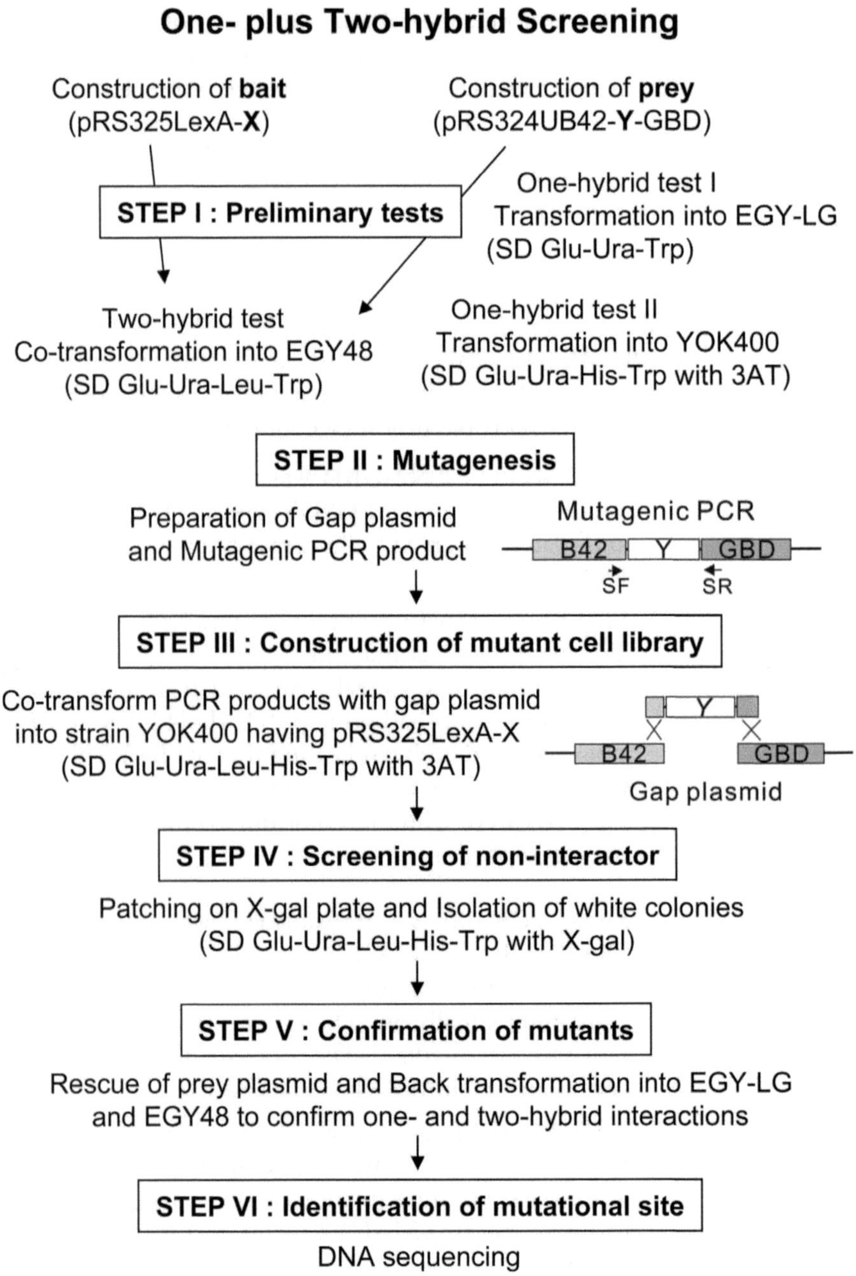

Fig. 3. Outline of the screening procedure of OPTHiS. See text for details.

or pRS324UB42-GBD, and then plated on SD Glu-Ura-Leu-Trp plate.

2. Five to six transformants were patched onto SD Glu-Ura-Leu-Trp plate media containing X-gal and vitamin D3 (10^{-7} M). After 12 h, colonies expressing LexA-VDR and B42-TN2-GBD fusions exhibited a blue color. By contrast, the colonies expressing

the LexA-VDR and B42-GBD fusions exhibited a white color, thereby indicating that the LexA-fused bait protein (LexA-VDR) has no autonomous activity and does not interact with B42 or GBD in yeast.

3.4. Mutagenic PCR

The Mn^{2+}-mediated PCR mutagenesis method (12) is utilized for the generation of a randomly mutated library of the prey gene, and two oligonucleotides are designed as primers for mutagenic PCR: forward primer (oligo-SF), matching the C-terminal region of the B42 activation domain, and reverse primer (oligo-SR), corresponding to an N-terminal portion of GBD. The mutagenic PCR products obtained with these primers commonly harbor approximately 100 bp of flanking regions at each end, with sequence identities to the gap plasmid prepared via EcoRI/BamHI digestions of pRS324UB42-GBD.

1. The PCR reaction was conducted in a 3 × 100-μL volume using *Taq* polymerase, two primers (oligo-SF and oligo-SR), and pRS324UB42-TN2-GBD as templates in the presence of 0.1 mM $MnCl_2$ (see Note 7). As a control, mutagenic PCR products were generated with or without 0.3 mM $MnCl_2$ in a reaction volume of 100 μL (for small-scale tests).
2. The program for the PCR reaction is as follows:

   Step 1: 94°C for 5 min

   Step 2: 94°C for 30 s

   Step 3: 55°C for 30 s

   Step 4: 72°C for 60 s

   Step 5: Repeat from steps 2–5 for 30 cycles

   Step 6: 72°C for 5 min
3. PCR products were purified and dissolved in distilled water at a concentration of 200–500 ng/μL.

3.5. Gap Plasmid Preparation

1. In order to generate the gap plasmid, pRS324UB42-TN2-GBD plasmid DNA was digested with EcoRI and BamHI restriction enzymes.
2. The gap plasmid was gel purified and dissolved in distilled water at a concentration of 200 ng/L.

3.6. One- Plus Two-Hybrid Screening

In order to construct the mutant cell library of the randomly mutated prey gene, we utilized a single-step method based on the in vivo gap repair (13).

3.6.1. One-Hybrid Screening for the Selection of Full-Length Mutant Allele of Prey

1. Strain YOK400 harboring the pSH18-34 reporter plasmid was transformed with the pRS325LexA-VDR bait plasmid and plated on SD Glu-Ura-Leu media.
2. Yeast cells harboring the bait plasmid were then cotransformed with 1 μg of gap plasmid and 4 μg of mutagenic PCR products

(0.1 mM $MnCl_2$) via high-efficiency yeast transformation methods (14).

3. Transformation mixtures were plated on ten plates of SD Glu-Ura-Leu-His-Trp media containing 10 mM 3AT to obtain singly isolated colonies.
4. For control tests, the YOK400 strain was cotransformed with 200 ng of the gap plasmid and 800 ng of mutagenic PCR products with 0 or 0.3 mM $MnCl_2$ and plated on SD Glu-Ura-Leu-His-Trp media containing 10 mM 3AT. Also, the pRS324UB42-TN2-GBD, pRS324UB42-FS-GBD, and gap plasmid were transformed individually into the YOK400 strain and plated on the same media (see Note 8).
5. After 4 days of incubation, the number of transformants with the His^+ phenotype was determined. Usually, the total number of transformants gradually decreased with the introduction of mutagenic PCR products generated under increased $MnCl_2$ concentration, thereby indicating the efficient elimination of truncation mutants in this step (see Note 9). The transformants harboring pRS324UB42-FS-GBD were then checked for growth on SD Glu-Ura-Leu-His-Trp plate containing 10 mM 3AT (Table 1).
6. The transformation of mutagenic PCR products (0.1 mM $MnCl_2$) with the gap plasmid was repeated until a sufficient number of mutant colonies was generated for the second screening (more than 2,000 colonies).

Table 1
Transformation and screening of interaction-defective mutants

	TRAP220-NID2	
Transforming DNAs[a]	**Total number of transformants**	**Mutation rate (%) (number of white/number picked)**
Gap only	22	0 (0/22)
Gap + PCR products (without $MnCl_2$)	1.47×10^3	0 (0/100)
Gap + PCR products (0.1 mM $MnCl_2$)	0.93×10^3	2.0 (50/2,465)[b]
Gap + PCR products (0.3 mM $MnCl_2$)	0.64×10^3	5.0 (5/100)
B42-TN2-GBD[c]	2.94×10^3	0 (0/100)
B42-FS-GBD[c]	0	ND

[a] For the small-scale test, 200 ng of gap plasmid and 800 ng of PCR products were generally used for cotransformation

[b] Data from the actual screening experiment which was scaled up fivefold than that of the small-scale test

[c] 1 μg of supercoiled prey plasmid was transformed for the control experiment

3.6.2. Two-Hybrid Screening for the Isolation of Interaction-Defective Mutants of Prey

1. 2,465 yeast colonies (generally, more than 1,000 colonies) exhibiting His+ phenotype were picked onto SD Glu-Ura-Leu-His-Trp plate media containing X-gal and vitamin D3 (10^{-7} M). Each of 100 yeast colonies from other transformation sets were also picked on the same plates and incubated at 30°C.
2. After 12 h, the yeast colonies evidencing a white or weak blue color were isolated as candidates for the interaction-defective mutants of TN2 with VDR (see Note 10).
3. The candidates were then retested for color phenotype by streaking on plates containing X-gal and vitamin D3.

3.7. Confirmation of Defective Interactions for the Isolated Mutants

3.7.1. One-Hybrid and Two-Hybrid Interactions in Yeast

1. Among the 50 mutant, isolated candidates evidencing a white color, the prey vector (p324UB42-TN2-GBD) was rescued from 30 colonies via yeast DNA preparation and *E. coli* (DH5α) transformation (15).
2. The rescued DNAs were transformed individually into the EGY48 strain expressing pEG202-VDR and plated on SD Glu-Ura-His-Trp media (for the two-hybrid interaction test). Prey vectors were also introduced into the EGY-LG strain harboring the pLGSD5 plasmid (UAS_{GAL}-*LacZ* reporter) and plated on SD Glu-Ura-Trp media to check for intact GBD (one-hybrid test).
3. Five to six transformants were then streaked onto X-gal plates and subjected to a liquid β-galactosidase assay for the selection of noninteracting mutants on the basis of quantitative data (for the two-hybrid interaction assay, 10^{-7} M of vitamin D3 was added in media). While all of the candidates evidenced a blue color in the one-hybrid test, 2 clones out of 30 candidates (less than 7%) exhibited a blue color in the two-hybrid test and turned out to be wild type (false positives). Finally, 28 clones that continued to exhibit a blue color in the one-hybrid test and a white color in the two-hybrid test were selected as final mutant candidates (see Note 11).
4. The prey plasmid DNAs were subjected to DNA sequencing in order to identify the mutational site(s) using oligo-ST as a primer (Fig. 4a).

3.7.2. In Vitro GST Pull-Down Assay

GST pull-down analysis was conducted to confirm in vitro interactions between isolated TRAP220-NID2 mutants and VDR.

1. pGEX4T-1 derivatives expressing the wild-type or mutants of TRAP220-NID2 were introduced into the DH5α *E. coli* strain.
2. One colony of transformants expressing GST derivatives (GST alone or GST-fused proteins) was individually inoculated in 10 mL of LB broth containing ampicillin and cultured overnight at 37°C with shaking.

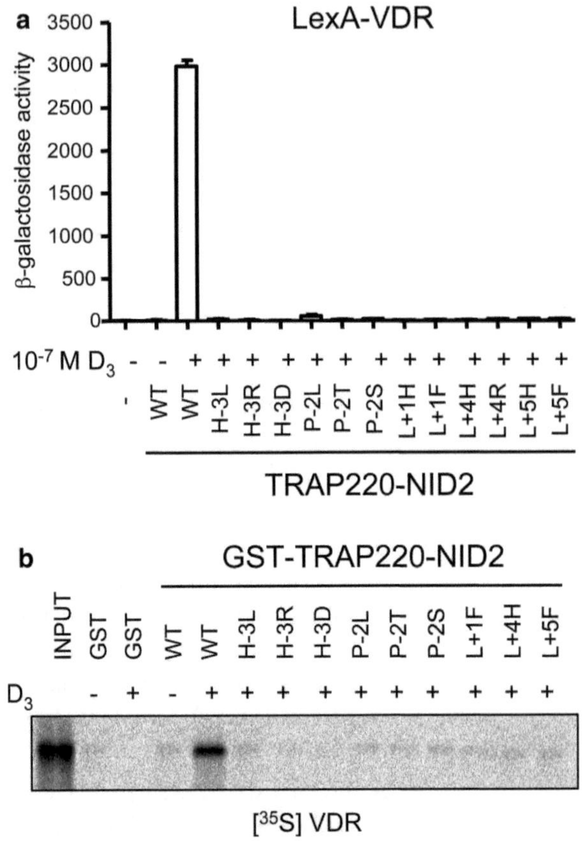

Fig. 4. Interactions of TRAP220-NID2 mutants with VDR. (**a**) Yeast two-hybrid assays were conducted using the EGY48 strains harboring the LexA*op*-*LacZ* reporter plasmid, the bait plasmid pEG202-VDR, and the pRS324UB42-TN2-GBD plasmid or its mutant derivatives. The pRS324UB42-GBD empty vector (–) was also introduced into the strains for use as negative controls. Transformants were cultivated in SD medium in the presence (+) or absence (–) of 10^{-7} M vitamin D3 (D_3). The mutational sites and the changed amino acids detected in the isolated TN2 mutants are shown, in which the first Leu of the LXXLL motif is denoted as +1. Liquid β-galactosidase assays were conducted with three or more transformants and the mean ± SE values are shown on the *Y*-axis. *WT* wild type. (**b**) GST pull-down assays for the interactions of VDR with the indicated TRAP220-NID2 mutants. The ^{35}S-labeled in vitro-translated VDR was incubated with GST-TRAP220-NID2 or its mutant derivatives in the presence (+) or absence (–) of 10^{-5} M vitamin D3. The bound proteins were analyzed via SDS-PAGE and autoradiography. INPUT indicates 10% of the in vitro-translated proteins utilized in the assays.

3. Cultured cells (5 mL) were then transferred into 200 mL of 2× yeast extract and tryptone (YT) medium containing ampicillin, and then incubated at 37°C with vigorous shaking until an OD_{600} of 0.5 was achieved.
4. GST proteins were overexpressed by the addition of 0.25 mM isopropyl-β-D-thiogalactopyranosine to the culture solution and 3 h of induction at 30°C.

5. The harvested cells were then suspended in 10 mL of HNE buffer [25 mM HEPES (pH 7.8), 0.2 mM EDTA, 20% glycerol] with protease inhibitor cocktails and disrupted via sonication.
6. Supernatants (cell lysates) were mixed with 200 μL of glutathione-agarose beads (50% slurry) overnight at 4°C with gentle agitation.
7. The beads were washed three times in 1 mL of PBS and suspended in 100 μL of PBS.
8. The quantity of purified GST-fused proteins was determined via SDS-PAGE and Coomassie staining.
9. VDR proteins were synthesized via the in vitro translation of pcDNA3HA-VDR construct using a TNT transcription-coupled translation system (Promega Corp., WI).
10. The radiolabeled VDR proteins were then added to similar quantities of GST or GST-fused proteins (2–4 μg) bound to glutathione-agarose beads preequilibrated with buffer A [50 mM Tris–HCl (pH 7.9), 5% glycerol, 1 mM EDTA, 1 mM dithiothreitol, 1× protease inhibitor, 0.01% NP-40, 150 mM KCl] at a final volume of 250 μL overnight at 4°C.
11. The beads were then washed three times in buffer A and the buffer solution was removed completely from beads.
12. The beads were mixed with sample buffer (20–30 μL) and boiled for 5 min at 100°C.
13. Bound proteins were analyzed via SDS-PAGE, followed by autoradiography using phospho-image reader (Fujix Bio Imaging Analyzer Bas 1000, Japan) (Fig. 4b).

4. Notes

1. The prey protein is expressed as a triple fusion between B42 and GBD; therefore, a complex protein that requires proper three-dimensional structure for interaction may not be an appropriate target for mutant screening.
2. There is a restriction in the length of the prey protein owing to the difficulty inherent to the maintenance of the optimal mutation rate with increasing prey size. Thus far, we have successfully analyzed prey proteins of up to 200 amino acids.
3. Other yeast strains can be utilized as host strains for OPTHiS. For example, the YLS500 strain (*MATα*, *ade2*, *leu2*, *lys2*, *trp1*, *ura3*, UAS_{GAL}-*HIS3*) was constructed via the integration of UAS_{GAL}-*HIS3* into the YPH500 strain. This strain has more available selection markers than the YOK400 strain, and can

therefore be utilized in more complex systems, such as 2.5-hybrid screening (16). With regard to the use of strain YLS500 in OPTHiS, SD Gal-Ura-His media without 3AT is generally utilized for the optimal growth of His^+ transformants.

4. YOK400 strain grows weakly on SD Glu-Ura-His plates and yields "false positives" during one-hybrid screening. However, these false positives are eliminated completely by the addition of low concentration of 3AT. 10 mM of 3AT is the optimal concentration for preventing false positives, but permits the optimal growth of His^+ transformants.
5. We recommend that the YOK400 strain be freshly prepared or carefully checked for the His^- phenotype prior to the initiation of each round of screening in order to minimize the generation of false positives in the first step of selection of the full-length alleles.
6. When the prey gene is cloned into the pRS324UB42-GBD vector, *E. coli* DH5α transformants appear as a mixture of small and large colonies. In the majority of cases, the small colonies proved to be the correct clone.
7. We selected 0.1 mM $MnCl_2$ for error-prone PCR for the actual screening because multiple-point mutations were generally introduced at higher concentrations of $MnCl_2$.
8. The individual transformation of the gap plasmid and the pRS324UB42-FS-GBD alone helps in determining whether the gap plasmid is digested fully by restriction enzymes and whether the chromosomal UAS_{GAL}-*HIS3* reporter gene of strain YOK400 functions properly under these selection conditions, respectively. The super-coiled form of pRS324UB42-TN2-GBD alone is also transformed in order to evaluate the transformation efficiency.
9. The validation experiment for the enrichment of full-length clones after one-hybrid selection is described in detail in our original article (17).
10. The blue color can become visible within 3–4 h after picking, depending on the binding strength of the proteins of interest. Usually, the percentage of white colonies (mutation rate) increased gradually with increased $MnCl_2$ concentrations in the PCR (0–0.3 mM) (Table 1). In the control tests, the His^+ colonies generated via individual transformations with pRS324UB42-TN2-GBD or gap plasmids alone were all blue in color on the X-gal plates.
11. In our OPTHiS, zero and low-level (less than 20%) backgrounds of false positives were generated in the first positive selection of the full-length alleles as well as in the second screening of the noninteractors, respectively. This result can be

contrasted with reports showing that reverse two-hybrid systems generally create a high background of false positive (more than 65%). This discrimination may be attributable to differences in the stabilities of the two-hybrid reporter systems utilized according to the methods described herein.

References

1. Chien, C. T., Bartel, P. L., Sternglanz, R., and Fields, S. (1991) The two-hybrid system: a method to identify and clone genes for proteins that interact with a protein of interest, *Proc Natl Acad Sci USA* *88*, 9578–9582.
2. Vidal, M., and Legrain, P. (1999) Yeast forward and reverse 'n'-hybrid systems, *Nucleic Acids Res* *27*, 919–929.
3. Leanna, C. A., and Hannink, M. (1996) The reverse two-hybrid system: a genetic scheme for selection against specific protein/protein interactions, *Nucleic Acids Res* *24*, 3341–3347.
4. Vidal, M., Brachmann, R. K., Fattaey, A., Harlow, E., and Boeke, J. D. (1996) Reverse two-hybrid and one-hybrid systems to detect dissociation of protein-protein and DNA-protein interactions, *Proc Natl Acad Sci USA* *93*, 10315–10320.
5. Vidal, M., Braun, P., Chen, E., Boeke, J. D., and Harlow, E. (1996) Genetic characterization of a mammalian protein-protein interaction domain by using a yeast reverse two-hybrid system, *Proc Natl Acad Sci USA* *93*, 10321–10326.
6. Shih, H. M., Goldman, P. S., DeMaggio, A. J., Hollenberg, S. M., Goodman, R. H., and Hoekstra, M. F. (1996) A positive genetic selection for disrupting protein-protein interactions: identification of CREB mutations that prevent association with the coactivator CBP, *Proc Natl Acad Sci USA* *93*, 13896–13901.
7. White, M. A. (1996) The yeast two-hybrid system: forward and reverse, *Proc Natl Acad Sci USA* *93*, 10001–10003.
8. Gray, P. N., Busser, K. J., and Chappell, T. G. (2006) A novel approach for generating full-length, high coverage allele libraries for the analysis of protein interactions, *Mol Cell Proteomics* *6*, 514–526.
9. Barr, R. K., Hopkins, R. M., Watt, P. M., and Bogoyevitch, M. A. (2004) Reverse two-hybrid screening identifies residues of JNK required for interaction with the kinase interaction motif of JNK-interacting protein-1, *J Biol Chem* *279*, 43178–43189.
10. Estojak, J., Brent, R., and Golemis, E. A. (1995) Correlation of two-hybrid affinity data with in vitro measurements, *Mol Cell Biol* *15*, 5820–5829.
11. Gyuris, J., Golemis, E., Chertkov, H., and Brent, R. (1993) Cdi1, a human G1 and S phase protein phosphatase that associates with Cdk2, *Cell* *75*, 791–803.
12. Cadwell, R. C., and Joyce, G. F. (1992) Randomization of genes by PCR mutagenesis, *PCR Methods Appl* *2*, 28–33.
13. Muhlrad, D., Hunter, R., and Parker, R. (1992) A rapid method for localized mutagenesis of yeast genes, *Yeast* *8*, 79–82.
14. Gietz, R. D., and Schiestl, R. H. (2007) Large-scale high-efficiency yeast transformation using the LiAc/SS carrier DNA/PEG method, *Nat Protoc* *2*, 38–41.
15. Hoffman, C. S., and Winston, F. (1987) A ten-minute DNA preparation from yeast efficiently releases autonomous plasmids for transformation of Escherichia coli, *Gene* *57*, 267–272.
16. Son, Y. L., Park, O. G., Kim, G. S., Lee, J. W., and Lee, Y. C. (2008) RXR heterodimerization allosterically activates LXR binding to the second NR box of activating signal co-integrator-2, *Biochem J* *410*, 319–330.
17. Kim, J. Y., Park, O. G., Lee, J. W., and Lee, Y. C. (2007) One- plus two hybrid system, a novel yeast genetic selection for specific missense mutations disrupting protein/protein interactions, *Mol Cell Proteomics* *6*, 1727–1740.

Chapter 13

Investigation of Membrane Protein Interactions Using the Split-Ubiquitin Membrane Yeast Two-Hybrid System

Julia Petschnigg, Victoria Wong, Jamie Snider, and Igor Stagljar

Abstract

Proteins are generally organized into molecular complexes, in which multiple interaction partners collaborate to carry out cellular processes. Thus, techniques to map protein–protein interactions have become pivotal for biological studies of as yet uncharacterized proteins. Investigation of interaction partners of membrane proteins is of special interest, as they play a major role in cellular processes and are often directly linked to human diseases. Owing to their hydrophobic nature, however, it has proven difficult to study their interaction partners. To circumvent this problem, a yeast-based genetic technology for the in vivo detection of membrane protein interactions, the split-ubiquitin membrane yeast two-hybrid (MYTH) system, has been developed. MYTH allows for detection of both stable and transient interactions and can be applied to large- and small-scale screens. It uses the split-ubiquitin approach, in which the reconstitution of two ubiquitin halves is mediated by a specific protein–protein interaction. Briefly, the bait membrane protein is fused to the C-terminal half of ubiquitin and an artificial transcription factor. The mutated N-terminal moiety of ubiquitin is fused to the prey protein. Upon interaction of bait and prey proteins, ubiquitin is reconstituted and further recognized by ubiquitin-specific proteases, which subsequently cleave off the transcription factor, thus resulting in reporter gene activation. To date, MYTH has been successfully applied to study interactions of membrane proteins from various organisms and has only recently been adapted for the identification of interaction partners of mammalian receptor tyrosine kinases.

Key words: Protein–protein interactions, Integral membrane proteins, Split-ubiquitin, Membrane yeast two hybrid (MYTH), cDNA screening

1. Introduction

Protein–protein interactions regulate nearly every cellular process, including cell signaling, regulation, and metabolic pathways, and furthermore, they can be predictive of functional relationships between interaction partners. Thus, the cellular milieu can be

Bernhard Suter and Erich E. Wanker (eds.), *Two Hybrid Technologies: Methods and Protocols*, Methods in Molecular Biology, vol. 812, DOI 10.1007/978-1-61779-455-1_13, © Springer Science+Business Media, LLC 2012

thought of as a complex network, or "interactome", of protein–protein interactions. Given that about a third of the whole proteome are membrane proteins and a vast number of them have disease-associated functions, the development of experimental approaches for identifying membrane protein interactors is of utmost importance for modern molecular biology. The necessity for the thorough mapping of membrane interactions is further corroborated by the fact that most drugs on the market are targeted toward membrane-resident proteins. Hence, protein–protein interactions can be used as targets for drug screens, with the goal of identifying molecules that can either restore or abolish protein–protein interactions (1).

To date, various methods have been established to characterize membrane protein interactions, such as tandem-affinity-purification (TAP), fluorescence-coupled technologies or the conventional yeast two-hybrid system (2–4). Despite the robustness of these techniques, membrane proteins have proven to be difficult for biochemical studies. Furthermore, methods such as the conventional two-hybrid are confined to monitor interactions solely within the nucleus, which is only partially feasible with membrane proteins (5, 6). An alternative in vivo detection method, called the split-ubiquitin membrane yeast two-hybrid (MYTH) method, circumvents these limitations. MYTH allows for identification of either membrane or cytosolic protein interactors of full-length integral membrane proteins, and also yields information about the protein topology in the membrane, with the only prerequisite being that the membrane proteins have their N- and/or C-terminus located in the cytosol (6–8).

The membrane yeast two-hybrid system is an adaptation of the split-ubiquitin assay (5, 9). Ubiquitin, an evolutionarily highly conserved 76 amino acid protein which serves as a tag for proteins destined for degradation by the 26S proteasome, can be split into two halves, the N-terminal Nub (aa 1–34) and the C-terminal Cub (aa 35–76). Upon reconstitution of these two halves into a functional ubiquitin protein, cytosolic ubiquitin specific proteases (UBPs) recognize this "pseudoubiquitin". In MYTH, the spontaneous association of Cub and Nub is inhibited by mutating isoleucine 13 to a glycine residue (NubG). The protein of interest (bait) is fused at either its N- or C-terminus to the C-terminal half of ubiquitin (Cub) and an artificial transcription factor (TF), consisting of the bacterial-derived LexA DNA binding domain and the *Herpes simplex* VP16 transactivator protein (LexA-VP16). A second protein (prey) is fused at its N- or C-terminus to the mutated amino-terminal half of ubiquitin (NubG). Since the mutated NubG is incapable of spontaneous reassociation with Cub, pseudoubiquitin is only reconstituted if the two halves are brought into proximity

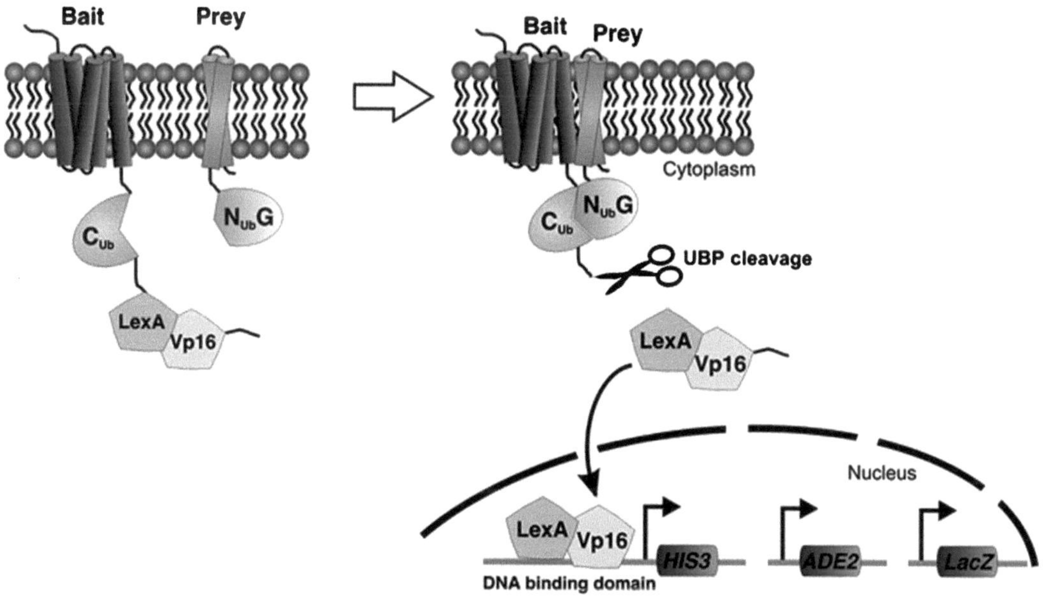

Fig. 1. Principle of MYTH. Ubiquitin is separated into a C-terminal half (Cub) and an N-terminal half (Nub). A point mutation in Nub at position 13 prevents spontaneous reassociation of Cub and Nub. Only an interaction between the bait and the prey proteins reconstitutes functional ubiquitin, which is recognized by UBPs. Upon cleavage, the transcription factor LexA-VP16 enters the nucleus and initiates reporter gene transcription.

through interaction of both bait and prey proteins. UBPs then recognize the functional ubiquitin and cleave between the bait protein and the C-terminal half of ubiquitin, thus releasing the LexA-VP16 transcription factor, which in turn activates reporter gene expression in the nucleus, such as *ADE2*, *HIS3*, or *LacZ* ((10–12); see Fig. 1).

MYTH can either be performed using a plasmid expressing the bait-Cub-TF construct (traditional MYTH, tMYTH), or by endogenous tagging of the bait with the Cub-TF sequence (integrated MYTH, iMYTH). The latter maintains expression of the bait protein at physiological levels and thus reduces potential perturbations caused by episomal overexpression (13).

Since its development, the split-ubiquitin membrane yeast two-hybrid system has been successfully used to identify both transient and stable interactions among various membrane proteins from yeast, plant, fly, worm, and humans (14–18). Uncovering novel protein interactions and dissecting their functions will provide more insight into membrane interactomes, thus being invaluable for the discovery of potential targets implicated in membrane protein-associated diseases.

2. Materials

2.1. Bait Construction and Validation

2.1.1. Bait Plasmid Construction by Gap Repair for tMYTH

1. Plasmid backbones for type I or type II baits (see Note 1).
2. DNA template to amplify the gene of interest.
3. Primers for PCR of gene of interest (harboring homologous sites to the vector for gap repair).
4. Restriction enzymes for the vector backbones.
5. PCR reagents and thermocycler.
6. THY.AP4, L40, or BY4741 yeast strains (see Note 2 and Subheading 2.6).
7. Selective yeast/*E. coli* media (solid and liquid) (see Subheading 2.7, items 3 and 5).
8. Competent *E. coli* cells (see Subheading 2.6).
9. Sequencing primers (see Note 3).

2.1.2. Bait Strain Construction for iMYTH

1. L2 or L3 plasmid template for amplification of the Cub-(YFP)-TF tagging cassette.
2. Primers for PCR of L2/L3 tagging cassette.
3. PCR reagents and thermocycler.
4. THY.AP4 or L40 yeast reporter strain (see Note 2 and Subheading 2.6).
5. YPAD (solid and liquid) (see Subheading 2.7, item 1).
6. YPAD + G418 (solid and liquid) (see Subheading 2.7, items 1 and 9).
7. Primers for PCR amplification and sequencing of tagged bait (see Note 3).

2.1.3. NubG/NubI Test

1. Bait plasmid/strain.
2. Control prey plasmids:
   (a) pFur4-NubG (negative control).
   (b) pFur4-NubI (positive control).
   (c) pOst1-NubG (negative control).
   (d) pOst1-NubI (positive control).
3. Selective media (solid) appropriate for strain harboring bait (SD-Trp or SD-Trp-Leu to select for transformation, and SD-Trp-Leu-His or SD-Trp-Leu-Ade-His to select for interaction) (see Note 2 and Subheading 2.7, item 3).
4. 3-AT (see Subheading 2.7, item 20).
5. Yeast transformation media and reagents (see Subheading 2.5.3).

2.1.4. Immuno-fluorescence/ Fluorescence Microscopy

1. Bait plasmid/strain.
2. YPAD (iMYTH bait) or selective medium (tMYTH bait) (see Subheading 2.7, items 1 and 3).
3. Anti-VP16 or anti-LexA primary antibody (if performing immunofluorescence).
4. FITC-coupled secondary antibodies (if performing immunofluorescence).
5. Standard immunofluorescence reagents (if performing immunofluorescence).
6. Fluorescence microscope.

2.2. Library Transformation

1. Plasmid libraries (NubG-X or X-NubG cDNA or genomic).
2. Bait plasmid/strain.
3. Liquid media appropriate for growth of yeast strain to be transformed; either YPAD (for iMYTH baits) or selective SD media appropriate for tMYTH bait plasmid (typically SD-Leu).
4. ssDNA, lithium acetate–TE buffer master mix, lithium acetate–PEG mix (see Subheading 2.7, items 11–18).
5. DMSO.
6. 0.9% NaCl (see Subheading 2.7, item 10).
7. 2× YPAD (see Subheading 2.7, item 2).
8. Selective media (solid) appropriate for strain harboring bait (SD-Trp or SD-Trp-Leu to select for transformation, and SD-Trp-Leu-His or SD-Trp-Leu-Ade-His to select for interaction) (see Note 2 and Subheading 2.7, item 3). For media used to select for interaction prepare both with and without X-Gal. Include 3-AT as established in the NubG/NubI test (see Subheading 2.1.3).

2.3. Prey Recovery and Sequencing

1. Positive colonies from X-Gal tests.
2. Yeast miniprep media and reagents (see Subheading 2.5.4).
3. *E. coli* transformation reagents (see Subheading 2.5.1).
4. *E. coli* media and miniprep reagents (see Subheading 2.5.2).
5. Sequencing primers (see Note 3).

2.4. Bait Dependency Test

1. Original bait plasmid/strain.
2. Unrelated bait plasmid/strain.
3. Plasmids harboring putative interactors.
4. Yeast transformation media and reagents (see Subheading 2.5.3).
5. Selective media (solid) appropriate for strain harboring bait/unrelated bait (SD-Trp or SD-Trp-Leu to select for transformation, and SD-Trp-Leu-His or SD-Trp-Leu-Ade-His to select for interaction) (see Note 2). For media used to select

for interaction prepare with X-Gal. Include 3-AT where required (see Subheading 2.1.3).

2.5. Basic Yeast/E. coli Techniques

2.5.1. E. coli Transformation

1. Chemically competent *E. coli* cells (see Subheading 2.6).
2. Plasmids to transform.
3. LB medium.
4. LB plates with antibiotics selecting for the plasmid marker (commonly used in MYTH: kanamycin or ampicillin) (see Subheading 2.7, items 5, 7 and 8).

2.5.2. E. coli Miniprep

1. Commercially available miniprep kit.
2. LB with antibiotics selecting for the plasmid marker (commonly used in MYTH: kanamycin or ampicillin) (see Subheading 2.7, items 5, 7 and 8).

2.5.3. Yeast Transformation

1. Yeast strain for transformation.
2. Plasmid DNA for transformation.
3. Liquid media appropriate for growth of yeast strain to be transformed; either YPAD or selective SD media appropriate for any plasmids being carried by the strain (e.g., SD-Leu for tMYTH bait strains).
4. Basic yeast transformation mix (TRAFO mix) (see Subheading 2.7, item 19).
5. Selective solid media for selection of transformed yeast.

2.5.4. Yeast Miniprep

1. Commercially available miniprep kits.
2. Selective liquid media (SD lacking the amino acid/nucleotide selecting for the plasmid marker).
3. 0.5 mm glass beads.

2.5.5. Colony PCR

1. Zymolyase (see Subheading 2.7, item 21).
2. Freshly streaked out colonies.
3. PCR reagents and thermocycler.
4. Primers for verification of insert.

2.6. Strains

1. *E. coli*: any competent *E. coli* cells suitable for plasmid propagation (DH5α, XL10 Gold etc.). Competent cells should have a high transformation efficiency (at least 10^7 transformants/microgram DNA) to ensure efficient recovery of yeast plasmids (see Note 4).
2. *Saccharomyces cerevisiae*: THY.AP4 [*MATa leu2 ura3 trp1:(lexAop)-lacZ (lexAop)-HIS3 (lexAop)-ADE2*] and L40

[*MATa HIS3 200 trp1-901 leu2-3, 112 ade2 LYS2:(lexAop)*^{4}-*HIS3 URA3:(lexAop)*^{8}-*lacZ GAL4*], any wild type strain for gap repair (e.g., BY4741). THY.AP4 and L40 are used for screens, BY4741 for gap repair of bait constructs (see Note 2). The best yeast strain to use depends on the bait and should be determined by the user.

2.7. Media and Solutions

1. YPAD: 1% bacto yeast extract, 2% bacto peptone, 2% glucose, 100 μM adenine sulfate, 2% agar (only for agar plates), dissolve in ddH_2O and autoclave at 121°C, 15 psi for 30 min.
2. 2× YPAD: 2% bacto yeast extract, 4% peptone, 4% glucose, 100 μM adenine sulfate, dissolve in ddH_2O and autoclave at 121°C, 15 psi for 30 min.
3. Selective/SD (synthetic drop-out) medium: 0.67% yeast nitrogen base (without amino acids, with ammonium sulfate), 2% glucose, 2% agar (only for agar plates), and 1× amino acid/nucleotide mix (see Subheading 2.7, item 4). Autoclave at 121°C, 15 psi for 30 min. If including 3-AT (*see* Subheading 2.7, item 20) in the medium, it must be added after autoclaving once the medium has cooled enough to be handled easily. If including X-Gal in the medium, add 100 ml sodium phosphate solution (see Subheading 2.7, item 22) and 0.8 ml X-Gal solution (see Subheading 2.7, item 6) per liter of media. Both solutions must be added after autoclaving when the media has cooled enough to handle easily. For storage, wrap X-Gal containing plates in aluminum foil. Store plates containing 3-AT and/or X-Gal at 4°C.
4. 10× amino acid drop-out mix (1 L): 400 mg adenine, 200 mg arginine, 200 mg histidine, 300 mg isoleucine, 1,000 mg leucine, 300 mg lysine, 1,500 mg methionine, 500 mg phenylalanine, 2,000 mg threonine, 400 mg tryptophan, 300 mg tyrosine, 200 mg uracil, 1,500 mg valine. Omit the desired amino acid(s)/nucleotide. Dissolve in ddH_2O, filter-sterilize, and store at 4°C.
5. LB medium: 1% bacto tryptone, 0.5% yeast extract, 0.5% NaCl, 2% agar (only for agar plates). Autoclave at 121°C, 15 psi, 30 min. Add the appropriate antibiotic once the medium has cooled enough to be handled easily.
6. X-Gal (5-bromo-4-chloro-3-indoyl-β-D-galactopyranoside) solution: 100 mg/ml stock in *N,N*-dimethyl formamide. Make fresh before use.
7. Kanamycin stock (50 mg/ml, 1,000×): dissolve 50 mg kanamycin in a final volume of 1 ml sterile ddH_2O. Filter-sterilize and store at −20°C until use. Final concentration in media: 50 μg/ml.

8. Ampicillin stock (100 mg/ml, 1,000×): dissolve 100 mg ampicillin in a final volume of 1 ml sterile ddH_2O. Filter-sterilize and store at −20°C until use. Final concentration in media: 100 μg/ml.
9. Geneticin (G418) stock (200 mg/ml, 1,000×): dissolve 200 mg G418 in a final volume of 1 ml sterile ddH_2O. Filter-sterilize and store at −20°C until use. Final concentration in media: 200 μg/ml.
10. 0.9% sodium chloride: dissolve 0.9 g NaCl in a final volume of 100 ml water and autoclave at 121°C, 15 psi, 30 min.
11. 1 M Lithium Acetate: Dissolve 10.2 g of lithium acetate in a final volume of 100 ml of ddH_2O and autoclave at 121°C, 15 psi, 30 min.
12. 1 M Tris, pH 7.5: dissolve 121.15 g Tris base in 800 ml ddH_2O, adjust the pH to 7.5 and fill up to 1 L and autoclave.
13. 0.5 M EDTA, pH 8.0: dissolve 93.05 g of EDTA in ddH_2O, adjust the pH to 8.0 using NaOH (EDTA does not dissolve unless a pH of 8.0 is reached),bring up to 500 ml with ddH_2O and autoclave at 121°C, 15 psi, 30 min.
14. 10× TE (Tris–EDTA) buffer: mix 20 ml of 0.5 M EDTA (pH 8.0), 100 ml of 1 M Tris pH 7.5, and 880 ml ddH_2O and autoclave at 121°C, 15 psi, 30 min.
15. 50% PEG (polyethylene-glycol): dissolve 50 g of PEG-3350 in a final volume of 100 ml and autoclave at 121°C, 15 psi, 30 min. 50% PEG should be tightly sealed and stored at room temperature.
16. ssDNA (single-stranded carrier DNA): Dissolve 200 mg of salmon sperm DNA in 100 ml sterile ddH_2O. Aliquot and boil at 100°C for 5 min, chill on ice and store at −20°C. Before every use, boil for 5 min and chill on ice.
17. Lithium acetate–Tris–EDTA master mix: mix 1.1 ml of 1 M lithium acetate, 1.1 ml of 10× TE buffer and fill up to 10 ml with ddH_2O. Vortex thoroughly.
18. Lithium acetate–PEG master mix: mix 1.5 ml of 1 M lithium acetate, 1.5 ml of 10× TE buffer and 12 ml 50% PEG. Vortex thoroughly.
19. Yeast TRAFO mix (per reaction): 240 μl 50% PEG, 36 μl 1 M lithium acetate, 24 μl ssDNA (boiled for 5 min and chilled on ice prior to use).
20. 1 M 3-AT (3-amino-1,2,4-triazole): dissolve 8.4 g of 3-AT in a final volume of 100 ml of ddH_2O. Filter-sterilize and store at −20°C.
21. Zymolyase (200 U/ml): dissolve zymolyase in sterile 1 M sorbitol and freeze at −20°C until use.

22. Sodium phosphate solution: Dissolve 30 g sodium phosphate monobasic and 70 g sodium phosphate dibasic per liter of ddH_2O and autoclave at 121°C, 15 psi, 30 min.

3. Methods

The entire MYTH screening procedure, from generation of bait to final identification of interactors can generally be completed in approximately 6 weeks. An overview of the major steps involved in a MYTH screen is provided in Fig. 2. These steps are described in detail below.

3.1. Bait Construction and Validation

Before constructing a MYTH bait, the user must first decide which variant of MYTH, either tMYTH or iMYTH, is most suitable for

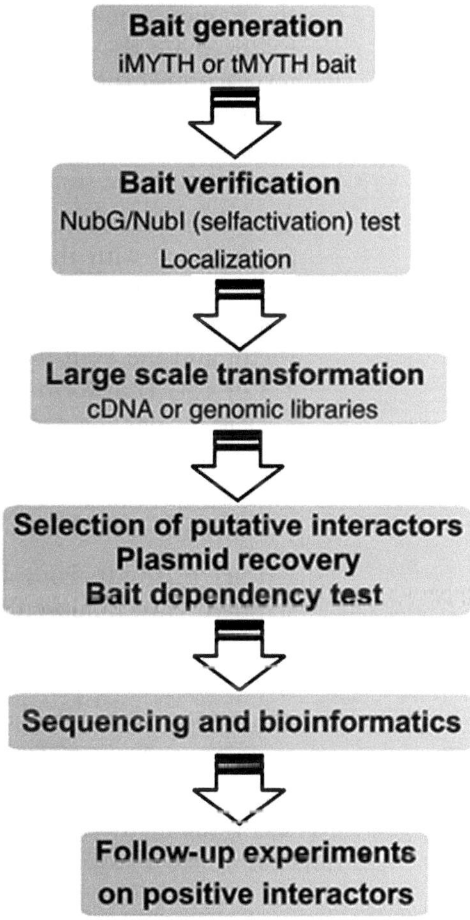

Fig. 2. Workflow of MYTH describing the general steps to identify putative interactors of a specific bait protein.

their protein of interest (see Note 5). While both MYTH variants are generally very similar, there are some differences in bait construction and the specific media used, as noted in the following protocol.

3.1.1. Bait Plasmid Construction by Gap Repair for tMYTH

While conventional restriction digest and ligation strategies can be used to generate tMYTH baits, we find that in vivo recombination methods such as gap repair are most convenient for tMYTH bait construction.

1. Choose the appropriate bait vector backbone (see Note 1) and cut the vector with an appropriate restriction enzyme (preferentially use two restriction enzymes) near Cub-TF/TF-Cub. Gel-purify the vector and store at −20°C until use. Cleavage should occur upstream of Cub-TF for C-terminal tagging or downstream of TF-Cub for N-terminal tagging.
2. Design primers for amplification of the bait gene. The forward primer should contain about 40 nucleotides homologous to the sequence upstream of the restriction site near Cub-TF, followed by the first 20 nucleotides of the bait gene. The reverse primer should contain about 40 nucleotides homologous to the sequence downstream of the restriction site, followed by the last 20 nucleotides of the bait gene. Note: C-terminal tagging requires that the stop codon of the bait gene is omitted.
3. Transform the PCR product (can be used directly without purification) with the digested (and gel-purified) bait vector into the target yeast strain using a standard yeast transformation protocol (see Subheading 3.5.3). Transform the digested vector and the PCR product for gap repair in a molar ratio of 1:3 or 1:4. Transform the digested vector (same amount used for gap repair) as a negative control and the undigested vector as a positive control.
4. Plate on appropriate selective media and incubate at 30°C until colonies are visible.
5. Extract plasmids from about three to five colonies using a standard yeast miniprep procedure (see Subheading 3.5.4).
6. Transform yeast miniprep DNA into highly competent *E. coli* and isolate DNA from transformants using a standard miniprep protocol (see Subheadings 3.5.1 and 3.5.2).
7. Verify the bait construct by sequencing (see Note 3).

3.1.2. Bait Strain Construction for iMYTH

iMYTH is the method of choice for studying yeast protein interactions, as it uses endogenously tagged bait proteins and hence represents a more wild-type like situation (see Note 5).

1. Choose the appropriate vector backbone, either L2 or L3, from which to amplify the tagging cassette. L3 has an additional YFP in the tag, thus allowing for localization of the bait.

Note, however, that addition of YFP may influence the screening procedures as it increases the size of the tag and could therefore lead to blockage of certain interactions due to steric hindrance. Important: To date, iMYTH has been primarily studied for C-terminal tagged baits, although it can be used for N-terminal tagging as well.

2. Design primers for amplification of the tagging cassette. The forward primer should contain about 40 nucleotides homologous to the end of the bait gene (omit the stop codon) and L2/L3 priming sequences. The reverse primer should contain about 40 nucleotides homologous to a region downstream of the bait gene and the L2/L3 reverse priming site.
3. Transform the amplified cassette into the target yeast strain according to the standard yeast transformation protocol (see Subheading 3.5.3) and plate cells on YPAD + G418 plates and grow at 30°C until colonies are visible.
4. Verify the integration by colony PCR and sequencing (see Note 3).

3.1.3. NubG/NubI Test

1. Transform the bait plasmid/strain with control plasmids according to the standard yeast transformation protocol (see Subheading 3.5.3).
2. Control plasmids: pOst1-NubG, pOst1-NubI; pFur4-NubG, pFur4-NubI (see Note 6).
3. Resuspend a single colony of each bait into sterile ddH_2O and make three tenfold serial dilutions.
4. Spot 2–5 μl volumes of the cells (altogether four concentrations) onto the following media:

   SD-Trp-Leu for tMYTH or SD-Trp for iMYTH (to ensure that transformation has occurred and that cells have similar growth) and on SD-Trp-Leu-Ade-His (when using THY.AP4 for tMYTH) and SD-Trp-Leu-His (when using L40 for tMYTH), or SD-Trp-Ade-His (when using THY.AP4 for iMYTH) or SD-Trp-His (when using L40 for iMYTH), with and without X-Gal to select for interacting bait-prey pairs. Important: Optimized 3-AT concentrations may be required to enhance the selection of bait-prey interactions.
5. Incubate at 30°C. An example of a NubG/I control test is given in Fig. 3a.

3.1.4. Immunofluorescence/Fluorescence Microscopy

In case of iMYTH, when using the L3 tagging cassette, bait proteins can be localized by fluorescence microscopy using direct visualization of live cells. Episomally tagged bait proteins or iMYTH bait strains using the L2 tagging cassette can be localized by immunofluorescence.

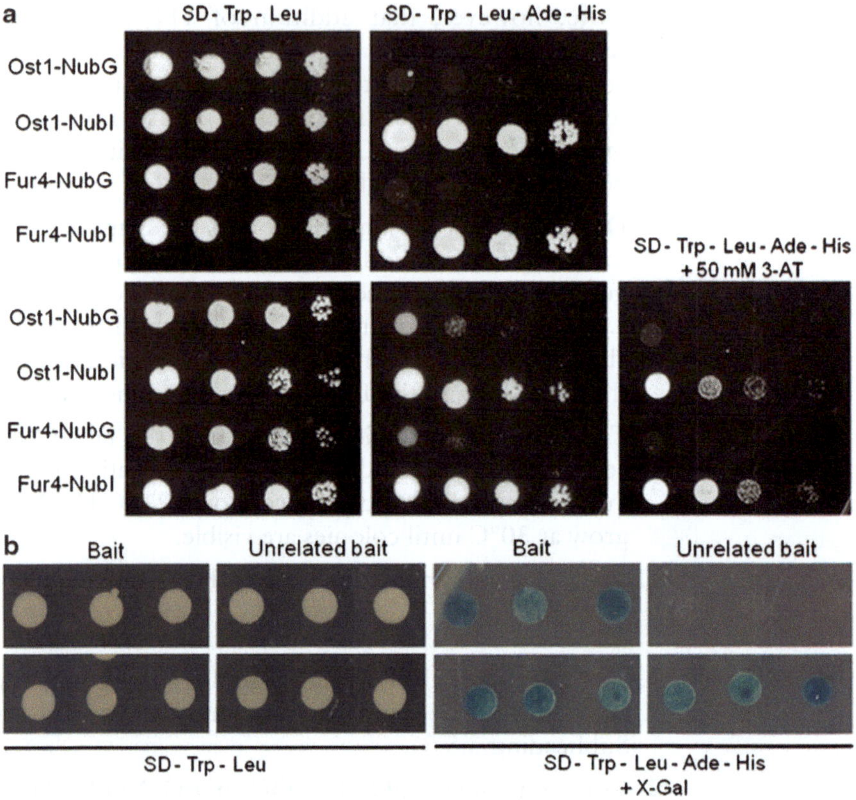

Fig. 3. Examples of MYTH validation experiments. (**a**) NubG/NubI control test. A bait protein is coexpressed with the control preys, which are NubG- and NubI-tagged Ost1p (ER protein) or Fur4p (plasma membrane protein). Serial dilutions are spotted onto selective media. The *panels* at the *far left* are transformation controls. Interaction (as indicated by growth of the transformed strain on selective media) of the bait only with NubI-controls, but not with NubG-controls, indicates that the bait is functional for MYTH and not self-activating (*top right panel*). In cases of self-activation (*bottom middle panel*), addition of 3-AT should be included to enhance the stringency of the bait-prey interaction (*bottom right panel*). (**b**) Bait dependency test (tMYTH). The original bait and an unrelated/artificial bait are transformed into the reporter strain together with putative prey interactors. Three individual colonies are spotted onto selective media. The four *panels* on the *left* show spotting onto SD-Trp-Leu media and serve as controls to ensure that both plasmids have been successfully transformed. The four *panels* on the *right* show spotting onto SD-Trp-Leu-Ade-His+X-Gal media and select only for cells in which bait-prey interaction occurs. Preys interacting with the original bait (as indicated by growth and *blue* coloration of transformed strains grown on selective media containing X-Gal), but not the unrelated/artificial bait, are considered real interactors (*right side, upper two panels*). The *lower two right panels* show an example of a failed bait dependency test as the prey interacts with both the original and the control baits.

1. For direct viewing of live cells, grow yeast strain in YPAD to log-phase, rinse the cells twice with 0.9% NaCl or minimal media and spot 1 μl of the cell suspension onto a microscope slide.
2. Visualize YFP using a fluorescence microscope.
3. For immunofluorescence, grow the strains in appropriate selective medium and perform standard immunofluorescence protocols, using primary antibodies against LexA or VP16 and fluorescence-coupled secondary antibodies. Visualize

using a fluorescence microscope, using settings suitable for the fluorophore.

3.2. Library Transformation (High-Throughput Screening)

1. Inoculate a single colony of MYTH strain in appropriate medium (selective SD-Leu medium for tMYTH or YPAD for iMYTH) and grow overnight at 30°C.
2. Inoculate 200 ml of selective medium (tMYTH) or YPAD (iMYTH) to an $OD_{600}=0.15$ and continue growing until an OD_{600} of 0.6–0.7 is reached (4–6 h).
3. Prepare the lithium acetate–PEG master mix and the lithium acetate–Tris–EDTA master mix.
4. Spin the 200 ml culture at 700×*g* for 5 min (divide into 4×50 ml tubes).
5. Wash the pellets with 30 ml ddH_2O.
6. Resuspend the pellets in 1 ml lithium acetate–Tris–EDTA mix.
7. Spin at 700×*g* for 5 min and resuspend pellets in 600 µl lithium acetate–Tris EDTA mix and transfer to 4×15 ml tubes.
8. Add the following components to each 15-ml tube:
   (a) 2.5 ml PEG–lithium acetate mix
   (b) 100 µl ssDNA (important: boiled for 5 min and chilled on ice)
   (c) 10 µg prey library DNA (see Note 7)
9. Vortex for 1 min.
10. Incubate at 30°C for 45 min and mix briefly every 15 min.
11. Add 160 µl of DMSO to each 15-ml tube and mix by inversion.
12. Incubate at 42°C for 20 min.
13. Spin at 700×*g* for 5 min and resuspend each cell pellet in 3 ml 2×YPAD pool the four samples.
14. Incubate at 30°C for 90 min.
15. Spin at 700×*g* for 5 min and resuspend the pellet in 4.9 ml sterile 0.9% NaCl.
16. Take 100 µl of the suspension and make 1:10, 1:100, and 1:1,000 serial dilutions.
17. Plate 100 µl of the dilutions (1:100 and 1:1,000) onto SD-Trp (iMYTH) or SD-Trp-Leu (tMYTH). Incubate at 30°C until colonies are visible. These plates are used to calculate the total number of transformants.
18. Divide the 4.8 ml cell suspension onto large plates of selective media (tMYTH: SD-Trp-Leu-His for L40 and SD-Trp-Leu-Ade-His for THY.AP4; iMYTH: SD-Trp-His for L40 and SD-Trp-Ade-His for THY.AP4) containing the appropriate

amount of 3-AT established in the NubG/NubI test (see Subheading 3.1.3). Incubate at 30°C until colonies are visible.

19. Perform the X-Gal test: Resuspend colonies in sterile 0.9% NaCl and plate aliquots onto selective media (for tMYTH: SD-Trp-Leu-His for L40 and SD-Trp-Leu-Ade-His for THY. AP4; for iMYTH: SD-Trp-His for L40 and SD-Trp-Ade-His for THY.AP4) + X-Gal (and plus 3-AT where required). Incubate at 30°C until colonies turn blue.

3.3. Prey Recovery, Sequencing, and Bioinformatic Analysis

1. Inoculate blue colonies from the X-Gal test in SD-Trp medium and grow overnight at 30°C. SD-Trp ensures the growth of prey plasmids, but not of bait plasmids.
2. Isolate yeast DNA using a commercial miniprep kit and glass beads (see Note 8 and Subheading 3.5.4).
3. Transform isolated yeast DNA into competent *E. coli* cells (see Subheading 3.5.1).
4. Isolate plasmid DNA from *E. coli* using commercial miniprep kits (see Subheading 3.5.2).
5. Sequence putative interactors using suitable sequencing primers, which are complementary to sequences within the NubG (see Note 3). Ensure that the identified interactor gene sequence is in frame with the NubG tag.
6. Identify preys by BLAST analysis and assemble a preliminary prey hit list.

3.4. Bait Dependency Test

1. Transform preliminary prey interactors into the original bait strain and an unrelated/artificial bait strain, using the standard yeast transformation protocol (see Note 9).
2. Resuspend multiple single colonies from the transformation into 100 μl of sterile ddH_2O and spot 3–5 μl onto selective media (for tMYTH: SD-Trp-Leu-His for L40 and SD-Trp-Leu-Ade-His for THY.AP4; for iMYTH: SD-Trp-His for L40 and SD-Trp-Ade-His for THY.AP4), including X-Gal (and 3-AT where required). Incubate plates at 30°C.
3. Preys that cause growth and blue color in the unrelated/artificial control and in the original bait strain are considered promiscuous.
4. Preys that cause growth and blue color with the original bait, but not the unrelated/artificial bait, are considered good hits and constitute interaction partners.
5. Remove promiscuous hits from the prey list and assemble the remaining list of preys, which are now considered as real interactors. An example of a bait dependency test is given in Fig. 3b.

3.5. Basic Yeast/E. coli Techniques

3.5.1. E. coli Transformation

1. Add 1–3 μl DNA to 10–30 μl of competent *E. coli* cells and mix gently.
2. Chill on ice for 30 min.
3. Heat-shock at 42°C for 45 s to 1 min.
4. Put samples on ice for 3 min.
5. Add 300 μl prewarmed (37°C) LB and incubate at 37°C for 1 h.
6. Spin down the cells (30 s at 16,000 × *g*), remove about 250 μl of the supernatant and plate the remaining cells on LB plus the antibiotic selecting for the plasmid marker.

3.5.2. E. coli Miniprep

1. Inoculate 3–5 ml of LB plus appropriate antibiotic with a single colony and grow overnight at 37°C.
2. Pellet the cells at 16,000 × *g* and extract the plasmid using any commercially available miniprep kit.

3.5.3. Yeast Transformation

1. Inoculate a single colony of your yeast strain into 5 ml appropriate media (e.g., SD or YPAD) depending on the strain being used and grow overnight at 30°C.
2. Inoculate fresh medium with overnight culture to a starting OD_{600} of ~0.15 and grow at 30°C for 3–4 h until an OD_{600} of ~0.6 is reached.
3. Harvest the cells at 700 × *g*, 5 min.
4. Wash the cells with 20 ml of sterile ddH_2O and spin at 700 × *g*, 5 min.
5. Resuspend the pellet in 1 ml of ddH_2O.
6. Add 300 μl of the TRAFO mix to 50–100 μl cells, add the DNA and mix thoroughly (either by vortexing or pipetting up and down).
7. Incubate at 30°C for 30 min.
8. Incubate at 42°C for 30 min.
9. Centrifuge at 700 × *g* for 5 min and remove the supernatant.
10. Add 200 μl YPAD (optional: incubate at 30°C for 1 h for recovery), spin cells down, resuspend pellet in sterile ddH_2O and plate on appropriate selective media. Let the plates dry and incubate at 30°C until colonies are visible.

3.5.4. Yeast Miniprep

1. Inoculate a single colony in 5 ml appropriate selective medium and grow at 30°C overnight.
2. Pellet the cells at 700 × *g* for 5 min and proceed with the plasmid extraction according to the manual of any commercially available miniprep kit. Important: To ensure efficient yeast lysis, add about 100 μl 0.5 mm glass beads to the pellet

(resuspended in resuspension buffer) and vortex vigorously for 10 min prior to proceeding with the miniprep protocol (see Note 8).

3. Elute the plasmid DNA in 20–30 μl of sterile ddH_2O.

3.5.5. Colony PCR

1. Resuspend a colony (critical: use freshly streaked out colonies) in 25 μl of zymolyase.
2. Incubate at 37°C for 15 min.
3. Directly use 1–2 μl as PCR template and perform standard PCR reactions.

3.6. Follow-Up Experiments

The outcome of a MYTH screen is typically a list of putative interactors, which can serve as the basis for further validation. The number of interactors may vary, dependent on the bait and the screening library, though, both cytosolic and membrane-bound interactors can be expected in the hit list. Commonly used follow-ups are techniques such as coimmunoprecipitation, fluorescence microscopy, knockdown, or overexpression studies in the original organism.

A recent example of characterization of interaction partners via MYTH is the identification of interaction partners of the human receptor tyrosine kinase EGFR. Receptor tyrosine kinases (RTKs) are cell surface receptors for many polypeptide growth factors, cytokines, and hormones. RTKs are crucial components of cell signaling cascades and also have a critical role in the development and progression of many types of cancer. Nevertheless, there is a lack of in-depth understanding of RTK networks because of their complex biochemical features and multiplicity (19). A modified MYTH was used to search for previously unknown epidermal growth factor receptor (EGFR)-interacting proteins. The yeast MFα signal sequence was added to EGFR devoid of its endogenous cleavable sequence, leading to stable protein expression. 87 proteins that bound to the EGFR in a ligand-independent fashion were identified. The complete interaction set was further validated by bioinformatics and a subset by immunoprecipitation. Among these EGFR-interacting proteins was the cytoplasmic lysine deacetylase HDAC6. HDAC6 deacetylates α-tubulin in stimulated cells – a process known to slow transport through the secretory pathway – and a feedback mechanism was identified where EGFR inactivates HDAC6 through phosphorylation, increasing α-tubulin acetylation. Together, this study shows that MYTH can now be extended as a novel tool to study receptor tyrosine kinases (14).

4. Notes

1. Based on the orientation of the C and N terminus, membrane proteins are classified as type I (cytosolic C-terminus) and type

II proteins (cytosolic N-terminus). It is important that the bait is tagged in such a way that the Cub-LexA-VP16 resides in the cytosol, as the UBPs are cytosolic enzymes. A membrane protein with both N- and C-termini in the cytosol can be tagged at either end. To date, various bait vector backbones for tMYTH as well as vectors from which cassettes can be amplified for use in iMYTH are available, allowing for N- and C-terminal tagging and different expression levels due to different promoters (see http://www.dualsystems.com for a complete vector list). Expression from the weak *CYC1* promoter leads to lower protein levels than expression driven by the strong *TEF1* or *ADH1* promoters. To circumvent mislocalization of heterologously expressed bait proteins, one might consider tagging the bait with yeast signal sequences to improve proper membrane localization (such as Ste2 or MFα). As both tag orientation and protein expression levels can affect the interactions found in the screen, the best vector has to be determined for each individual gene. It is also important to ensure that gene sequences are in frame with the MYTH tag, and, for C-terminal bait tagging, that the native stop codon is omitted.

2. Yeast strains for MYTH are THY.AP4 and L40, which both contain the *LacZ* reporter gene in addition to auxotrophic markers. THY.AP4 contains three selectable markers (*ADE2*, *HIS3*, and *lacZ*) and grows faster, whereas L40 contains two selectable markers (*HIS3* and *lacZ*) and is sometimes better for screening results. For bait generation, any yeast strain (e.g., BY4741) can be used. In iMYTH, baits are integrated and do not need specific selection. Thus, different media have to be used according to the strain (L40 or THY.AP4) and MYTH method (tMYTH or iMYTH) and are indicated in the Subheadings 2 and 3.

3. For verification of tMYTH bait constructs, commonly available primers such as T7 or M13 (forward and reverse) can be used. If the gene of interest is long (>2,000 bp), additional sequencing primers within the gene should be designed. For iMYTH bait constructs, primers must be generated to allow for PCR amplification and sequencing of the junction between the bait and the integrated tagging cassette. To this end, a gene-specific primer should be designed for each bait, either in the forward direction corresponding to sequences near the 3′ end of the gene (when performing C-terminal tagging), or in the reverse direction corresponding to sequences near the 5′ end of the gene (when performing N-terminal tagging). The gene-specific primer must be paired with a primer internal to the KanMX cassette (in the reverse direction for C-terminal tagging, or the forward direction for N-terminal tagging). The PCR product amplified using such a primer pair can then be sequenced to verify proper integration and the integrity of the bait–tagging

cassette junction. Sequencing of preys derived from NubG-prey libraries (http://www.dualsystems.com) is performed using the following primers. For NubG-X libraries: 5′ CCGATACCATCGACAACGTTAAGTCG 3′ and for X-NubG libraries: 5′ CGACTTAACGTTGTCGATGGTATCGG 3′.

4. Efficient transformation of yeast plasmid miniprep DNA requires the use of competent *E. coli* cells with high transformation efficiency. Low temperature methods, such as that of Inoue et al. (20), are recommended to produce cells of sufficient competency.
5. Before starting a MYTH screen, it has to be determined on a case by case basis which MYTH method to use. There are two different types of MYTH screens, the traditional MYTH (tMYTH) and the integrated MYTH (iMYTH). In tMYTH, the bait protein is expressed on a plasmid, which is the method of choice for heterologously expressed proteins in yeast. In iMYTH, the bait is endogenously tagged; hence, this method is only suitable for native yeast baits. The advantage of iMYTH is that the bait is expressed under the native promoter, which helps to eliminate problems such as overexpression artifacts or mislocalization.
6. The NubG/NubI test gives information about the expression/functionality and the self-activating potential of bait proteins (see Fig. 3a). The bait is cotransformed with an interacting (NubI) and non-interacting (NubG) control prey. Optimally, the bait protein should grow on selective media in the presence of the NubI-control prey, but not the NubG-control prey. Commonly used control preys are Ost1, an ER-resident protein, and Fur4, a plasma-membrane localized protein. If the bait is self-activating (corresponding to growth on selective media with the NubG-prey), the stringency of the selective media using 3-AT in increasing concentrations can be improved (typical concentrations include 10, 25, 50 and 100 mM). When 3-AT is included in the NubG/NubI test, it is crucial to apply the same concentration in the subsequent screening process. Lack of growth in the presence of the NubI-control prey might indicate low levels of bait expression, which can be increased using bait vectors with strong promoters (*ADH1*, *TEF1*).
7. cDNA libraries containing N- and C-terminally tagged NubG preys are available from various organisms and tissues (http://www.dualsystems.com). Depending on the individual bait expression profiles, prey cDNA libraries from specific tissues or organisms may be required.
8. Treatment of yeast cells with zymolyase and incubation at 37°C for 30–60 min can also result in higher plasmid yields prior to glass bead disruption.

9. Performing the bait dependency test is necessary to determine real interactors of the bait and get rid of spurious preys. Thus, an unrelated bait, which is likely not to interact with the preys, has to be chosen carefully. Alternatively, an "artificial" bait is successfully used in our laboratory, which consists of the Matα signal sequence, the transmembrane domain of CD4 and Cub-TF (see Fig. 3b).

Acknowledgments

The Stagljar lab is supported by grants from the Canadian Foundation for Innovation (CFI), the Canadian Institute for Health Research (CIHR), the Canadian Cancer Society Research Institute (CCSRI), the Heart and Stroke Foundation, the Cystic Fibrosis Foundation, the Ontario Genomics Institute, and Novartis. J.P. is a recipient of an FWF-Erwin-Schrödinger postdoctoral fellowship.

References

1. Fetchko, M., Auerbach, D., and Stagljar, I. (2003) Yeast genetic methods for the detection of membrane protein interactions: potential use in drug discovery, *BioDrugs* **17**, 413–424.
2. Auerbach, D., Thaminy, S., Hottiger, M. O., and Stagljar, I. (2002) The post-genomic era of interactive proteomics: facts and perspectives, *Proteomics* **2**, 611–623.
3. Suter, B., Kittanakom, S., and Stagljar, I. (2008) Two-hybrid technologies in proteomics research, *Curr Opin Biotechnol* **19**, 316–323.
4. Suter, B., Kittanakom, S., and Stagljar, I. (2008) Interactive proteomics: what lies ahead?, *Biotechniques* **44**, 681–691.
5. Fields, S., and Song, O. (1989) A novel genetic system to detect protein-protein interactions, *Nature* **340**, 245–246.
6. Stagljar, I., and Fields, S. (2002) Analysis of membrane protein interactions using yeast-based technologies, *Trends Biochem Sci* **27**, 559–563.
7. Iyer, K., Burkle, L., Auerbach, D., Thaminy, S., Dinkel, M., Engels, K., and Stagljar, I. (2005) Utilizing the split-ubiquitin membrane yeast two-hybrid system to identify protein-protein interactions of integral membrane proteins, *Sci STKE* **2005**, pl3.
8. Stagljar, I., Korostensky, C., Johnsson, N., and te Heesen, S. (1998) A genetic system based on split-ubiquitin for the analysis of interactions between membrane proteins in vivo, *Proc Natl Acad Sci USA* **95**, 5187–5192.
9. Johnsson, N., and Varshavsky, A. (1994) Split ubiquitin as a sensor of protein interactions in vivo, *Proc Natl Acad Sci USA* **91**, 10340–10344.
10. Kittanakom, S., Chuk, M., Wong, V., Snyder, J., Edmonds, D., Lydakis, A., Zhang, Z., Auerbach, D., and Stagljar, I. (2009) Analysis of membrane protein complexes using the split-ubiquitin membrane yeast two-hybrid (MYTH) system, *Methods Mol Biol* **548**, 247–271.
11. Snider, J., Kittanakom, S., Curak, J., and Stagljar, I. (2010) Split-ubiquitin based membrane yeast two-hybrid (MYTH) system: a powerful tool for identifying protein-protein interactions, *J Vis Exp*. **36**. pii: 1698
12. Thaminy, S., Miller, J., and Stagljar, I. (2004) The split-ubiquitin membrane-based yeast two hybrid system, *Methods Mol Biol* **261**, 297–312.
13. Paumi, C. M., Menendez, J., Arnoldo, A., Engels, K., Iyer, K. R., Thaminy, S., Georgiev, O., Barral, Y., Michaelis, S., and Stagljar, I. (2007) Mapping protein-protein interactions for the yeast ABC transporter Ycf1p by integrated split-ubiquitin membrane yeast two-hybrid analysis, *Mol Cell* **26**, 15–25.
14. Deribe, Y. L., Wild, P., Chandrashaker, A., Curak, J., Schmidt, M. H., Kalaidzidis, Y., Milutinovic, N., Kratchmarova, I., Buerkle, L., Fetchko, M. J., Schmidt, P., Kittanakom, S.,

Brown, K. R., Jurisica, I., Blagoev, B., Zerial, M., Stagljar, I., and Dikic, I. (2009) Regulation of epidermal growth factor receptor trafficking by lysine deacetylase HDAC6, *Sci Signal* **2** (102), p. ra84.

15. Fetchko, M., and Stagljar, I. (2004) Application of the split-ubiquitin membrane yeast two-hybrid system to investigate membrane protein interactions, *Methods* **32**, 349–362.
16. Gisler, S. M., Kittanakom, S., Fuster, D., Wong, V., Bertic, M., Radanovic, T., Hall, R. A., Murer, H., Biber, J., Markovich, D., Moe, O. W., and Stagljar, I. (2008) Monitoring protein-protein interactions between the mammalian integral membrane transporters and PDZ-interacting partners using a modified split-ubiquitin membrane yeast two-hybrid system, *Mol Cell Proteomics* 7, 1362–1377.
17. Paumi, C. M., Chuk, M., Chevelev, I., Stagljar, I., and Michaelis, S. (2008) Negative regulation of the yeast ABC transporter Ycf1p by phosphorylation within its N-terminal extension, *J Biol Chem* **283**, 27079–27088.
18. Thaminy, S., Auerbach, D., Arnoldo, A., and Stagljar, I. (2003) Identification of novel ErbB3-interacting factors using the split-ubiquitin membrane yeast two-hybrid system, *Genome Res* **13**, 1744–1753.
19. Ullrich, A., and Schlessinger, J. (1990) Signal transduction by receptors with tyrosine kinase activity, *Cell* **61**, 203–212.
20. Inoue, H., Nojima, H., and Okayama, H. (1990) High efficiency transformation of Escherichia coli with plasmids. *Gene* **96**, 23–28.

Chapter 14

Application of the Split-Protein Sensor Trp1 to Protein Interaction Discovery in the Yeast *Saccharomyces cerevisiae*

Mandana Rezwan, Nicolas Lentze, Lukas Baumann, and Daniel Auerbach

Abstract

Yeast two-hybrid based systems are powerful tools for the detection and characterization of protein–protein interactions (PPIs). However, some important protein classes, e.g., integral membrane proteins and transcription factors, are difficult to study using these technologies.

To overcome these limitations, we have employed a novel protein complementation screening platform. Protein interactions are detected by reconstitution of the split-protein sensor TRP1, enabling trp1 cells to grow on medium lacking tryptophan. Since the interaction readout is direct and independent of transcriptional reporter activation the rate of false positives is lowered. Furthermore, the technology allows for detection of protein interactions in their natural setting, e.g., the cytosol, the nucleus, and at cellular or organellar membranes. The protocols used for screening are explained in detail and as an example we describe the isolation of novel binding partners found with APP screened against a human cDNA library.

Key words: Split-protein sensor, Protein–protein interaction, Yeast-two hybrid, Membrane protein, Transcription factors, Screening, Tryptophan, cDNA library, Amyloid precursor protein (APP)

1. Introduction

Protein–protein interactions are at the heart of most biological processes and various methods have been developed to identify and study protein–protein interactions over the last years. Despite impressive achievements both in method development as well as applications of these methods to complex biological problems, there is still a generally acknowledged need for new methods to identify the interacting partners of a given protein.

The most widely used methods for the identification of new protein–protein interactions rely either on the isolation of native protein complexes and subsequent identification by mass spectrometry or on

Bernhard Suter and Erich E. Wanker (eds.), *Two Hybrid Technologies: Methods and Protocols*, Methods in Molecular Biology, vol. 812, DOI 10.1007/978-1-61779-455-1_14, © Springer Science+Business Media, LLC 2012

the screening of a protein of interest against a cDNA library using a suitable selection or screening system. The most popular example for the latter approach is the yeast two-hybrid system (1, 2).

However, many proteins are difficult to investigate using conventional yeast two-hybrid systems, including integral membrane proteins, membrane-associated proteins, and nuclear proteins, such as transcription factors (3). To address these limitations, several split-protein sensor systems have been developed (4, 5). Here, a reporter protein is expressed as two separated fragments, which by themselves do not reassemble into a functional protein. If the two fragments are expressed as fusions to two interacting proteins, the induced proximity of the two fragments results in the formation of quasi-native split reporter protein, whose activity can be measured. The activity of the reporter protein indicates an interaction between the two proteins. Commonly used split-protein assays include the split ubiquitin assay (6, 7), the split dihydrofolate reductase system (8), assays based on split autofluorescent proteins (9, 10), and the split β-lactamase assay (11). Here, we describe a split-protein sensor (called split-Trp) based on *N*-(5′-phosphoribosyl)-anthranilate isomerase Trp1p from *Saccharomyces cerevisiae* (12).

The *S. cerevisiae* enzyme *N*-(5′-phosphoribosyl)-anthranilate isomerase (encoded by the *TRP1* gene) can be split into an N-terminal (NTrp) and a C-terminal (CTrp) fragment (13). The fragments have very little affinity for each other and therefore do not spontaneously reassociate when coexpressed (Fig. 1a). NTrp

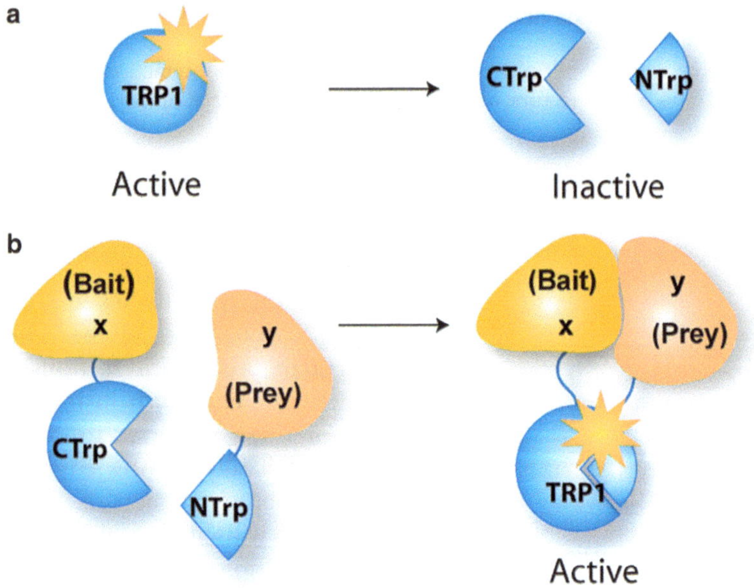

Fig. 1. The SplitTrp mechanism. (**a**) The TRP1 gene is split into a N-terminal (NTrp) and a C-terminal (CTrp) fragment which cannot reassociate spontaneously. (**b**) By fusing NTrp and CTrp to two interacting proteins, TRP1 protein is produced and allows grow on medium lacking tryptophan.

and CTrp are then fused to two interacting proteins X and Y. Upon interaction of X and Y the NTrp and CTrp fragments are forced into close proximity and quasi-native split-Trp is reconstituted (Fig. 1b). Reconstitution of the split-Trp sensor enables trp1 deficient yeast cells to grow on medium lacking tryptophan.

For screening, the cDNA encoding a protein of interest is cloned into the plasmid p-bait-CTrp and is transformed into the yeast strain CRY1. A cDNA library encoding NTrp-prey fusion proteins is transformed into yeast bearing the bait plasmid and the cotransformants are plated on selective medium lacking tryptophan. After incubation for 10 days, colonies are picked up and analyzed by isolation of the library plasmid and sequencing of the cDNAs. Subsequently, a bait dependency test is performed to validate the interactors.

2. Materials

2.1. Plasmid Construction

2.1.1. p-Bait-CTrp Bait Vector

The bait vector is a shuttle plasmid containing a URA3 marker and a CEN/ARS origin of replication for propagation in yeast. Expression of the bait-CTrp fusion protein is driven by a copper inducible promoter pCup. Two *SfiI* sites allow in-frame cloning of the bait cDNA, together with an N-terminal myc epitope tag and a C-terminal V5 epitope tag (see Fig. 2).

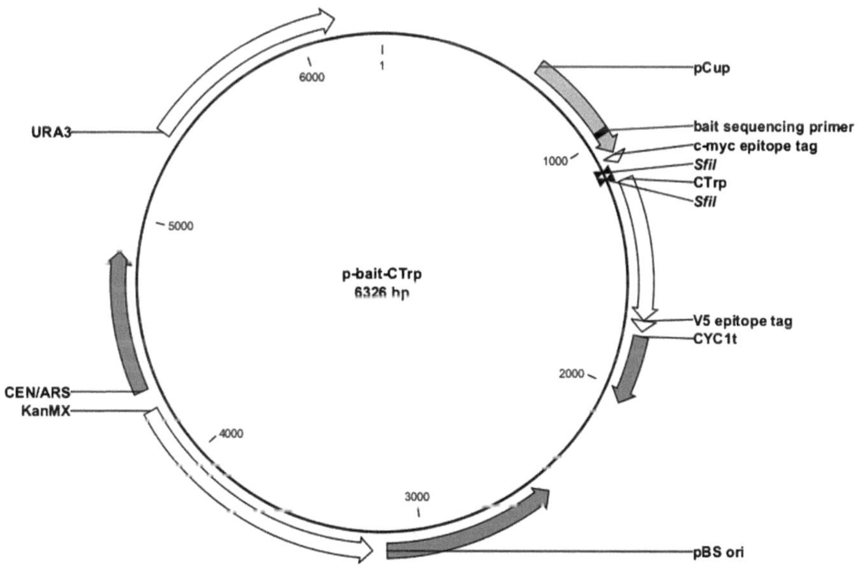

Fig. 2. The p-bait-CTrp bait vector. The bait cDNA is subcloned into the *SfiI* sites and expression of the fusion protein is driven by the copper inducible promoter pCup.

2.1.2. Primers for Bait Amplification

1. Forward bait primer: 5′-gatc**g gcc att acg gcc** (NNN)$_5$-3′.
2. Reverse bait primer: 5′-gatc **ggc cga ggc ggc c**gc (NNN)$_5$-3′ *SfiI* restriction sites are bold and underlined. The proper reading frame is indicated.
3. Bait sequencing primer: 5′-ccttgtcttgtatcaattgc-3′.

2.1.3. Library Vector

The library vector is a shuttle plasmid containing a LEU2 marker and a CEN/ARS origin of replication for propagation in yeast. Expression of the NTrp-prey fusion protein is driven by a methionine inducible promoter pMET25. Two *SfiI* sites allow in-frame cloning of the prey cDNA, together with an N-terminal HA epitope tag (see Fig. 3).

2.1.4. Control Vectors

1. p-NTrp-empty: The empty library vector serves as a background control in the functional assay.
2. p-NTrp135-Alg5: This vector contains an extended version on the NTrp domain (135 aa instead of 44 aa). The NTrp135 fragment reassociates spontaneously with CTrp and serves as positive control in the functional assay (see Subheading 3.2).

2.2. Functional Assay

2.2.1. Small Scale Yeast Transformation

1. Strain CRY1: *Mat*a *ura3-1 trp1-1 his3-11, 15 leu2-3,112 ade2-1 can1-100 GAL.*
2. YPAD: Mix 10 g Bacto yeast extract, 20 g Bacto peptone, 40 mg Adenine sulfate, and 20 g Bacto agar (for plates only) and fill up to 1 l with H_2O (see Note 1). Autoclave the medium and let the

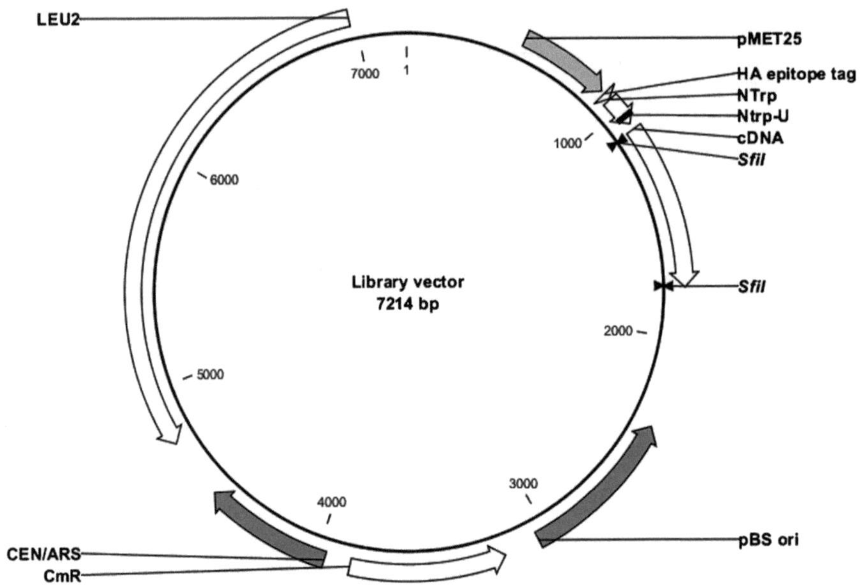

Fig. 3. The library vector. cDNAs are subcloned into *SfiI* sites and expression is driven by the methionine inducible promoter pMET25.

medium cool down to 50°C. Add 50 ml sterile 40% glucose solution. Store for up to 6 months at room temperature.

3. 50% PEG: Combine in a glass beaker 50 g PEG 4000 and 80 ml H_2O. Stir until completely dissolved and adjust the total volume to 100 ml with water. Sterilize by passing through a 0.22-μm pore filter (see Note 2). Store for up to 1 year at room temperature.
4. 1 M LiOAc, pH 7.5: Combine in a glass beaker 10.2 g LiOAc and H_2O to 100 ml. Dissolve by stirring. Sterilize by passing through a 0.22-μm pore filter. Store for up to 1 year at room temperature.
5. Salmon sperm DNA, sodium salt (e.g., deoxyribonucleic acid sodium salt from salmon testes, Sigma) (2 mg/ml) (Recipe see Note 3).
6. 0.9% NaCl: Add 0.9 g NaCl to 100 ml H_2O. Autoclave and store at room temperature up to 1 year.
7. SD-Ura-Leu: mix 0.6 g Dropout mix (-Ura-His-Trp-Leu), 6.7 g Bacto yeast nitrogen base, and 20 g Bacto agar (for plates only) and fill up to 1 l with H_2O. Autoclave the medium and let the medium cool down to 50°C. Add 50 ml sterile 40% glucose solution. Add appropriate amino acids to final concentration as follows: Histidine 20 mg/l and Tryptophan 20 mg/l. Store at room temperature up to 6 months.

2.2.2. Yeast Spotting

1. SD-Ura-Leu, see Subheading 2.2.1.
2. 0.9% NaCl, see Subheading 2.2.1.
3. SD-Ura-Leu-Trp-Met: mix 0.6 g Dropout mix (-Ura-Ade-His-Trp-Leu-Met), 6.7 g Bacto yeast nitrogen base and 20 g Bacto agar (for plates only) and fill up to 1 l with H_2O. Autoclave the medium and let the medium cool down to 50°C. Add 50 ml sterile 40% glucose solution. Add appropriate amino acids to final concentration as follows: Histidine 20 mg/l and Adenine 10 mg/l. Store at room temperature for up to 6 months.
4. SD-Ura-Leu-Trp-Met + $CuSO_4$: Prepare as SD-Ura-Leu-Trp-Met. Add $CuSO_4$ to a final concentration of 100 μM. Store at room temperature up to 6 months.
5. SD-Ura-Trp-Leu: Mix 0.6 g Dropout mix (-Ura-His-Trp-Leu), 6.7 g Bacto yeast nitrogen base, and 20 g Bacto agar (for plates only) and fill up to 1 l with H_2O. Autoclave the medium and let the medium cool down to 50°C. Add 50 ml sterile 40% glucose solution. Add Histidine to a final concentration of 20 mg/l. Store at room temperature for up to 6 months.

2.3. cDNA Library Screening

2.3.1. Bait Culture Preparation

1. For yeast small scale transformation, see Subheading 2.2.1.
2. SD-Ura: Mix 0.6 g Dropout mix (-Ura-His-Trp-Leu), 6.7 g Bacto yeast nitrogen base, and 20 g Bacto agar (for plates only) and fill up to 1 l with H_2O. Autoclave the medium and let the medium cool down to 50°C. Add 50 ml sterile 40% glucose solution. Add appropriate amino acids to final concentration as follows: Histidine 20 mg/l, Tryptophan 20 mg/l, and Leucine 100 mg/l. Store at room temperature up to 6 months.

2.3.2. Library Scale Yeast Transformation

1. 10× TE buffer: 100 mM Tris–Cl, pH 7.5, 10 mM EDTA, store at room temperature up to 1 year.
2. SD-Ura, see Subheading 2.3.1.
3. SD-Ura-Leu-Trp-Met + $CuSO_4$, see Subheading 2.2.2.
4. For all other reagents, see Subheading 2.2.1.

2.4. Clone Analysis

2.4.1. Yeast Plasmid Isolation

1. Glass beads, acid washed (Sigma G1277).
2. Plasmid isolation kit (Macherey-Nagel), also available for 96-well plates.

2.4.2. PCR

1. Forward colony primer: 5′-atgtacccatacgatgttccagattacg-3′.
2. Reverse colony primer: 5′-gttagttaaagcttaggcctggctaagg-3′.
3. Sequencing primer NTrp-U: 5′-agattccgatgctgacttg-3′.
4. EasyClone PCR kit (Dualsystems Biotech).

2.5. Bait Dependency Test

2.5.1. Retransformation into Yeast

1. Linearized library vector: Incubate library vector with *SfiI* (1 U/µg DNA) for 2 h at 50°C. Load the sample on an agarose gel and isolate the linearized DNA. Purify with a standard DNA extraction kit. Measure DNA concentration. 35 ng is needed for each retransformation.
2. For all other reagents, see Subheading 2.2.1.

2.5.2. Yeast Spotting

See Subheading 2.2.2.

3. Methods

3.1. Bait Plasmid Construction

The cDNA sequence encoding a bait protein of interest is cloned into the vector p-bait-CTrp. The cDNA is amplified by PCR and subcloned into the *SfiI* sites present in the bait vector. Primers have to be designed in a way that the bait is in frame with the downstream CTrp sequence.

3.1.1. p-Bait-CTrp

1. Design primers for bait amplification using the primer sequences given in Subheading 2.1.2.
2. Perform PCR using an appropriate proof reading polymerase or the EasyClone PCR kit (see Subheading 2.4.2).
3. Load the PCR fragment on an agarose gel. Isolate the DNA band, purify using a standard gel purification kit and digest with *SfiI* (1 U/μg DNA) at 50°C for 2 h.
4. Ligate the insert into *SfiI* linearized p-bait-CTrp using standard cloning technique.
5. Sequence bait construct using the bait sequencing primer (see Subheading 2.1.2).

3.2. Functional Assay

The bait plasmid is cotransformed into yeast with control vectors (see Subheading 2.1.4) to confirm that the newly constructed fusion protein is expressed properly. The functional assay is also used to determine whether the bait interacts nonspecifically with the empty library vector, which would result in nonspecific growth in the library screening procedure.

3.2.1. Yeast Small Scale Transformation

1. Grow CRY1 in 2 ml YPAD per transformation at 30°C, shaking at 200 rpm to an OD_{546} of 0.8.
2. Pellet 2 ml of yeast culture per transformation at 700 g for 5 min and resuspend in 100 μl H_2O.
3. Prepare the following microfuge tubes:

	Tube 1 (μg)	Tube 2 (μg)
p-Bait-CTrp	0.5	0.5
p-NTrp135-Alg5	0.5	–
p-NTrp-empty	–	0.5

4. Prepare 300 μl PEG–LiOAc master mix per transformation (see Note 4).
5. Add 300 μl PEG–LiOAc master mix to each tube and vortex briefly.
6. Add 100 μl yeast cells (from step 2) to each tube and vortex for 1 min to thoroughly mix all components.
7. Incubate in a 42°C water bath for 45 min.
8. Pellet yeast for 5 min at 700 g at room temperature.
9. Resuspend each yeast pellet in 100 μl of 0.9% NaCl and plate 100 μl of each transformation onto one SD-Ura-Leu plate.
10. Incubate plates at 30°C until colonies are visible (3–4 days).

3.2.2. Yeast Spotting

1. From each plate prepared in Subheading 3.2.1 pick several colonies and inoculate into 1 ml SD-Ura-Leu medium. Grow overnight at 30°C with shaking.
2. Measure the OD_{546} and adjust ODs to 1. Prepare 1:10, 1:100 and 1:1,000 dilutions and spot 5 μl drops on SD-Ura-Leu, SD-Ura-Leu-Trp, SD-Ura-Leu-Trp-Met, and SD-Ura-Leu-Trp-Met + $CuSO_4$ medium. Incubate plates at 27°C for up to 10 days (see Note 5).
3. Examine the plates. Growth of all strains should be visible on SD-Ura-Leu until the 1:1,000 dilution. On SD-Ura-Leu-Trp, SD-Ura-Leu-Trp-Met, and SD-Ura-Leu-Trp-Met + $CuSO_4$ plates containing the transformants of the bait construct and the positive control p-NTrp135-Alg5 colonies should be visible until the 1:100 dilution, depending on the expression level of the bait. See also Note 6.

3.3. cDNA Library Screening

Library screening is performed by sequential transformation. First, the bait construct is transformed into CRY1 and is plated on selective medium. In the second step, the strain bearing the bait plasmid is transformed with the cDNA library.

3.3.1. Bait Culture Preparation

1. Grow CRY1 in 2 ml YPAD per transformation at 30°C, shaking at 200 rpm, to an OD_{546} of 0.8.
2. Pellet 2 ml of yeast culture per transformation at 700 g for 5 min and resuspend in 100 μl H_2O.
3. Add 1 μg of each bait construct to separate microfuge tubes.
4. Prepare 300 μl PEG–LiOAc Master mix per transformation.
5. Add 300 μl PEG–LiOAc master mix to each tube of DNA and vortex briefly.
6. Add 100 μl yeast cells (from step 2) to each tube and vortex for 1 min to thoroughly mix all components.
7. Incubate in a 42°C water bath for 45 min.
8. Pellet yeast for 5 min at 700 g at room temperature.
9. Resuspend each yeast pellet in 100 μl of 0.9% NaCl and plate 100 μl of each transformation onto one SD-Ura plate.
10. Incubate plates at 30°C until colonies are visible (3–4 days).

3.3.2. Library Scale Transformation

1. Inoculate a single colony of the bait plasmid bearing strain into 10 ml SD-Ura medium and grow for 8 h at 30°C with shaking.
2. Inoculate 100 ml SD-Ura medium with the entire 10 ml culture and grow overnight at 30°C with shaking.
3. Transfer the amount corresponding to 30 OD units of overnight culture into 50-ml Falcon tubes and collect the cells by centrifugation at 700 g for 5 min at room temperature.

4. Resuspend all pellets in a total of 200 ml of 2 × YPAD medium and grow to an OD_{546} of 0.6 (two cell doublings) in a 1-l Erlenmeyer flask at 30°C with shaking at 200 rpm.
5. Split the culture into four 50-ml Falcon tubes and centrifuge at 700 g for 5 min.
6. Resuspend each pellet in 30 ml of H_2O by vortexing.
7. Centrifuge at 700 g for 5 min at room temperature.
8. Prepare 1.1 × TE/LiOAc by combining 0.88 ml 10× TE buffer, 0.88 ml 1 M LiOAc and 6.24 ml H_2O.
9. Wash the cell pellets once with 1 ml 1.1 × TE/LiOAc.
10. Remove the supernatants and resuspend the cells in 600 μl of 1.1 × TE/LiOAc.
11. Prepare four 50-ml Falcon tubes with 7 μg library plasmid and 250 μl single-stranded carrier DNA (2 mg/ml) in each tube (see Note 7).
12. Prepare PEG–LiOAc Master Mix: 12 ml 50% PEG, 1.5 ml 10× TE buffer, 1.5 ml 1 M LiOAc.
13. Add 600 μl cells from step 10 to each DNA-containing Falcon tube from step 11.
14. Vortex briefly to mix.
15. Add 2.5 ml PEG–LiOAc Master Mix to each tube.
16. Vortex for 1 min to thoroughly mix all components.
17. Incubate at 30°C for 45 min (see Note 8).
18. Add 160 μl DMSO (Sigma) to each Falcon tube.
19. Incubate at 42°C for 20 min (see Note 8).
20. Pellet cells at 700 g for 5 min at room temperature.
21. Resuspend each pellet in 3 ml of 2× YPAD and pool the contents of all four tubes into one Falcon tube.
22. Incubate at 30°C for 90 min with gentle shaking (200 rpm).
23. Pellet the cells by centrifugation at 700 g for 5 min at room temperature.
24. Resuspend the pellet in 4.7 ml of 0.9% NaCl.
25. Plate 300 μl suspension per plate onto 15 big (14 cm diameter) SD ULTM + $CuSO_4$ plates.
26. Use the remaining resuspension to prepare 1:10, 1:100, and 1:1,000 dilutions in 0.9% NaCl, and plate 100 μl aliquots onto two SD-Ura-Leu plates for each dilution. Use these plates to calculate the transformation efficiency (number of transformants per μg cDNA = (dilution factor × 4.7 × 10)/28). The transformation efficiency should be greater than 1×10^5 cfu/μg cDNA.

27. Transformants on the selection plates should appear after 7–10 days at 27°C (see Note 5) (between 10 and 100; the number of colonies obtained varies for different bait proteins).

3.4. Clone Analysis

Plasmids are isolated using a standard plasmid isolation kit together with a modified purification procedure and cDNAs are amplified by PCR and sequenced.

3.4.1. Yeast Plasmid Isolation

1. Pick all colonies from the selection plates (Subheading 3.3.2, step 27) and inoculate into 1 ml SD-Ura-Leu medium. Grow overnight at 30°C with shaking.
2. Harvest the cells by centrifugation at 700 g for 5 min. Resuspend the pellets in resuspension buffer from a standard plasmid isolation kit. Add 300 μl glass beads (Sigma) and vortex vigorously for 5 min.
3. Proceed with the samples according to the plasmid isolation kit manual (see Note 9).

3.4.2. PCR

1. Use the yeast plasmid extracts from Subheading 3.4.1 as template for PCR amplification of the prey plasmid inserts.
2. Amounts for one PCR are indicated below. Add an additional 10% when preparing the master mix.
3. Aliquot 9 μl master mix into thin wall PCR tubes or a 96-well PCR plate.

Amount (μl)	Component
1	Yeast plasmid extract
1	10× EasyClone buffer
0.1	Forward primer
0.1	Reverse primer
0.8	dNTPs (2.5 mM)
0.05	50× EasyClone mix
7.0	H_2O
10	Total

4. Perform a PCR with the following conditions:

Temperature (°C)	Time	Cycles
95	1 min	1×
95	15 s	
63	30s	35×
72	3 min	
72	7 min	1×

5. Check 20 PCR reactions on a 1% agarose gel, loading 1.5 μl PCR reaction per well. The success rate of prey amplification should be 80 %, i.e., the majority of the PCR reactions should have yielded one or several distinct bands.
6. Purify the PCR products using a PCR purification kit (NucleoSpin Extract II, Macherey-Nagel or NucleoSpin 96 Extract II, Macherey-Nagel). Elute with 18 μl (Nucleospin Extract II) or 80 μl (NucleoSpin 96 Extract II), respectively.
7. Sequence the purified PCR fragments with the forward sequencing primer NTrp-U.

3.5. Bait Dependency Test

The PCR fragments are retransformed together with the linearized library vector into the bait bearing strain. Due to the homology between the ends of the PCR fragments and the library vector ends, the PCR fragments are recombined into the library vectors to yield the original library plasmid. Spotting of the transformants on selective media confirms bait-dependent protein–protein interactions. An example is given in Fig. 4.

3.5.1. Retransformation into Yeast

1. Cotransform the PCR products from Subheading 3.4.2 together with the linearized library vector (*SfiI* digested pNTrp-empty) into the yeast strain CRY1 bearing the bait vector.

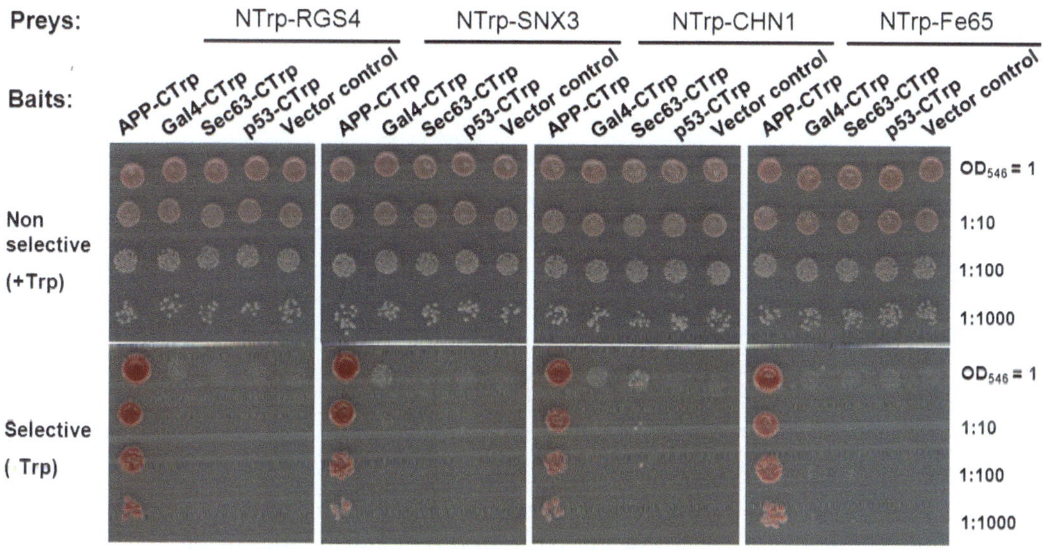

Fig. 4. Spotting bait dependency test APP screen. To investigate PPIs involving APP, the Split-Trp APP bait was screened against a total human brain NTrp-cDNA library. The screen revealed the known APP interactor Fe65, as well as several novel putative interaction partners of APP (RGS4, SNX3, and CHN1). Both integral membrane proteins as well as soluble proteins that locate to different cellular compartments are represented among the interactors of APP. All interactors were bait dependent. Yeast bearing bait plasmid and library plasmids expressing control proteins (Gal4, Sec63, p53) or empty library vector did not grow on medium lacking tryptophan.

2. Perform small scale transformation as described in Subheading 3.2.1. Use 2 µl of the PCR extracts and 35 ng of the linearized library vector. Perform the transformation in a 0.5 ml 96-deep-well masterblock.
3. Harvest the cells, resuspend in 50 µl 0.9 % NaCl, spot 5 µl 1:1 and of a 1:30 dilution onto SD-Ura-Leu. Incubate plates at 30°C until colonies are visible (3–4 days).

3.5.2. Yeast Spotting

1. Inoculate the strains from Subheading 3.5.1 in 1 ml SD-Ura-Leu overnight at 30°C with shaking (see Note 10).
2. Measure the OD_{546} the next day and adjust ODs to 1.
3. Prepare 1:10, 1:100, and 1:1,000 dilutions and spot 5 µl drops on SD-Ura-Leu, SD-Ura-Leu-Trp, SD-Ura-Leu-Trp-Met, and SD-Ura-Leu-Trp-Met + $CuSO_4$. Incubate plates at 27°C for up to 10 days (see Note 5).
4. Real interactors will grow on plates lacking tryptophan (see Note 11).

4. Notes

1. All solutions are prepared in water that has 18.2 MΩ-cm resistivity and that is sterilized by autoclaving. This standard is referred to as H_2O in the text.
2. Preparation and handling of PEG: 50% PEG is very viscous and therefore, pipetting has to be done very carefully. During sterilization, it is crucial to avoid evaporation, as the concentration of PEG is a critical parameter for achieving good transformation efficiency.
3. Preparation of salmon sperm DNA:
   - Weigh 200 mg salmon sperm DNA sodium salt into the sterile glass beaker
   - Add a sterile stir bar
   - Add 100 ml sterile H_2O
   - Dissolve the large chunks of DNA by drawing the suspension up and down a few times with a 25-ml sterile plastic pipet
   - Place the glass beaker into an ice–water bath and place the ice-water bath on top of a magnetic stirrer
   - Stir for 6–12 h to dissolve the DNA completely
   - Dispense the DNA in 1 ml aliquots into sterile microcentrifuge tubes
   - Denature the DNA by placing the microcentrifuge tubes into a boiling water bath or a heating block set to 99°C for 5 min

- Immediately transfer the tubes to an ice–water bath
- Store the denatured single-stranded DNA up to 1 year at –20°C

4. Prepare all master mixes for the yeast transformation freshly, on the day that it is needed. Discard unused solutions. Sterilized stock solutions can be stored up to 1 year at room temperature.
5. Sensitivity of split Trp system is increased by incubating yeast at 27°C instead of 30°C.
6. If the bait is not positive in the functional assay, switch the orientation by ligating the cDNA into the p-CTrp-bait vector and repeat the functional assay. The p-CTrp-bait construct expresses the fusion protein in the opposite orientation (p-CTrp-bait) and may circumvent steric hindrance observed with the p-bait-CTrp construct.
7. ssDNA should be boiled prior to large-scale cDNA library transformation. Boil the ssDNA at 95°C for 10 min and transfer immediately to ice for 5 min. Repeat one more time. Store on ice until use.
8. The cell suspension should be mixed manually by shaking for several seconds every 10 min.
9. Yeast plasmid isolation efficiency can be increased by incubating the suspension at room temperature for 10 min after Buffer 2 (lysis buffer) was added. Use Nucleo Spin Multi-96 Plus Plasmid kit (Macherey-Nagel) or GeneJET™ Plasmid Miniprep kit (Fermentas).
10. It is recommended to use the strains generated for the functional assay in the bait dependency test as positive and negative controls.
11. Positive interactors found in the split-Trp assay should be confirmed in an independent experiment, such as copurification or coimmunoprecipitation.

References

1. Fields, S. and Song, O. (1989) A novel genetic system to detect protein-protein interactions. *Nature*. **340**, 245–246.
2. Brückner, A., Polge, C., Lentze, N., Auerbach, D. and Schlattner U. (2009) Yeast two-hybrid, a powerful tool for Systems Biology. *Int J Mol Sci*. **10** (6), 2763–2788.
3. Auerbach, D., Thaminy, S., Hottiger, M.O. and Stagljar I. (2002) The post-genomic era of interactive proteomics: facts and perspectives. *Proteomics*. **2** (6), 611–23.
4. Snider, J., Kittanakom, S., Curak, J. and Stagljar, I. (2010) Split-Ubiquitin based membrane yeast two-hybrid (MYTH) system: a powerful tool for identifying protein-protein interactions. *J Vis Exp*. **1** (36), 1698.
5. Ozawa, T. (2009) Protein reconstitution methods for visualizing biomolecular function in living cells. *Yakugaku Zasshi*. **129** (3), 289–95.
6. Johnsson, N. and Varshavsky, A. (1994) Split ubiquitin as a sensor of protein interactions in vivo. *Proc. Natl. Acad. Sci USA*. **91** (22), 10340–4.
7. Stagljar, I., Korostensky, C., Johnsson, N. and te Heesen, S. (1998) A genetic system based on split-ubiquitin for the analysis of interactions

between membrane proteins in vivo. *Proc. Natl. Acad. Sci USA.* **95** (9), 5187–92.

8. Remy, I., Campbell-Valois, F. X. and Michnick, S. W. (2007) Detection of protein-protein interactions using a simple survival protein-fragment complementation assay based on the enzyme dihydrofolate reductase. *Nature Publishing Group.* **2** (9), 2120–2125.
9. Jackrel, M.E., Cortajarena, A.L., Liu, T.Y. and Regan, L. (2009) Screening libraries to identify proteins with desired binding activities using a split-GFP reassembly assay. *ACS Chem. Biol.* **Dec 28.**
10. Misawa, N., Kafi, A.K., Hattori, M., Miura, K., Masuda, K. and Ozawa T. (2010) Rapid and high-sensitivity cell-based assays of protein-protein interactions using split click beetle Luciferase complementation: an approach to the study of G-Protein-coupled receptors. *Anal.Chem.* **82** (6), 2552–60.
11. Remy, I., Ghaddar, G. and Michnick, S.W. (2007) Using the beta-lactamase protein-fragment complementation assay to probe dynamic protein-protein interactions. *Nat Protoc.* **2** (9), 2302–6.
12. O'Hare, H., Juillerat, A., Dianisková, P. and Johnsson, K. (2008) A split-protein sensor for studying protein-protein interaction in mycobacteria. *Journal of Microbiological Methods.* **73** (2), 79–84.
13. Tafelmeyer, P., Johnsson, N. and Johnsson, K. (2004) Transforming a ((beta)/(alpha))8-barrel enzyme into a split-protein sensor through directed evolution. *Chemistry & Biology.* **11** (5), 681–689.

Chapter 15

Tetracycline Repressor-Based Mammalian Two-Hybrid Systems

Kathryn Moncivais and Zhiwen Jonathan Zhang

Abstract

The study of protein–protein interactions is critical for the understanding and regulation of biological systems. To that end, yeast two-hybrid systems have been used to study protein–protein interactions in vivo, but they frequently suffer from a high incidence of false positives when applied to mammalian systems. A novel mammalian two-hybrid system has recently been developed which exhibits lower background and higher sensitivity than earlier mammalian two-hybrid systems. It has successfully detected interactions with dissociation constants ranging from 0.99 nM to 55 μM. The system was built upon the tetracycline repressor–tetracycline operator interaction and is suitable for use in the study of most, if not all, mammalian protein–protein interactions.

Key words: Two-hybrid, Tetracycline repressor, Protein–protein interactions

1. Introduction

Protein–protein interactions mediate most cellular functions at some point (1). It is therefore imperative that biomedical scientists come to an understanding of in vivo protein–protein interactions in order to manipulate them in the lab and/or correct those which have been altered by disease or genetic disorder. Without this understanding, scientists can have no hope of developing effective treatments for a wide range of medical disorders. While many in vitro techniques exist for the study of protein–protein interactions, they share the common problem of not mimicking the environment in a living cell or human being. The most logical remedy for this problem is to study protein–protein interactions within a living organism or cell, but there are few technologies capable of accomplishing this. Those that do accomplish this are frequently plagued by high background and a lack of sensitivity.

Bernhard Suter and Erich E. Wanker (eds.), *Two Hybrid Technologies: Methods and Protocols*, Methods in Molecular Biology, vol. 812, DOI 10.1007/978-1-61779-455-1_15,

The first system to successfully study protein–protein interactions in vivo was yeast two-hybrid system. Two-hybrid systems convert bait–prey interactions to "on" signals for a chosen reporter protein. A DNA binding domain is fused to a bait protein, and a prey protein is fused to a transcriptional activator. If the bait and prey interact, the DNA binding domain's interaction with the upstream activator sequence will bring the transcriptional activator into proximity with the reporter protein gene, and the reporter protein will be expressed. While these systems are relatively simple to apply in a yeast system, they do not necessarily provide accurate information regarding a mammalian system (as they are employed in yeast). For this reason, mammalian two-hybrid systems were developed. Mammalian two-hybrid systems utilize the GAL4 transcriptional activator, making them basically a transfer of the yeast two-hybrid system into a mammalian environment (2). This system works the best for the detection of strong protein–protein interactions in vivo.

The Zhang lab has developed a novel tetracycline-repressor based mammalian two-hybrid system (trM2H) capable of studying protein–protein interactions in vivo, in mammalian cells, exhibiting low background, and detecting interactions with high sensitivity (3). The trM2H system, shown in Fig. 1, is based upon the tetracycline repressor (TetR) to tetracycline operator (TetO) interaction, which is a tightly regulated interaction used in mammalian protein expression systems to achieve highly sensitive regulation of gene expression. By harnessing the tight regulation of this TetR–TetO interaction and employing it in the trM2H system, both

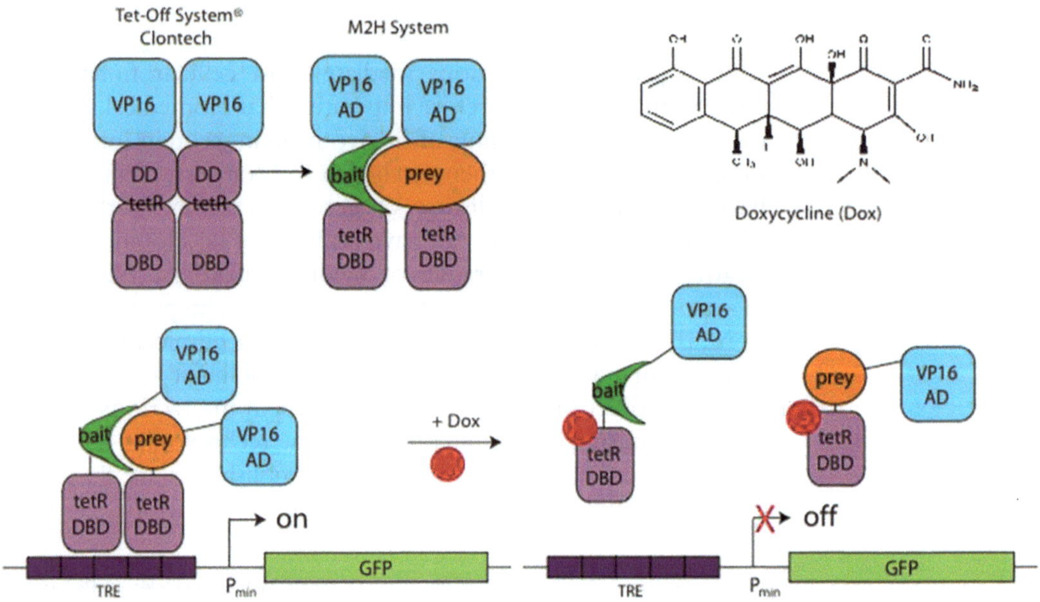

Fig. 1. Schematic of trM2H system.

homo-dimerization and hetero-dimerization of proteins with dissociation constants ranging from 0.99 nM to 55 μM were able to be detected. Other two-hybrid systems have not yet detected interactions as weak as the trM2H system, nor have they achieved background signal as minimal as the trM2H system (4). Researchers can apply the trM2H system to the study of a wide range of protein–protein interactions with the advantage of producing reliable in vivo information in a relatively short period of time.

From the construction of bait–prey plasmids to final fluorescence microscopy determination of interaction, the entire process can be completed in less than 2 weeks. The following protocols provide step-by-step instructions for each facet of the trM2H system and will enable researchers of all levels of experience to successfully use the trM2H system for the study of the protein–protein interactions of their choice.

While the trM2H system has proven useful for the study of individual protein–protein interactions, this technology could be used for several potentially high impact applications. Since the trM2H system is highly sensitive it could be used to study more sensitive protein–protein interactions like those mediated through posttranslational modifications. Because the trM2H system has a built-in negative selection tool via the dox disruption assay, it could be an ideal platform for cell-based high-throughput screening of therapeutic small molecule/peptide protein–protein interaction disruptors. The trM2H could also be used to create a standard curve for protein–protein interaction characterization. Interactions with known dissociation constants could be tested with the trM2H and subsequently quantified using FACS. A standard curve could be created from that data, relating dissociation constant to fluorescence intensity. In this way, the trM2H system could be used to not only confirm/deny interactions, but to characterize the strength of interaction as well.

2. Materials

2.1. Cloning and Construction of Two-Hybrid Bait–Prey Plasmids

While some researchers may find that simple cut/paste cloning is sufficient for their chosen bait and prey, others may have to use polymerase chain reaction (PCR) to create their gene with compatible restriction sites for the empty trM2H bait–prey vector, called the trM2H base plasmid. Supplies required for only PCR-based cloning will be denoted by "PCR" and are not necessary for simple cut/paste cloning.

1. Plasmids containing the desired bait and prey genes (in an appropriate expression cassette).
2. trM2H base plasmid (Fig. 2).

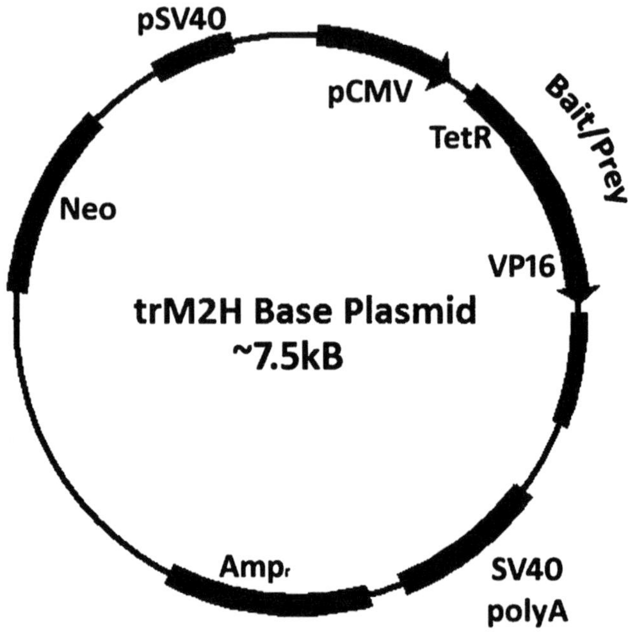

Fig. 2. The trM2H Base Plasmid confers ampicillin resistance to bacterial hosts and neomycin resistance to mammalian hosts; the bait–prey segment is flanked by the restriction sites AgeI and KpnI on the N and C termini, respectively.

3. Appropriate restriction enzymes and buffers: Age1, KpnI, NEB Buffer 3, and bovine serum albumin (BSA) for the trM2H base plasmid.
4. Agarose gel and DNA loading buffer.
5. Qiagen or similar gel extraction kit.
6. Ligase and compatible buffer.
7. Competent *Escherichia coli* cells.
8. Luria–Bertani (LB) medium.
9. LB agar plates with appropriate selective antibiotic(s).
10. Qiagen or similar plasmid purification kit of desired scale.
11. Forward and reverse sequencing primers.
12. Forward and reverse primers introducing the desired restriction sites (PCR).
13. dNTPs (PCR).
14. Thermostable DNA polymerase (PCR).

2.2. Cell Culture and Transfection

1. Tissue culture treated T-75 flasks and desired well plates for transfection (24-well plates are recommended).
2. HEK293 cells (Invitrogen).
3. Complete HEK medium: Combine and sterile filter Dulbecco's modified Eagle's medium (DMEM) with 10% (v/v) fetal bovine

serum (FBS), 1% (v/v) MEM nonessential amino acid solution, and 1% (v/v) L-glutamine.

4. 0.25% Trypsin–EDTA (Invitrogen).
5. Fugene 6 (Roche), FugeneHD (Roche), or similar transfection agent.
6. Opti-MEM (Gibco/BRL) or equivalent reduced serum medium.
7. pTRETight AcGFP plasmid (Invitrogen) or equivalent fluorescent reporter plasmid with tetracycline response element.
8. trM2H bait–prey plasmids.

2.3. Doxycycline Disruption Assay

Doxycycline (dox) (MP Biomedicals).

2.4. Fluorescent Imaging

1. Nikon Eclipse TE2000-S microscope with a FITC HyQ filter (Chroma, Rockingham, VT) or equivalent.
2. Image capture software with manual capture mode.

2.5. FACS Sorting

4% Paraformaldehyde in PBS (Sigma–Aldrich).

2.6. Cell Lysis and SDS-Polyacrylamide Gel Electrophoresis

1. 5× Passive lysis buffer (Promega).
2. Complete Mini Protease inhibitor tablet (Roche).
3. 6× SDS-polyacrylamide gel electrophoresis (SDS-PAGE) loading buffer: Combine 7 mL 4× Tris–HCl/SDS, pH 6.8, 3.8 g (3 mL) glycerol (30% final), 1 g SDS (10% final) (weigh in hood), 0.93 g DTT (0.6 M) or 5% βME (0.5 mL), and 1.2 mg bromophenol blue (0.012%), add H_2O to 10 mL. Store in 0.5 mL aliquots at −20°C.
4. Running buffer: 25 mM Tris–HCl, 200 mM glycine, and 0.1% SDS (w/v). Store at 4°C.
5. SDS-PAGE precast gel(s) (Invitrogen).
6. Prestained protein molecular weight markers.

2.7. Western Blotting for Reporter Protein Expression

1. Transfer Buffer: 48 mM Tris, 39 mM glycine, 20% methanol v/v, 0.037% SDS (w/v), pH 8.3.
2. Nitrocellulose membrane (Millipore).
3. Chromatography paper (Whatman).
4. Blocking buffer: 4% w/v dry milk in 10 mM Tris–HCL pH 8.0, 150 mM NaCl, 0.1% v/v Tween20.
5. Anti-GFP Tag antibody (Applied Biological Materials) or other primary antibody against your chosen reporter protein.
6. Rabbit anti-goat IgG (H+L)-AP conjugated secondary antibody (Santa Cruz) or other secondary antibody against your chosen primary antibody.

7. PhosphaGlo AP (KPL, Gaithersburg, MD) or other developing solution.
8. Kodak BioMax Light Film or other film responsive to chosen developing solutions/secondary antibody.

3. Methods

3.1. Cloning and Construction of Two-Hybrid Bait–Prey Plasmids

Bait–Prey inserts can be prepared either by restriction enzyme digest to remove from the original vector with restriction sites compatible with the trM2H base plasmid, or they can be prepared by PCR to introduce compatible restriction sites to the gene. The trM2H base vector utilizes the restriction enzymes AgeI and KpnI, with AgeI located on the N terminus of the bait–prey and KpnI on the C terminus. In the protocol, AgeI and KpnI are referred to as RE1 and RE2, respectively. The protocols have been generalized to RE1 and RE2 so that the reader may apply them to other mammalian two-hybrid systems or a second-generation trM2H, which would not necessarily use AgeI and KpnI. A single extra or missing base in the insert region will cause a frame-shift in the insert as well as the C terminus of the trM2H base vector, rendering the protein nonfunctional. Therefore, it is imperative that each new plasmid be sequenced to confirm the correct amino acid sequence of the intended bait and prey before they are used for transfection. Some inserts may prove unusually challenging to insert into the trM2H base plasmid. In such cases, the reader is advised to consult one of the other books in this series such as Methods in Molecular Biology: PCR Cloning Protocols.

3.1.1. Plasmid Purification

Commercially available kits should be used according to the manufacturer's instructions. Qiagen kits can be used to purify bacterial cultures of a wide range of volumes and purify enough DNA for multiple transfection or cloning experiments at once. Table 1 shows

Table 1
Recommended bacterial plasmid propagation culture volumes for mammalian transfections of various sizes

Well size	Recommended culture size for transfection of single well (mL)	DNA yield (μg)
T-75 flask	50	≤200
6-Well plate	10	≤40
12-Well plate	5	≤20
24-Well plate	2	≤16
96-Well plate	2	≤16

the recommended DNA amounts and culture sizes for experiments of various sizes. After determining the amount of DNA necessary for a particular experiment, choose the corresponding commercial kit and follow the manufacturer's instructions.

3.1.2. Preparing Bait–Prey Insert by Restriction Enzyme Cutting

1. Combine 1 μg insert carrying vector with 1 μL RE1, 1 μL RE2, 5 μL 10× buffer, 5 μL 10× BSA (optional), and sterile water up to 50 μL.
2. Incubate at 37°C for 1 h.
3. Add 6 μL 10× DNA loading buffer and proceed to run an agarose gel when the trM2H base plasmid has been digested as well (see Subheading 3.1.4).

3.1.3. Preparing Bait–Prey Insert by PCR

1. Set up and perform PCR reactions as suggested by the polymerase manufacturer.
2. According to the manufacturer's instructions, set up a double digest of the PCR product and incubate overnight. Check manufacturer's temperature suggestion for overnight incubations to ensure longevity of the enzyme.
3. Add 6 μL of 10× DNA loading buffer and proceed to run an agarose gel when the trM2H base plasmid has been digested as well (see Subheading 3.1.4).

3.1.4. Prepare the trM2H Backbone by Restriction Digest

1. Combine 1 μg insert carrying vector with 1 μL RE1, 1 μL RE2, 5 μL 10× buffer, 5 μL 10× BSA (optional), and sterile water up to 50 μL.
2. Incubate at 37°C for 1 h.
3. Directly after incubation for 1 h, use an alkaline phosphatase, according to manufacturer's directions, to dephosphorylate the sticky ends of the vector. This will prevent backbone ligations later in the protocol. This usually takes an additional hour.
4. After the phosphatase incubation is completed, add 6 μL of 10× DNA loading dye.

3.1.5. Prepare and Run the Agarose Gel

Steps 1 and 2 should be completed while the backbone is digesting.

1. Pour a 1% agarose gel with ethidium bromide (EtBr) in an appropriate-sized casting mold (reference).
2. Use a comb with wells large enough to accommodate the entire 50 μL of each digest + 6 μL of loading dye.
3. Once the backbone and insert digests are completed, load the backbone digest into one well and the insert digest into another.
4. Load the recommended amount of 1 Kb DNA ladder into another well.

5. Based upon the size of the gel, run the samples at 10 V/cm gel for approximately 20–30 min, monitoring the progress of the loading dye as it runs.
6. Stop the gel when the fastest moving band of the loading dye comes within 1 cm of the end of the gel. See Note 1 if your insert is smaller than 500 bp.

3.1.6. Cut Out and Purify the Appropriate Bands

1. Using an ultraviolet gel imager, locate the appropriate bands in each of the lanes. The trM2H base vector digest band of interest is the band with slow mobility (i.e., the band corresponding to a larger size), while the insert carrying vector band of interest has fast mobility (i.e., the band corresponding to a smaller size).
2. Using a clean razor blade, remove each of the bands of interest from the gel and place them in preweighted microcentrifuge tubes. Be sure to label them properly.
3. Use a commercial gel extraction kit to extract DNA from the gel slices. Qiagen gel extraction kits are easy and routinely achieve high yield.

3.1.7. Ligate and Trasform into Competent E. coli Cells

1. Prepare ligation reactions according to the instructions provided for your particular brand of ligase. Each commercially available ligase comes with a very detailed sample protocol which will yield the best results for that particular ligase.
2. As soon as the ligation is completed, place the reaction on ice with two vials of chemically competent *E. coli* cells. When the cells have thawed, add 2–10 μL of ligation reaction to one vial of the TOP10 cells and label it.
3. Place both vials of cells in a 42°C water bath for 45 s.
4. Remove the cells from the water bath and place them on ice for 2–3 min.
5. Add 1 mL LB medium to each vial of cells, then place in a 37°C shaker for 45 min–1 h.
6. While the transformations are shaking, place the desired number of LB + antibiotic plates in a 37°C incubator to warm.
7. After the 45 min to 1 h incubation, plate various amounts of each transformation on the prewarmed plates.
8. Incubate plates at 37°C for 10–14 h, then remove from incubator and store at 4°C until ready to use.

3.1.8. Screen Colonies for Correct Sequence

1. Pick 2–5 colonies from the ligated plasmid plate and use to inoculate small volumes of LB medium containing the appropriate selective antibiotic.
2. Grow cultures 12–14 h at 37°C with shaking.

3. Before spinning down cultures, prepare glycerol stocks of each colony by combining equal volumes of culture and sterile 50% glycerol. These can be stored at −80°C and used to propagate the correct plasmid after sequencing.
4. Follow manufacturer's instructions to purify DNA from cultures.
5. Using the forward and reverse sequencing primers, sequence the plasmids from each colony to determine which, if any, has the correct sequence.

3.1.9. Propagate Correct Plasmid

1. Using the glycerol stock carrying the plasmid(s) with the correct sequence, inoculate and grow-up a culture large enough to provide DNA for several transfection experiments (see Table 1).
2. Purify DNA using one of the previously suggested kits.

3.2. Cell Culture and Transfection

1. Unless otherwise stated, cells should be maintained in complete HEK medium in a humidified incubator at 37°C with 5% CO_2.
2. HEK293 cells should be washed once with PBS and passaged with trypsin when approaching 80–90% confluence.
3. Cells should be passaged and seeded into 6-, 12-, 24-, or 96-well plates for transfection and transfected at 60–80% confluency.
4. Table 2 shows the conditions required for a single bait–prey heterodimer experiment (see Note 2). Although many researchers

Table 2
Conditions for a single bait–prey interaction experiment

	Negative control	Transfection negative control	Transfection positive control	Bait homodimer control	Prey homodimer control	Bait–prey interaction determina-tion
Tet-responsive reporter protein plasmid		✓	✓	✓	✓	✓
trM2Hbait plasmid				✓		✓
tlM2H prey plasm id					✓	✓
pTet-Off plasmid			✓			

The top row names the experimental condition; the far left column indicates the plasmid corresponding to each row. A check mark under a condition column indicates that the plasmid in that row should be included in the transfection for that condition

would instinctively include the "Negative Control" column in Table 2, it is included explicitly in this table for the benefit of the mammalian transfection novice. This negative control (untransfected cells) must be included so that auto-fluorescence of mammalian cells can be distinguished from background expression of acGFP (which can be seen in the "Transfection Negative Control" condition).

5. Transfection should be completed according to the transfection agent manufacturer's instructions. It may be necessary to optimize transfection conditions for the best results (see Note 3).
6. After transfection, HEK medium should be replaced at least every 24 h; cells can be assayed for reporter gene expression 48–72 h posttransfection.

3.3. Tetracycline Disruption Assay

Genetic assays are always subject to artifacts. This is especially true in mammalian cells. One way to eliminate the false positives is to have a negative control, in this case a disruption assay. This type of negative selection is especially useful if one wants to use trM2H for a high throughput screening.

1. 72 h posttransfection replace complete HEK medium with complete HEK medium + 1 mM doxycycline.
2. Complete HEK medium with 1 mM doxycycline should be replaced every 24 h, and cells can be assayed (with fluorescence microscopy or FACS) 24–48 h after the addition of doxycycline. Figure 3 shows an example result of this assay.

3.4. Fluorescent Imaging of Transfected Cells

1. Using a standard fluorescence microscope and image capture software, take images of each condition using the same exposure, gain, and excitation light.
2. It is imperative that researchers take representative images of each condition. Some weak binding pairs may produce very few fluorescent cells, but if those cells happen to be in the same

Fig. 3. Doxycycline disruption assay. Visualization of GFP expressed in HEK293H cells by transfecting with a homodimerizing trM2H-bait plasmid and pTRETightAcGFP. Pictures were taken 72 h posttransfection. (**a**) Doxycycline absent. (**b**) 1 mM doxycycline was added 24 h posttransfection. (**c**) 1 mM doxycycline was added 48 h posttransfection.

viewing window, the well as a whole can be easily misrepresented by unscrupulous image capture. Always capture an image that is representative of the well as a whole and not the best or only cluster of fluorescence.

3. Figure 4 shows example pictures with their respective experimental conditions.

3.5. FACS (Optional)

1. After 48–72 h, follow the passaging instructions in Subheading 3.2 but stop before resuspending cells in complete HEK medium.
2. Resuspend the cells in sterile PBS.
3. Centrifuge cells again at 2,000 rpm for 5 min.
4. If using unfixed cells, consult the manufacturer's instructions for the FACS machine you will be using to find the optimal cell density.
5. If using fixed cells, resuspend cells in an appropriate volume of 4% paraformaldehyde in PBS.
6. Incubate cells at least 15 min at room temperature.
7. Cells can now be stored up to 2 weeks at 4°C until ready to be used.
8. Consult manufacturer's instructions to find optimal cell density for FACS sorting.
9. Use a hemocytometer or other counting method to determine the cell density of the cells in 4% paraformaldehyde in PBS.
10. Use 4% paraformaldehyde in PBS to dilute cells to the density found in step 8.
11. Follow manufacturer's instructions for running samples on a FACS machine.

3.6. Cell Lysis and SDS-PAGE

These instructions assume the use of an Invitrogen XCell SureLock® electrophoresis system, but can be easily adapted to other systems.

1. After 48–72 h, follow the passaging instructions in Subheading 3.2 but stop before resuspending cells in complete HEK medium.
2. Resuspend the cells in 2.5× passive lysis buffer (stock comes in 5×, so it must be diluted 1:1 with sterile water before resuspending cells) supplemented with Complete Mini protease inhibitor tablet(s).
3. Incubate cells on ice 30 min–1 h.
4. While cells are lysing, set up the PAGE electrophoresis unit: open the commercial gel, mark the lanes with a permanent marker for easier sample loading, remove the comb, and rinse the wells with running buffer. Lock the gel in place in the XCell SureLock® unit, and fill both sides of the unit with running buffer.

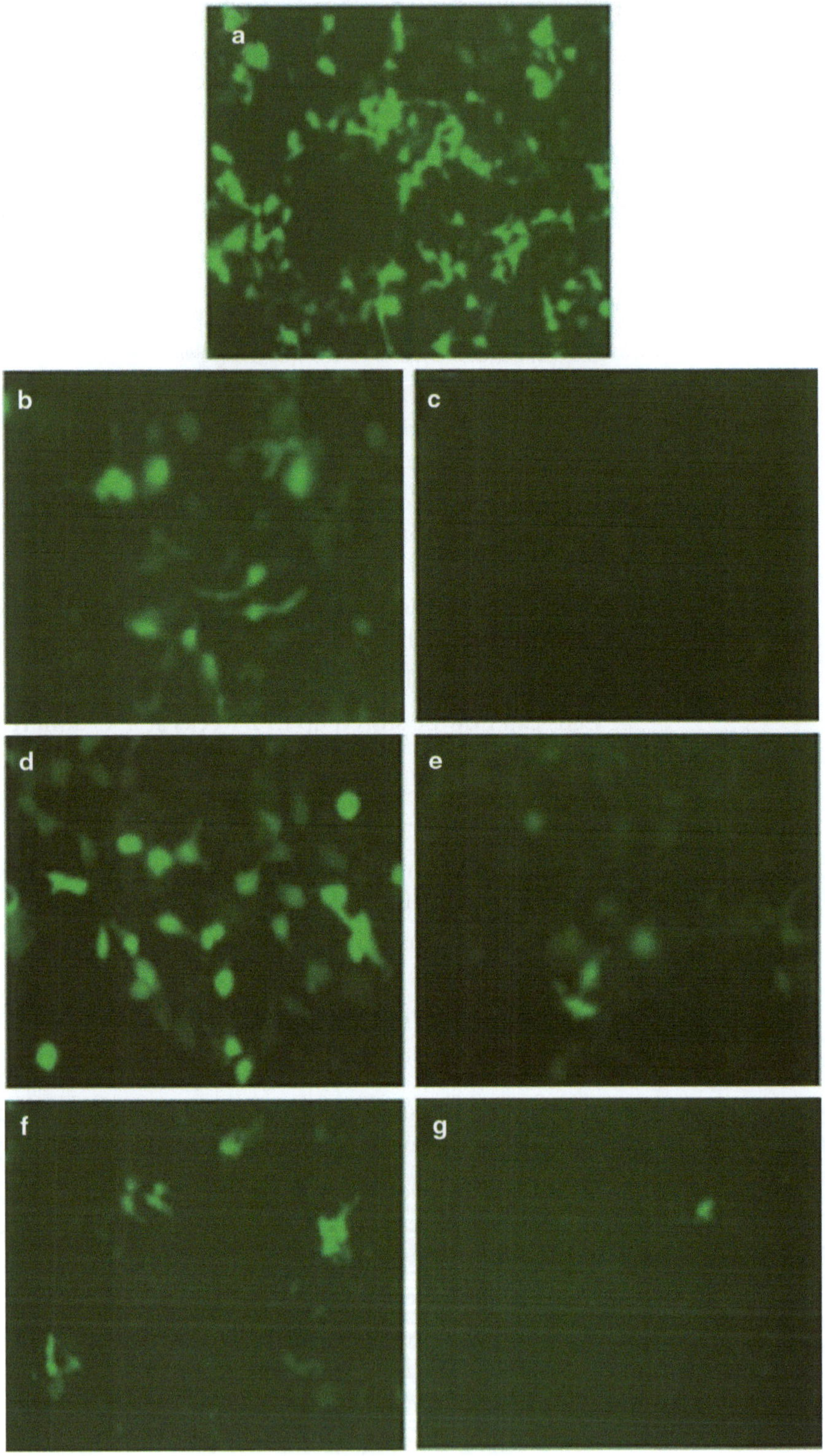

Fig. 4. Fluorescent images of GFP expression in HEK293H cells transfected with (**a**) pTRETightAcGFP (Tet-responsive reporter plasmid) + pTetOff (*transfection positive control*). (**b**) pTRETightAcGFP + trM2H-GCN4LZ (homodimerizing bait protein). (**c**) pTRETightAcGFP + trM2H-GCN4LZMU (nondimerizing bait protein). (**d**) pTRETightAcGFP + trM2H-Fos + trM2H-Jun (*interacting bait and prey proteins*). (**e**) pTRETightAcGFP + trM2H-Fosmu + trM2H-Jun (*noninteracting bait and prey proteins; GFP expression is activated by weak dimerization of trM2H-Jun protein*). (**f**) pTRETightAcGFP + trM2H-SrtA (*homodimerizing bait protein*). (**g**) pTRETightAcGFP + trM2H-SrtA (*nondimerizing bait protein*).

5. Determine the protein concentration in each cell sample using a NanoDrop 2000 (Thermo Scientific, Wilmington, DE) and use PBS to bring all cell samples to the same protein concentration.
6. Add loading buffer to all cell samples and boil for 10 min.
7. Load samples and prestained molecular weight markers into gel (consult gel manufacturer's instructions for maximum loading volume per well).
8. Run the gel at 100–150 V until the desired resolution of bands is achieved.
9. Turn off the power supply and disconnect the electrophoresis system from it.

3.7. Western Blotting for Reporter Protein Expression

This protocol assumes the use of Trans-Blot SD® semidry transfer cell but can be adapted for other units.

1. Carefully remove the gel from its casting plates and cut off the lanes/stacking portion of the gel.
2. Wash the gel in deionized water.
3. Incubate the gel in transfer buffer for 5–10 min.
4. While the gel is incubating, wet the membrane and six layers of filter paper in transfer buffer.
5. Set up the transfer sandwich: place three layers of filter paper on the blotter followed by the nitrocellulose membrane. Place the protein gel on top of the nitrocellulose membrane followed by the remaining three layers of filter paper. Use a roller to remove any air bubbles.
6. Assemble the unit and transfer at 15 V for 10 min or 10 V for 15–30 min. You may need to increase this time depending on the chosen protein, transfer buffer, etc. (see Note 4).
7. After turning off the power supply, disassemble the unit and remove the nitroceullulose membrane to a clean dish filled with blocking buffer.
8. Incubate the membrane with shaking for 1 h at room temperature, or 4°C overnight.
9. Pour off the blocking buffer, briefly rinse the membrane with TBST (see Note **5**), and add primary antibody in the dilution recommended by the manufacturer (2 μL in 10 mL blocking buffer for anti-GFP Tag antibody) (see Note 6). Incubate the membrane with shaking for 1 h at room temperature or 4°C overnight.
10. Wash the membrane three times for 10 min in blocking buffer without antibody.
11. Incubate the membrane in blocking buffer with secondary antibody at the manufacturer's recommended dilution (2 μL

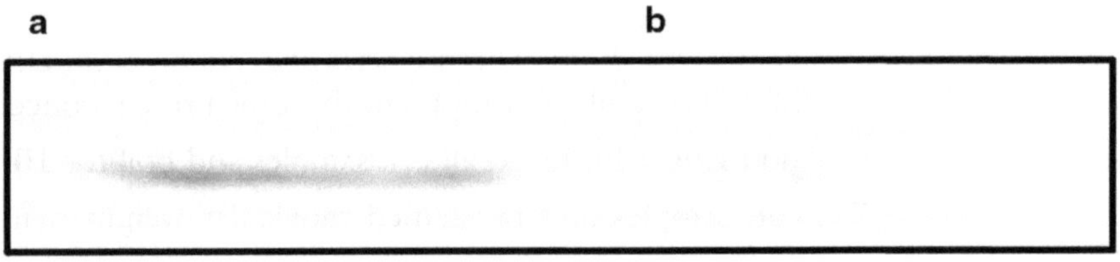

Fig. 5. Example Western blot of (**a**) a homodimerizing trM2H-bait protein and (**b**) a nondimerizing trM2H-bait protein.

in 10 mL blocking buffer for rabbit anti-goat IgG (H+L)-AP antibody) for 1 h at room temperature or 4°C overnight.

12. Wash the membrane three times for 10 min in blocking buffer without antibody.
13. Briefly rinse the membrane with TBST.
14. Follow the developing reagent manufacturer's instructions for developing the membrane. A sample Western blot is shown in Fig. 5.

4. Notes

1. For inserts smaller than 500 bp, the gel should be stopped earlier than suggested, and they may require more plasmid to be digested in order to show up on the gel.
2. For heterodimer experiments, it is imperative that controls are used for each individual bait–prey plasmid. If either bait or prey molecule forms a tight homodimer, the trM2H system may not be the optimal method of assessing interaction of the heterodimer between the bait and prey.
3. Each transfection reagent will come with detailed instructions for optimization of the transfection procedure. We have found that optimal conditions vary widely between plasmids as well as cell lines. Therefore, when researchers encounter low or no expression of reporter protein after transfection, it is strongly suggested that the optimization protocol for the individual transfection reagent be employed. In many cases, this will solve the problem.
4. Incomplete protein transfer is frequently the culprit in Western blot failure. If a blot returns negative results, the gel should be Coomassie blue stained to ascertain the amount of protein

retained in the gel. If this reveals that the transfer was incomplete, the transfer times and voltages should be adjusted until a successful transfer is achieved before proceeding to the rest of the Western blot protocol.

5. When washing or applying antibody during Western blotting, the buffers should never be applied directly to the membrane as this may dislodge proteins. Buffers should always be added to an area of the dish where the membrane is not located so that they can slowly flow onto the membrane without disturbing the proteins on it.
6. Antibody dilutions may need to be optimized, depending on the quality of the antibody and the choice of developing reagent. If membrane transfer was successful and the blot still does not develop well, the primary and secondary antibody dilutions should be optimized, as the manufacturer's instructions are not always optimized for this application.

Acknowledgments

The authors would like to thank Gabrielle Thibodeaux and Roshani Cowmeadow for initial development of the trM2H system, and Aiko Umeda, Jie Zhu, Liang Xiang, YungAh Lee, Wonjae Lee, Jay Xiao, Kristi Kim, Young Park, and Sarah Chun for technical assistance.

References

1. Phizicky, E.M. and S. Fields (1995) Protein-protein interactions: methods for detection and analysis *Microbiol Rev.* 59, 94–123
2. Luo, Y., et al. (1997) Mammalian two-hybrid system: A complementary approach to the yeast two-hybrid system *Biotechniques* 22, 350–352
3. Thibodeaux, G.N., et al. (2009) A tetracycline repressor-based mammalian two hybrid system to detect protein-protein interactions in vivo *Anal Biochem.* 386, 129–31
4. Huang H., J.B.M., Bader J.S. (2007) Where have all the interactions gone? Estimating the coverage of two-hybrid protein interaction maps *PLoS Computational Biology* 3, 2155–2174

Chapter 16

The Fluorescent Two-Hybrid (F2H) Assay for Direct Analysis of Protein–Protein Interactions in Living Cells

Kourosh Zolghadr, Ulrich Rothbauer, and Heinrich Leonhardt

Abstract

Information about protein interactions is crucial for the understanding of cellular processes. Current methods for the investigation of protein–protein interactions (PPIs) require either removal of the proteins from their normal cellular environment, perturbation of the cells or costly instrumentation and advanced technical expertise (Fields and Song, Nature 340:245–246, 1989; Deane et al., Mol Cell Proteomics 1:349–356, 2002; Kerppola, Nat Rev Mol Cell Biol 7:449–456, 2006; Blanchard et al., Mol Cell Proteomics 5:2175–2184, 2006; Miller et al., Mol Cell Proteomics 6:1027–1038, 2007; Miyawaki, Dev Cell 4:295–305, 2003; Parrish et al., Curr Opin Biotechnol 17:387–393, 2006; Sekar and Periasamy, J Cell Biol 160:629–633, 2003). Here, we describe a simple assay to directly visualize and analyze PPIs in single living cells. By adapting a *lac* operator/repressor system, we generated a stable nuclear interaction platform. A fluorescent bait protein is tethered to the interaction platform and assayed for co-localization of fluorescent prey fusion proteins. This fluorescent two-hybrid (F2H) assay allows the investigation of cell cycle dependent PPIs. With this cell based assay protein interactions even from different subcellular compartments can be visualized in real time (Zolghadr et al., Mol Cell Proteomics 7:2279–2287, 2008). The simple optical readout enables automated imaging systems to segment and analyze the acquired data for high-throughput screening of PPIs in living cells in response to external stimuli and chemical compounds.

Key words: Protein–protein interaction, Cell based assay, Fluorescent protein, High-throughput screen, Live cell studies, End-point assay, Compound screen

1. Introduction

The analysis of interacting networks of genes, proteins and metabolites is essential to understand the difference between normal and abnormal cell functions. In this context, protein–protein interactions (PPIs) are of major interest in various fields of life sciences ranging from basic research to pharmaceutical compound screening. A multitude of different cell based approaches

Bernhard Suter and Erich E. Wanker (eds.), *Two Hybrid Technologies: Methods and Protocols*, Methods in Molecular Biology, vol. 812, DOI 10.1007/978-1-61779-455-1_16, © Springer Science+Business Media, LLC 2012

to screen and analyze PPIs has been described (1–8). Fluorescent fusion proteins became excellent biosensors of cellular functions and have been used to analyze the dynamics of cellular processes now for many years. Here, we present our novel fluorescent two-hybrid (F2H) assay for the direct visualization and analysis of PPIs in the native environment of living mammalian cells (9–11). The rationale for the F2H assay is based on the observation that proteins are freely roaming the cell unless interactions with certain cellular components transiently immobilize and enrich them at specific structures (12). We take advantage of previously described cell lines, which harbor a stable integration of about 200–1,000 copies of a plasmid each carrying 256 copies of the *lac* operator sequence, such as BHK clone #2 and U2OS clone 2-6-3 (13, 14). These cells are co-transfected with two expression constructs encoding the fluorescent fusion proteins of interest.

The first expression construct encodes a fluorescent bait protein consisting of a red fluorescent protein (mCherry, mCh), the lac repressor (LacI), and the protein X to be tested for interactions (bait) resulting in the triple fusion protein *mCh-LacI-X* (Fig. 1a). This fluorescent fusion protein binds to the chromosomal *lac* operator array generating a fluorescence based protein–protein interaction biosensor (PPIB). The interaction platform becomes visible due to the focal enrichment of the red fluorescent protein signal as a bright red spot in the nucleus of the cell.

The second expression construct encodes the green fluorescent protein (GFP) fused to the potential interaction partner protein Y (GFP-Y, prey). The prey protein Y may either interact with the bait protein X leading to co-localization of the green and red fluorescent signals (Fig. 1b) or may not interact, resulting in a dispersed/ diffuse distribution of the prey protein fluorescence within the cell (Fig. 1c) (9).

This simple read-out of the F2H assay based on the optical co-localization of two distinguishable fluorophores (Fig. 2) enables the analysis of mutation and deletion constructs to map PPI domains. Likewise, the effect of external biochemical or physical stimuli on PPI can be studied in the cellular environment in high-throughput without the requirement of specialized technical expertise. In combination with automated image acquisition and analysis, the F2H assay opens new possibilities to screen for molecules and drugs inducing or disrupting PPI in living cells and in real time.

2. Materials

2.1. Molecular Cloning

1. F2H bait and prey expression vectors pF2H-bait and pF2H-prey (Fig. 1a).
2. pF2H_forward and pF2H_reverse primers.

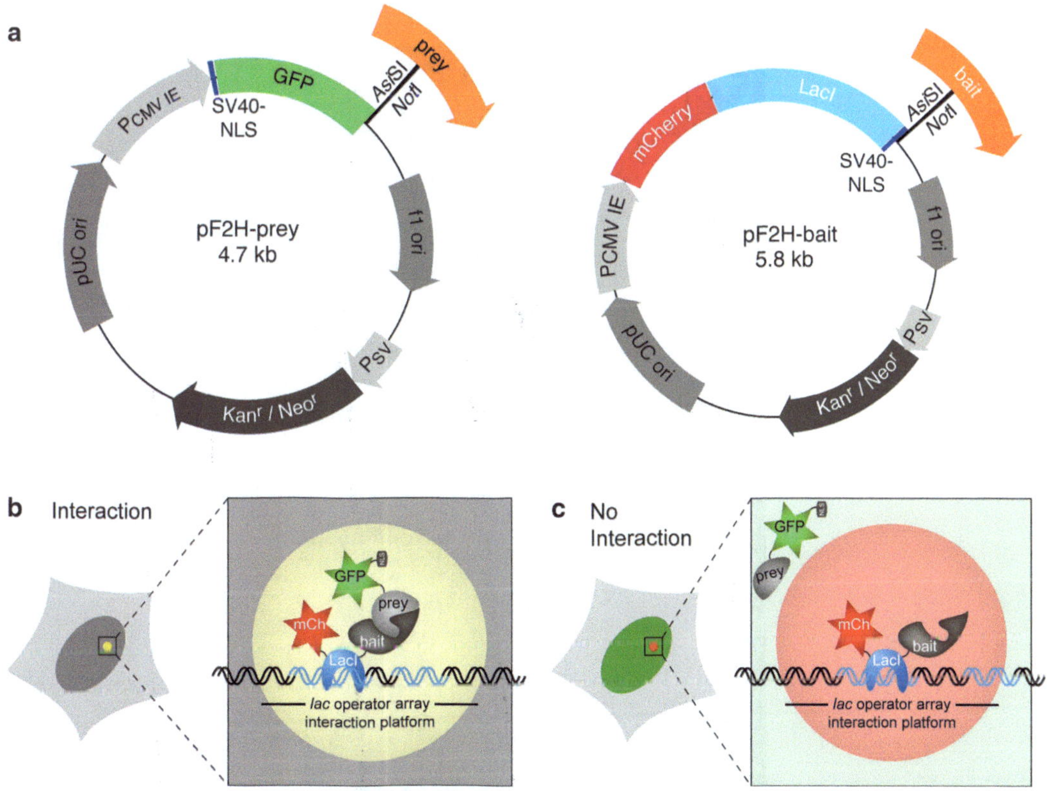

Fig. 1. Schematic outline of the fluorescent two-hybrid (F2H) assay. (**a**) pF2H-prey and pF2H-bait expression vectors coding for GFP labeled prey proteins and RFP labeled bait proteins used for the F2H assay (**b**) The LacI domain of the bait protein mediates binding to the interaction platform, the chromosomally integrated *lac* operator array, which is visible as a *red fluorescent spot* in nuclei of transfected cells. In case of interaction with the bait protein, the prey becomes enriched at the same spot resulting in co-localization of *red* and *green* fluorescent signals. (**c**) If the prey does not interact with the bait protein, it remains dispersed in the nucleus and the *lac* operator array is only visible through the bound bait protein (*red spot*).

3. Restriction enzymes *AsiSI* and *NotI* and respective buffers (New England Biolabs, NEB).
4. For all other methods and reagents for PCR amplification, restriction enzyme digestion, ligation, transformation, and DNA preparation please refer to classical molecular cloning protocols.

2.2. Cell Culture and Transfection

1. Transgenic BHK cells (clone #2) and U2OS cells (clone 2-6-3) containing *lac* operator repeats. Alternatively, any other cell line containing *lac* operator arrays can be used (see Note 1).
2. Dulbecco's modified Eagle's medium (DMEM) (all cell culture solutions from PAA Laboratories, Germany) supplemented with 10% fetal bovine serum (FBS Gold) and 150 μg/ml hygromycin B.

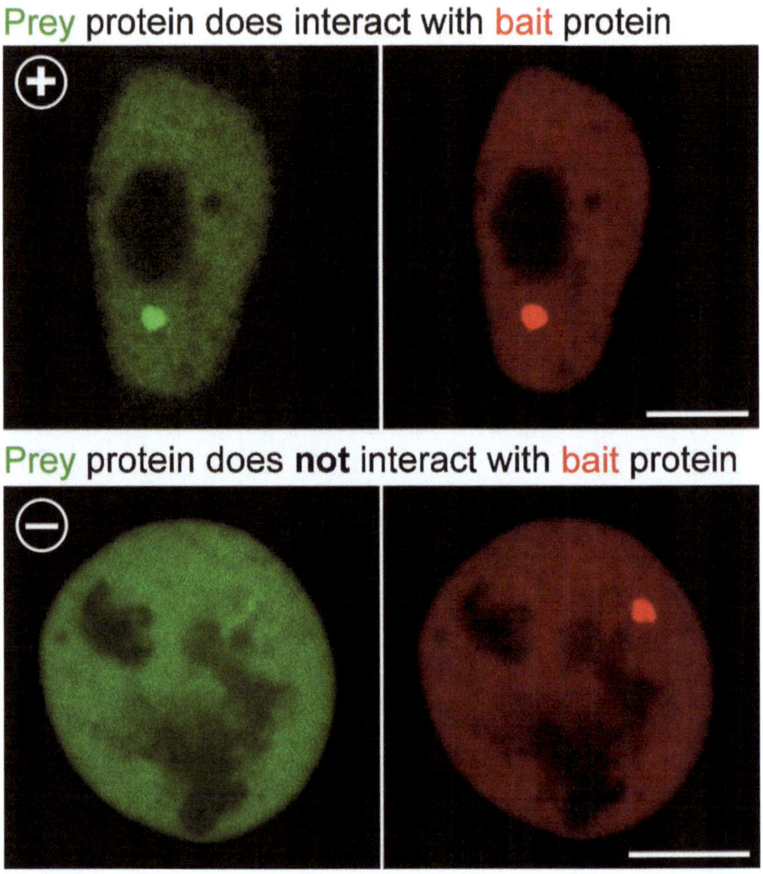

Fig. 2. Application example of the F2H assay. Selected confocal images of a cell showing interaction (*top row*) between DNA Ligase III (*green*) and XRCC1 (*red*) and a cell showing no interaction (*bottom row*) between DNA Ligase I (*green*) and XRCC1 (*red*). Bars: 5 µm.

3. Phosphate-buffered saline (PBS), 0.5 mM ethylenediamine tetraacetic acid (EDTA) and solution of 0.25% trypsin.
4. Polyplus transfection reagent jetPEI™ (BIOMOL, Hamburg, Germany).

2.3. Sample Preparation

1. For fixed samples: Microscope cover slips (18 × 18 × 0.17 mm; Carl Roth, Karlsruhe, Germany); for live cell experiments: ibidi µ-slide I or µ-slide IV (ibidi, Munich, Germany); for high-throughput screens: µClear 96-well Microplate (Greiner Bio-One, Frickenhausen, Germany).
2. PBS.
3. 36.5% Formaldehyde: Prepare a 3.7% (w/v) solution in PBS fresh for each experiment.
4. Wash solution: 0.02% Tween-20 in PBS (PBS-T).
5. Permeabilization solution: 0.2% Triton X-100 in PBS-T.

6. Nuclear stain: 200 nM DAPI (4,6-diamidino-2-phenylindole) in PBS-T.
7. Mounting medium: VectaShield (Vector Laboratories, CA, USA).

3. Methods

3.1. Molecular Cloning

For a detailed description of a classical molecular cloning procedure (restriction digest, ligation, transformation, and DNA preparation from bacteria) please refer to a dedicated protocol (e.g., Molecular Cloning: A Laboratory Manual or http://cshprotocols.cshlp.org/)

In brief:

1. PCR amplify coding cDNAs of the proteins of interest using primers pF2H_forward (5′-CCCCC*GCGATCGC*XXXXXXXXXXXXX-XX 3′) and pF2H_reverse (5′-ATTCTTAT*GCGGCCGC*TCAXXXXXXXXXXXXX-3′) containing an *AsiSI* restriction site as forward primer and an *NotI* as reverse primer (restriction sites are in italic, stop-codon in bold, X represents nucleotides corresponding to the cDNA sequence of the protein of interest).
2. Digest PCR fragments and expression vectors pF2H-bait and pF2H-prey with restriction enzymes *AsiS*I/*Not*I according to manufacturer's instructions.
3. Ligation of fragments into vectors.
4. Transformation (e.g., in *Escherichia coli* XL1-Blue) and growth under Kanamycin resistance.
5. Mini-/Maxi-Preparation of DNA from bacteria.
6. Sequencing of the obtained constructs (see Note 2).

3.2. Cell Culture and Transfection

1. Transgenic BHK cells (clone #2) or U2OS cells (clone 2-6-3) containing *lac* operator repeats are cultured under selective conditions in DMEM supplemented with 10% fetal calf serum and 150 μg/ml hygromycin B (PAA Laboratories) as described (13, 14). Subconfluently grown cells are passaged every 2–3 days with trypsin/EDTA.
2. For microscopy cells are seeded at appropriate density (40–60% confluence) either on 18 × 18 glass cover slips (for fixed samples), for live cell imaging in μ-slides (ibidi) or for high-throughput screens in μClear 96-multiwell plates (Greiner).
3. 4–6 h after seeding, adhered cells are then co-transfected with expression constructs pF2H-bait and pF2H-prey using Polyplus transfection reagent jetPEI™ according to the manufacturer's instructions (see Notes 3 and 4).

4. After incubation over night at 37°C with 5% CO_2, the transfection medium is exchanged to fresh culture medium.
5. Live cells can either be observed immediately or after incubation for another 12–24 h before fixation with 3.7% formaldehyde in PBS for 15 min at room temperature.
6. The fixation solution is discarded (into a hazardous waste container) and exchanged with PBS-T.
7. Cells are permeabilized and counterstained for 5–10 min with 0.2% Triton X-100 in PBS-T and DAPI.
8. Wash three times with PBS-T.
9. Cover slips and μ-slides (ibidi): Mount in Vectashield (Vector Laboratories, CA, USA), μClear plates (Greiner): Store plate with 100 μl PBS-T per well.

3.3. Microscopy

Live or fixed cells expressing fluorescent fusion proteins can be analyzed by either wide-field epifluorescence, confocal laser-scanning, or spinning disk microscopes. For high-content analyses (HCA), a high-throughput screening microscope may be used. The only technical prerequisite for the F2H assay is appropriately equipped microscopes to distinguish green and red fluorescent proteins, e.g.:

GFP	Excitation filter HQ480/40X	Emission filter HQ535/50M
RFP	Excitation filter HQ565/30X	Emission filter HQ620/60M
DAPI	Excitation filter D360/40X	Emission filter HQ460/40M

For the detection of the subcellular focal signal (interaction platform) with epifluorescent wide-field systems or with high-throughput microscopes (e.g., INCell Analyzer 1000, GE Healthcare) air objectives with 20×/0.45 NA ELWD Plan Fluor, 40×/0.60 NA ELWD Plan Fluor are sufficient. Images of living and fixed cells are acquired with exposure times typically in the range of 200 ms for GFP fusion proteins and 600 ms for RFP fusion proteins.

Alternatively, a laser scanning confocal microscope (e.g., Leica TCS SP5) equipped with a 63×/1.4 NA Plan-Apochromat oil immersion objective can be used to obtain higher resolution data. Here, fluorophores are typically excited with a 405 nm laser-line (for DAPI), a 488 nm laser-line (for GFP), and a 561 nm laser-line (for RFP). Confocal image stacks of living or fixed cells are typically recorded with a frame size of 512 × 512 pixels, a pixel size of 50–100 nm, a *z*-step size of 250 nm, and the pinhole opened to 1 Airy unit.

4. Notes

1. The F2H assay takes advantage of cell lines with a stable integration of a *lac* operator array to immobilize a lac repressor fused to fluorescently labeled proteins of interest (bait) (9). For optimal signal-to-noise ratio, we suggest working with the BHK clone #2 cell line (13). However, other readily usable human, mouse, and *Drosophila* cell lines have been described (14–18).
2. The F2H assay works independent on the subcellular localization of the proteins of interest. The SV40-NLS (nuclear localization signal) in the F2H expression vectors directs the proteins to be analyzed for interaction to the nucleus of the cell and in case of the bait protein to the interaction platform (9).
3. The F2H assay may yield false positive results, which need to be controlled for. An initial screen can identify prey proteins that bind to the *lac* operator array in the absence of a bait protein. In such cases, these proteins can then be only used as baits.
4. To screen compound libraries for induction or disruption of a specific PPI, stable cell lines should be generated to obtain constant and reproducible expression levels of bait and prey fusion proteins.

Acknowledgments

We thank Lothar Schermelleh for comments on the manuscript. We also thank D.L. Spector for providing BHK clone#2 and U2OS.2-6-3 cells containing a *lac* operator array. This work was supported by the Center for NanoScience (CeNS), the Nanosystems Initiative Munich (NIM), the BioImaging Network Munich (BIN), and by grants from the Deutsche Forschungsgemeinschaft (DFG).

References

1. Fields, S., and Song, O. (1989) A novel genetic system to detect protein-protein interactions, *Nature* **340**, 245–6.
2. Deane, C. M., Salwinski, L., Xenarios, I., and Eisenberg, D. (2002) Protein interactions: two methods for assessment of the reliability of high throughput observations, *Mol Cell Proteomics* **1**, 349–56.
3. Kerppola, T. K. (2006) Visualization of molecular interactions by fluorescence complementation, *Nat Rev Mol Cell Biol* **7**, 449–56.
4. Blanchard, D., Hutter, H., Fleenor, J., and Fire, A. (2006) A differential cytolocalization assay for analysis of macromolecular assemblies in the eukaryotic cytoplasm, *Mol Cell Proteomics* **5**, 2175–84.

5. Miller, C. L., Arnold, M. M., Broering, T. J., Eichwald, C., Kim, J., Dinoso, J. B., and Nibert, M. L. (2007) Virus-derived platforms for visualizing protein associations inside cells, *Mol Cell Proteomics* **6**, 1027–38.
6. Miyawaki, A. (2003) Visualization of the spatial and temporal dynamics of intracellular signaling, *Dev Cell* **4**, 295–305.
7. Parrish, J. R., Gulyas, K. D., and Finley, R. L., Jr. (2006) Yeast two-hybrid contributions to interactome mapping, *Curr Opin Biotechnol* **17**, 387–93.
8. Sekar, R. B., and Periasamy, A. (2003) Fluorescence resonance energy transfer (FRET) microscopy imaging of live cell protein localizations, *J Cell Biol* **160**, 629–33.
9. Zolghadr, K., Mortusewicz, O., Rothbauer, U., Kleinhans, R., Goehler, H., Wanker, E. E., Cardoso, M. C., and Leonhardt, H. (2008) A fluorescent two-hybrid assay for direct visualization of protein interactions in living cells, *Mol Cell Proteomics* 7, 2279–87.
10. Meilinger, D., Fellinger, K., Bultmann, S., Rothbauer, U., Bonapace, I. M., Klinkert, W. E., Spada, F., and Leonhardt, H. (2009) Np95 interacts with de novo DNA methyltransferases, Dnmt3a and Dnmt3b, and mediates epigenetic silencing of the viral CMV promoter in embryonic stem cells, *EMBO Rep* **10**, 1259–64.
11. Fellinger, K., Rothbauer, U., Felle, M., Langst, G., and Leonhardt, H. (2009) Dimerization of DNA methyltransferase 1 is mediated by its regulatory domain, *J Cell Biochem* **106**, 521–8.
12. Phair, R. D., and Misteli, T. (2000) High mobility of proteins in the mammalian cell nucleus, *Nature* **404**, 604–9.
13. Tsukamoto, T., Hashiguchi, N., Janicki, S. M., Tumbar, T., Belmont, A. S., and Spector, D. L. (2000) Visualization of gene activity in living cells, *Nat Cell Biol* **2**, 871–8.
14. Janicki, S. M., Tsukamoto, T., Salghetti, S. E., Tansey, W. P., Sachidanandam, R., Prasanth, K. V., Ried, T., Shav-Tal, Y., Bertrand, E., Singer, R. H., and Spector, D. L. (2004) From silencing to gene expression: real-time analysis in single cells, *Cell* **116**, 683–98.
15. Robinett, C. C., Straight, A., Li, G., Willhelm, C., Sudlow, G., Murray, A., and Belmont, A. S. (1996) In vivo localization of DNA sequences and visualization of large-scale chromatin organization using lac operator/repressor recognition, *J Cell Biol* **135**, 1685–700.
16. Tumbar, T., Sudlow, G., and Belmont, A. S. (1999) Large-scale chromatin unfolding and remodeling induced by VP16 acidic activation domain, *J Cell Biol* **145**, 1341–54.
17. Dietzel, S., Zolghadr, K., Hepperger, C., and Belmont, A. S. (2004) Differential large-scale chromatin compaction and intranuclear positioning of transcribed versus non-transcribed transgene arrays containing {beta}-globin regulatory sequences, *J Cell Sci* **117**, 4603–14.
18. Vazquez, J., Belmont, A. S., and Sedat, J. W. (2001) Multiple regimes of constrained chromosome motion are regulated in the interphase Drosophila nucleus, *Curr Biol* **11**, 1227–39.

Chapter 17

ArrayMAPPIT: A Screening Platform for Human Protein Interactome Analysis

Sam Lievens, Nele Vanderroost, Dieter Defever, José Van der Heyden, and Jan Tavernier

Abstract

Mammalian protein–protein interaction trap (MAPPIT) is a two-hybrid technology to identify and characterize interactions of proteins with other proteins or organic molecules in living mammalian (human) cells. The method relies on complementation of a modified cytokine receptor complex. Protein interaction restores the signalling competence of the complex, which is monitored through the activation of a reporter gene. Here, we describe a protocol that has been recently developed to increase the utility of MAPPIT as a tool to identify novel interactions. In the ArrayMAPPIT assay, a collection of prey proteins which is arrayed in high-density microtiter plates is efficiently screened for interaction partners using reverse transfection into a bait-expressing cell pool.

Key words: Interactomics, Protein–protein interaction, High-throughput screening, Two-hybrid, MAPPIT

1. Introduction

Protein–protein interaction analysis aims at connecting proteins into functional assemblies, ranging from stable multi-protein machines, such as proteasomes, to transient protein encounters as observed in the context of a signaling cascade. Since protein function is essentially mediated by interactions with other (extra-) cellular molecules, most importantly other proteins, this knowledge is an important starting point for understanding the complexity of biological systems at the molecular level. Large-scale protein interaction mapping, or (protein) interactomics, is a relatively young "omics" field that has been expanding at a rapid pace during the last 5 years, both at the level of the available tools for data generation and analysis and with regard to the amount and the quality of the data produced (1–3).

Bernhard Suter and Erich E. Wanker (eds.), *Two Hybrid Technologies: Methods and Protocols*, Methods in Molecular Biology, vol. 812, DOI 10.1007/978-1-61779-455-1_17, © Springer Science+Business Media, LLC 2012

The bulk of the interactome data is being produced using yeast two-hybrid, the classical genetic complementation strategy (4), and biochemical (tandem) affinity purification methods (5). However, recently also a number of other approaches, including protein complementation assay (PCA) (6), luminescence-based mammalian interactome (LUMIER) (7), and protein chips (8), have been employed in large-scale protein network charting projects.

Our group developed mammalian protein–protein interaction trap (MAPPIT), a two-hybrid approach based on complementation of a cytokine receptor complex that operates in mammalian cells, allowing mammalian proteins to be analyzed in their native context (9). In this strategy, a cytokine receptor mutant that has been rendered signalling-deficient by mutating its recruitment sites for signal transducer and activator of transcription (STAT) signalling molecules is genetically fused to a protein (fragment) of interest (the "bait"; Fig. 1). The putative binding partner (the "prey" protein) is tethered to a fragment of a cytokine receptor that does contain functional STAT docking sites. Interaction between bait and prey proteins restores a functional cytokine receptor complex, leading to STAT-mediated activation of a reporter gene. One of the main advantages of MAPPIT is that reporter activation through the complemented receptor additionally requires stimulation by the appropriate cytokine ligand, which adds an extra level of control to the system. This feature, together with the simplicity of the assay readout, makes the technology particularly useful toward high-throughput applications. The basic binary MAPPIT setup that assays the interaction between designated protein pairs is being applied at large scale for the validation of yeast two-hybrid data in proteome-wide interactome mapping projects (10–13). To enable complex prey cDNA libraries to be screened for novel interaction partners of a particular bait protein, an FACS-based protocol was designed that has been successfully applied to identify novel subunits of E3 ubiquitin ligase complexes (14, 15). Here, we describe ArrayMAPPIT, an efficient screening approach based on reverse transfection that enables rapid interrogation of an array of prey proteins for potential binding partners of a designated bait (15).

In ArrayMAPPIT, a collection of prey encoding plasmids are individually mixed with a luciferase reporter plasmid and complexed with a transfection reagent and a number of stabilizing agents, and these complexes are then spotted into microtiter plates (Fig. 2). Stacks of these "prey arrays" are prepared in parallel and after drying, these can be stored for an extended period of time. To screen such an array, a set of plates is filled with a suspension of bait-expressing cells, resulting in their "reverse" transfection with the spotted prey (and reporter) plasmid (16). The cells of the resulting "reverse transfected cell array" contain a different bait-prey combination in each well of the array. Upon stimulation with the appropriate cytokine ligand, the luciferase reporter gene is activated

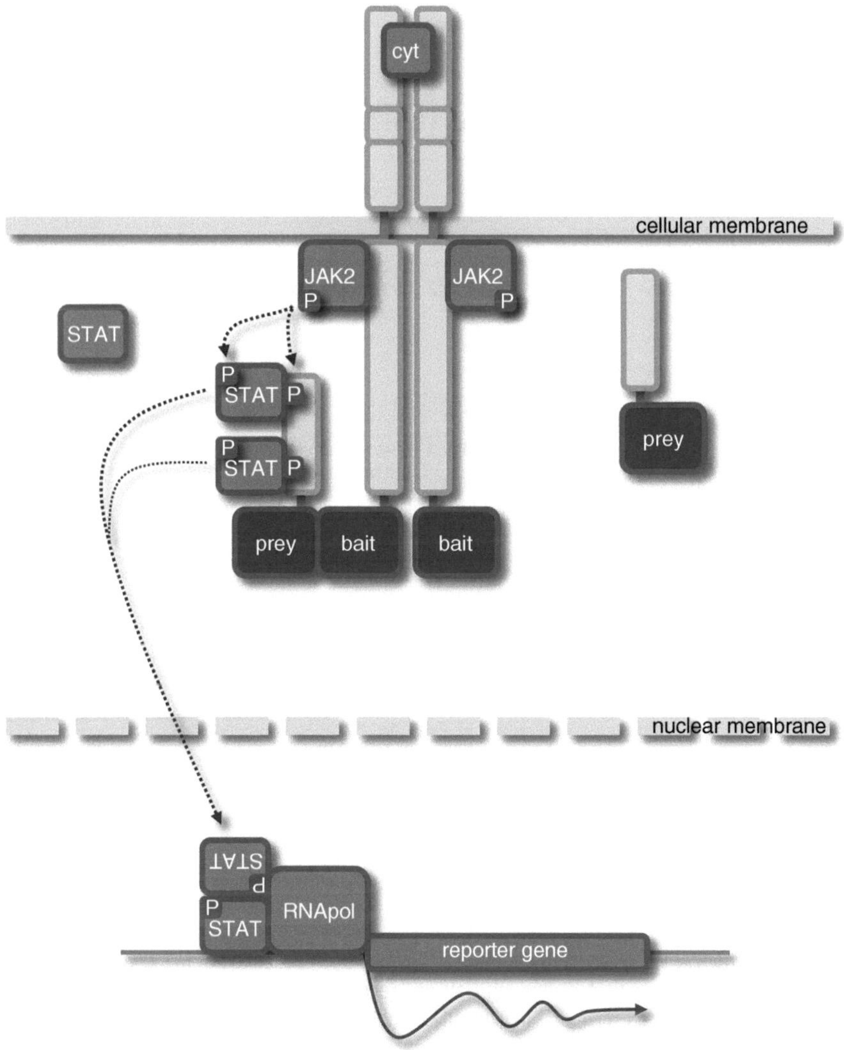

Fig. 1. Concept of the MAPPIT two-hybrid approach. The bait protein (fragment) is genetically fused to a cytokine receptor which is made signalling-deficient by mutation of signal transducer and activator of transcription (STAT) recruitment sites, whereas the prey is tethered to a portion of a receptor containing intact STAT docking sites. Binding of the appropriate cytokine ligand to the chimeric bait receptor leads to activation of the constitutively associated Janus kinases (JAKs). Interaction between bait and prey brings both receptor fragments into proximity, reconstituting a functional receptor complex. The activated JAKs phosphorylate the tyrosine motifs of the prey chimaera in trans, leading to STAT recruitment. These transcription factors are in turn phosphorylated by the JAK kinases, resulting in their activation and subsequent dissociation and translocation to the nucleus. In the nucleus, STAT dimers induce STAT-dependent reporter gene transcription.

in those cells harbouring an interacting bait and prey pair, which is detected using a bioluminescence plate reader. The system is fast: the screening itself takes only 5 days from seeding the cells for transient bait transfection up to signal readout; it is robust: in the assay setup described in this protocol multiple replicates for every bait–prey combination are tested in each screen and signals for ligand-stimulated and unstimulated cells are combined which eliminates

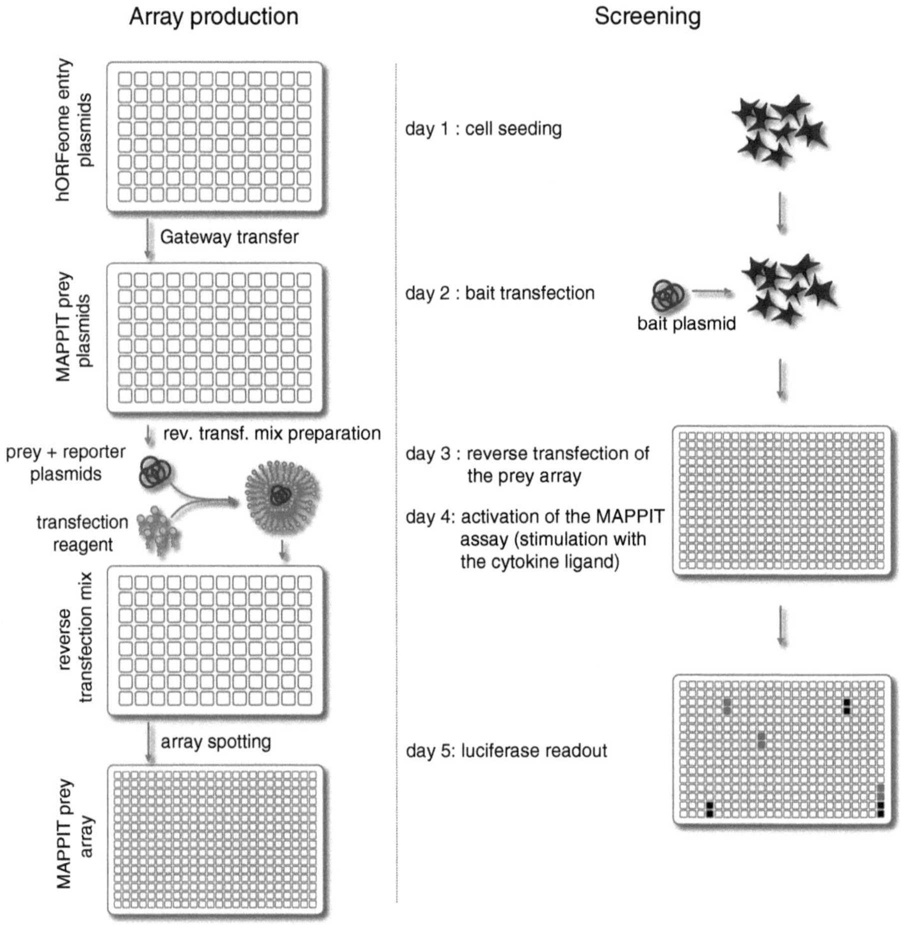

Fig. 2. Overview of the ArrayMAPPIT assay. Array production (*left panel*). A collection of full length human ORFeome entry clones is transferred to MAPPIT prey vectors through Gateway recombinatorial cloning. The resulting prey plasmids are used to produce reverse transfection mixes by individually combining them with a suitable transfection reagent and a number of additional stabilizing agents. Also the luciferase reporter plasmid used for the MAPPIT readout is included in these mixtures. The reverse transfection mixes are then spotted in 384-well microtiter plates, dried and stored until needed for screening. Screening (*right panel*). Cells are seeded and transfected in bulk with the desired MAPPIT bait construct. The resulting pool of bait-expressing cells is seeded in the MAPPIT prey array, leading to their additional (reverse) transfection with the spotted prey and reporter plasmids. The MAPPIT assay is activated by stimulating the cells with the appropriate cytokine ligand, and signals are read out using a luciferase assay.

important sources of technical error; and it can be automated: array preparation and most steps of the screening process can be programmed on a robotic platform.

2. Materials

2.1. Array Production

1. LB medium: Add 10 g bacto-tryptone, 5 g yeast extract, and 10 g NaCl to 1 l distilled water, bring to pH 7.5 with 5N NaOH, sterilize by autoclaving.

2. 2× YT medium: Add 16 g bacto-tryptone, 10 g bacto-yeast extract, and 5 g NaCl to 1 l distilled water, bring to pH 7.0 with 5N NaOH, sterilize by autoclaving.
3. Hepes/EtOH washing buffer: 10 mM Hepes in 80% EtOH (store at room temperature).
4. EC/sucrose buffer: Dissolve sucrose in EC buffer (Effectene Transfection Reagent kit, Qiagen) at a concentration of 0.4 M, and filter-sterilize (store at 4°C).
5. Enhancer/EC/sucrose buffer: Mix Enhancer (Effectene Transfection Reagent kit, Qiagen) with EC/sucrose buffer at a ratio 3/2 (store at 4°C).
6. Effectene: From the Effectene Transfection Reagent kit (Qiagen; store at 4°C).
7. 0.2% Gelatin stock: Add 0.1 g gelatin (Type B: 225 Bloom; Sigma) to 50 ml of double distilled water, keep in waterbath at maximum 60°C until completely dissolved, filter-sterilize while still warm (store at 4°C for up to 1 month).
8. Diluted gelatin solution: Warm the 0.2% gelatin stock in waterbath (maximum 60°C) and dilute 25 times (make fresh before each use).

2.2. Screening

1. Culture medium: Dulbecco's modified Eagle's medium (DMEM) supplemented with 10% foetal calf serum and antibiotics (1/100 dilution of penicillin/streptomycin solution; Invitrogen).
2. $CaCl_2$ buffer: Prepare a 2.5 M $CaCl_2$ stock and filter-sterilize (store at −20°C).
3. 2× Hepes-buffered saline (HeBS) buffer: 280 mM NaCl, 1.5 mM Na_2HPO_4, and 50 mM Hepes in double distilled water, adjust the pH to 7.05 with 1 M NaOH (an exact pH is critical for transfection efficiency; the optimal range is very narrow: from 7.05 to 7.12), and filter-sterilize (store at −20°C, prevent repeated freeze–thaw cycles).
4. Human epo stock: Reconstitute lyophilized stock (R&D systems) in growth medium at a concentration of 5 μg/ml (store at −20°C for long term, keep at 4°C for weekly use, and prevent repeated freeze–thaw cycles).
5. Mouse leptin stock: Reconstitute lyophilized stock (R&D systems #498-OB-01 M) in growth medium at a concentration of 20 μg/ml (store at −20°C for long term, keep at 4°C for weekly use, and prevent repeated freeze–thaw cycles).
6. 5× lysis buffer stock solution: 125 mM Tris–phosphate (pH 7.8), 10 mM DTT, 10 mM CDTA (*trans*-1,2-diaminocyclohexane-*N,N,N′,N′*-tetra acetic acid), 50% glycerol, 5% Triton X-100 in double distilled water (store at −20°C).

7. Luciferin substrate buffer: 40 mM Tricine, 2.14 mM$(MgCO_3)_4Mg(OH)_2 \cdot 5H_2O$, 5.34 mM $MgSO_4$, 66.6 mM DTT, 0.2 mM EDTA, 509 μM Coenzyme A, 734 μM ATP, and 940 μM d-luciferin (store at −20°C).
8. Lysis buffer/luciferin mix: Mix 5× lysis buffer stock solution with luciferin substrate buffer and double distilled water in a ratio 10/4.2/2.8 (make fresh before each use).

3. Methods

The protocol is organized in two sections, the first part dealing with the production of the arrays starting from the plasmid preparation and the second part describing their use in the actual screening. The procedure described is largely automated and in our setup makes use of two separate robotic platforms, one for plasmid preparation (Tecan Freedom Evo 150 equiped with a four-channel liquid handling arm, a gripper arm, and an integrated Tecan Infinite 200 plate reader for spectrophotometry) and a second one for the production of the arrays and the luciferase readout at the end of the screening (Tecan Freedom Evo 200 with an eight-channel liquid handling arm, a 96-pin multi-channel arm and a gripper; this robot is coupled to a Perkin-Elmer Envision plate reader for bioluminescence detection). The formats applied are 96-well microtiter plates for preparation of prey plasmids and reverse transfection mixes, which are then spotted into 384-well plates for screening (Fig. 3). As a consequence, every array plate contains four replicates of each prey transfection mix. Two of these are left unstimulated and two are stimulated with the cytokine that activates the MAPPIT signal. Two wells of each 96-well plate are reserved for controls, two preys corresponding to JAK2-binding proteins which serve as universal positive controls as they will generate a signal in combination with any bait receptor through their interaction with JAK2, the kinase that is constitutively associated with the modified cytokine receptor used in MAPPIT (Fig. 3).

3.1. Array Production

3.1.1. Prey Plasmid Preparation

Although the way of purifying plasmid DNA is not particular to this method and any plasmid preparation method that produces transfection-grade quality DNA is fine, we include this part of our protocol because it contains a number of tricks to obtain a fair amount of high-quality plasmid DNA (about 8 μg of high-copy plasmid on average starting from a 1.5 ml culture) using standard silica-based columns. The preys are derived from the human ORFeome collection (17) (http://horfdb.dfci.harvard.edu) and have been transferred to the pMG1 prey vector (Fig. 4) (15) using Gateway recombinatorial cloning.

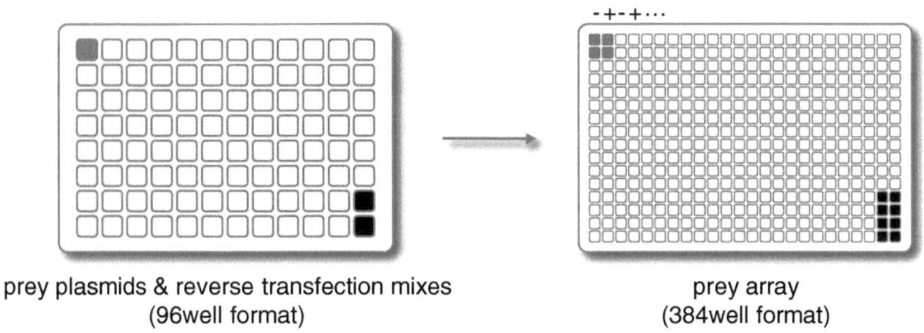

Fig. 3. Layout of the ArrayMAPPIT plate setup. Prey plasmids and reverse transfection mixes are prepared in 96-well microtiter plates. Two wells (positions G12 and H12) are reserved for universal positive controls, corresponding to preys that bind to JAK2. The reverse transfection mixes are spotted into 384-well plates in 4 squarewise neighbouring replicates. Odd columns are left unstimulated, whereas even columns are stimulated with cytokine ligand. As a result, for each bait–prey combination the prey array contains two unstimulated and two stimulated replicates which are used to calculate an induction factor which, when higher than 1, point to a potential protein–protein interaction.

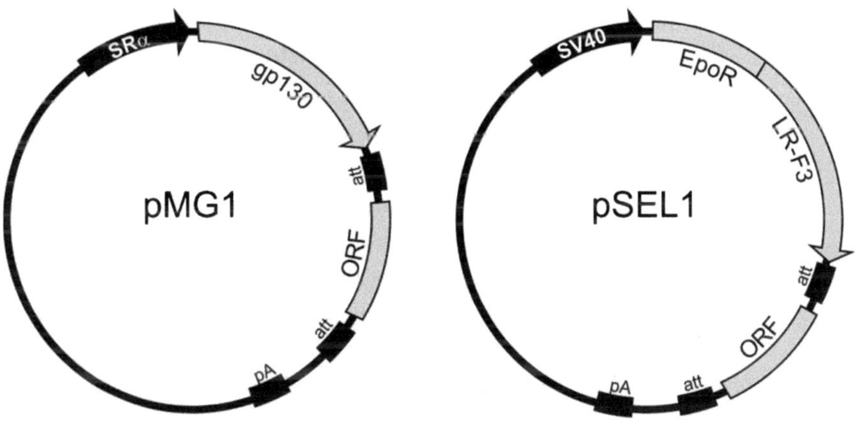

Fig. 4. MAPPIT prey and bait vectors. Human ORFs are transferred into the MAPPIT prey (pMG1) and bait (pSEL1) destination vectors using Gateway cloning through recombination between "att" sites. The resulting prey construct is a chimaera of the ORF and a portion of the gp130 cytokine receptor, and expression is driven by the synthetic SRα promoter. Baits expressed from the pSEL1 vector are fusions with an EpoR/LR chimaera containing 3 Y-F mutations in the cytoplasmic tail ("LR-F3"), and expression originates from the SV40 promoter. Both vectors contain an SV40 polyadenylation sequence ("pA").

1. Using a 96-pin replicator, inoculate a pre-culture from a 96-well glycerol slant into a 96-well microtiter plate containing 150 μl of LB medium containing selective antibiotic per well and incubate overnight in an incubator shaker at 37°C (see Note 1).
2. Using a 96-pin replicator, start from the pre-culture to inoculate a 96-deepwell block containing 1.5 ml of 2× YT medium supplemented with selective antibiotic per well and incubate for 24 h in an incubator shaker at 37° (see Note 2).
3. Prepare plasmid DNA using a silica-based 96-well format plasmid preparation kit (e.g. NucleoSpin Robot-96 Plasmid from

Macherey-Nagel) according to the guidelines of the manufacturer. After the final washing step of the standard protocol, an extra washing step consisting of 600 µl of Hepes/EtOH buffer should be included to remove excessive salt in order to obtain the purity desired for (reverse) transfection applications (see Note 3).

4. Measure the DNA concentration by spectrophotometry of the OD at 260 nm.
5. Transfer 1 µg aliquots of each prey plasmid to a U-bottom 96-well plate and dry in a speedvac (see Note 4). Seal the plate and store at –20°C.

3.1.2. Reverse Transfection Mix Preparation and Spotting the Arrays

The procedure described here essentially follows the original "lipid method" described by the Sabatini group, using Effectene as a transfection reagent (16). The volumes mentioned below correspond to what is needed to make a reverse transfection mix for one prey and can serve as a starting point to programme a robot pipetting multiple 96-well plates in parallel.

1. To the 1 µg of dried prey plasmid DNA in each well of the 96-well plate prepared in Subheading 3.1.1., add 11.5 µl of a solution consisting of 0.5 µg pXP2d2-rPAP1-luciferase reporter plasmid (9) in EC/sucrose buffer. Shake for 8 min to dissolve the prey plasmid DNA (see Note 5).
2. Add 5 µl of enhancer/EC/sucrose mix and shake for 1 min. Incubate the mixture at room temperature for 5 min to facilitate condensation of the plasmid DNA.
3. Add 5 µl of Effectene and shake for 1 min. Incubate the mixture at room temperature for 10 min to allow Effectene-DNA complexes to form.
4. Add 300 µl of a diluted gelatin solution and mix by pipetting up and down (see Note 6).
5. Using the 96-pin multi-channel arm, spot 4 µl into 4 adjacent wells of white 384-well microtiter plates (Fig. 3). With the volumes used here, 18 plates can be spotted in parallel.
6. Dry the plates overnight at room temperature (see Note 6) and store them dark, dry, and cool (see Note 7).

3.2. Screening

The screening process comprises a 5-day protocol and implies the transient bulk transfection of a cell pool with the bait encoding construct using the classical calcium phosphate method, followed by reverse transfection of the resulting bait-expressing cells with individual preys by seeding them in the wells of the 384-well prey array plates containing the dried transfection mixes. As a result, every bait-prey combination is expressed in a different well. Stimulation with the appropriate cytokine ligand activates the

MAPPIT system, resulting in the synthesis of the firefly luciferase reporter enzyme in those cells harbouring an interacting protein pair, which is measured using a standard firefly luciferase assay.

3.2.1. Bait Transfection (Day 1–2)

1. Seed HEK293-T cells at a density of 7×10^6 cells per 175-cm^2 flask in 35 ml of culture medium (see Note 8).
2. Incubate overnight at 37°C, 8% CO_2.
3. For each flask to be transfected, prepare a mix containing 35 μg of bait plasmid DNA (e.g. cloned in the pSEL bait vector; Fig. 4) (9, 14) and 175 μl 2.5 M $CaCl_2$, supplemented with double distilled water to a final volume of 1,750 μl.
4. Add this mix dropwise to a vial containing 1,750 μl 2× HeBS buffer while vigourously vortexing the vial or bubbling air into it through a fine pipette tip (see Note 9).
5. Incubate 15 min at room temperature to allow DNA precipitates to form.
6. Briefly vortex the solution and add it to the cells.
7. Incubate overnight at 37°C, 8% CO_2.

3.2.2. Reverse Transfection of the Prey Array (Day 3)

1. Remove the medium, wash the cells with PBS, detach with 7 ml of trypsin and resuspend with culture medium.
2. Spin down, remove the supernatant, resuspend the cells in culture medium at a density of 350,000/ml and pipette through a 70-μm cell strainer to remove any cell clusters (see Note 10).
3. Seed the cells on to the 384-well prey array plates at 15 μl per well (which corresponds to around 5,000 cells per well) (see Notes 11 and 12). In addition, seed a couple of rows in untreated transparent 384-well plates to be able to track cell growth microscopically.
4. Incubate overnight at 37°C, 8% CO_2.

3.2.3. Activation of the MAPPIT System (Day 4)

1. Add 15 μl of culture medium without or with the appropriate cytokine ligand (10 ng/ml epo or 100 ng/ml leptin, depending on the extracellular domain utilized in the chimeric MAPPIT bait receptor construct (9, 14)) to the odd or even columns of the 384 well prey array plates, respectively (see Note 13).

3.2.4. Luciferase Read-Out (Day 5)

1. Remove the culture medium to a similar level in all wells (see Note 14).
2. Add 17 μl of lysis buffer/luciferin mix to each well.
3. Measure luciferase activity in a bioluminescence plate reader.

3.2.5. Data Analysis

For every bait–prey combination, calculate the ratio between the average of the luciferase counts of the two stimulated samples and the average of the luciferase counts of the two unstimulated samples.

If this ratio or induction factor is higher than 1, it points to a potential interaction between that particular bait and prey protein. Depending on the level of confidence required, a different cut-off can be set to select candidates for retesting (for an example, see ref. 15). The candidate interaction partners are then first re-evaluated in MAPPIT binary assays, including a negative control bait (constructs where no bait or an irrelevant bait is fused to the MAPPIT receptor) to discriminate between preys that bind specifically to the fused bait protein and preys that bind to the MAPPIT receptor itself. The control preys which are present on every plate serve as positive controls for the quality of the individual plates and the screening process in general.

4. Notes

1. Including a pre-culture step is not essential but generally results in more evenly grown cultures and hence more consistent and evenly distributed plasmid yields.
2. The use of 2× YT medium will significantly increase the yield compared to LB or even TB medium, and in our hands 24 h incubation further increases the yield compared to an overnight incubation. Additionally, sufficient aeration of the cultures is of critical importance to obtain high yields, which depends both on a sufficiently high shaking speed and a small orbital diameter of the incubator shaker.
3. Although "transfection grade" ion-exchange resin-based plasmid kits are marketed specifically for the purpose of mammalian cell transfection, we experienced that at least in the protocol described here, plasmid DNA prepared using the much cheaper silica-based columns in combination with the extra washing step described gives similar results.
4. In our setup, the DNA is eluted from the silica columns directly in thin-bottom 96-well plates which are compatible with UV spectrophotometry. Before robotic transfer of the 1 μg aliquots, it is advised to transfer the eluted DNA to V-shaped PCR-type microtiter plates because the liquid column will be significantly higher in these plates than in the flat-bottom spectrophotometry plates, reducing robot pipetting errors.
5. It is not essential to start from dried prey plasmid DNA to make the reverse transfection mixes; however, it is a simple way of equalizing the volume of prey plasmid DNA across the different wells of the plate.
6. Compared to the original Sabatini protocol we add a larger volume of a less concentrated gelatin solution in order to

increase the volumes for spotting, which should result in lower variation of the pippeted volumes. As the dilution of the reverse transfection mixes does not favour the stability of the Effectene-DNA complexes, they should be spotted and subsequently dried immediately after preparation.

7. One of the main advantages of reverse transfection arrays is that they can be stored at room temperature for a long time. In our hands, arrays have shown to be functional after storage at room temperature for up to 6 months. The original Sabatini protocol mentions that the quality of the arrays can be further improved by storing the slides at 4°C; however, we prefer to store them at temperatures not below 16°C to prevent condensation on the arrays upon transfer to room temperature.
8. In order to obtain high signals, and thus high transfection efficiencies of both bait and prey, it is absolutely critical to start from cells that are in "good shape", meaning they are in logarithmic growth phase and have been kept subconfluent at all times. In order to maintain subconfluency, we never allow the cells to reach densities higher than around 20×10^6 per 175-cm^2 flask.
9. Vigourous agitation of the buffer is critical to obtain fine DNA precipitates which result in high transfection efficiencies. The method we experienced to be most efficient is to blow air into the solution using an automatic pipettor with a 5-ml plastic pipette fitted at its end with a 20-μl pipette tip.
10. Removal of cell clumps is essential to obtain evenly distributed amounts of cells across the different wells of the prey array plates, which in turn is important to reduce signal variability.
11. It is critical to keep the amount of medium as low as possible during reverse transfection in order to obtain high signals, most likely due to the higher transfection efficiency obtained in conditions where the redissolved Effectene–DNA complexes are more concentrated.
12. For dispension of cell suspensions into microtiter plates, we use microplate dispenser devices (e.g. Thermo Scientific Multidrop Combi). During dispension, it is important to stir the cell suspension from time to time to ensure even distribution of the cells across the plates.
13. As for the bait-expressing cell suspension, this dispension step is most efficiently done using an automated microplate dispenser.
14. Because it is difficult to completely remove all medium with robot pins without disturbing the cells, we leave some medium in the wells (around 10 μl). The presence of phenol red in the medium results in slightly lower absolute luciferase counts; however, since in MAPPIT we evaluate the ratio between the luciferase activity measured in the stimulated versus the unstimulated cells, this does not affect the outcome.

References

1. Figeys, D. (2008) Mapping the human protein interactome *Cell Res* **18**, 716–24.
2. Russell, R. B., and Aloy, P. (2008) Targeting and tinkering with interaction networks *Nat Chem Biol* **4**, 666–73.
3. Sanderson, C. M. (2009) The Cartographers toolbox: building bigger and better human protein interaction networks *Brief Funct Genomic Proteomic* **8**, 1–11.
4. Fields, S. (2009) Interactive learning: Lessons from two hybrids over two decades *Proteomics.*
5. Burckstummer, T., Bennett, K. L., Preradovic, A., Schutze, G., Hantschel, O., Superti-Furga, G., and Bauch, A. (2006) An efficient tandem affinity purification procedure for interaction proteomics in mammalian cells *Nat Methods* **3**, 1013–9.
6. Tarassov, K., Messier, V., Landry, C. R., Radinovic, S., Serna Molina, M. M., Shames, I., Malitskaya, Y., Vogel, J., Bussey, H., and Michnick, S. W. (2008) An in vivo map of the yeast protein interactome *Science* **320**, 1465–70.
7. Barrios-Rodiles, M., Brown, K. R., Ozdamar, B., Bose, R., Liu, Z., Donovan, R. S., Shinjo, F., Liu, Y., Dembowy, J., Taylor, I. W., Luga, V., Przulj, N., Robinson, M., Suzuki, H., Hayashizaki, Y., Jurisica, I., and Wrana, J. L. (2005) High-throughput mapping of a dynamic signaling network in mammalian cells *Science* **307**, 1621–5.
8. Jones, R. B., Gordus, A., Krall, J. A., and MacBeath, G. (2006) A quantitative protein interaction network for the ErbB receptors using protein microarrays *Nature* **439**, 168–74.
9. Eyckerman, S., Verhee, A., Van der Heyden, J., Lemmens, I., Van Ostade, X., Vandekerckhove, J., and Tavernier, J. (2001) Design and application of a cytokine-receptor-based interaction trap *Nature Cell Biology* **3**, 1114–19.
10. Boxem, M., Maliga, Z., Klitgord, N., Li, N., Lemmens, I., Mana, M., de Lichtervelde, L., Mul, J. D., van de Peut, D., Devos, M., Simonis, N., Yildirim, M. A., Cokol, M., Kao, H. L., de Smet, A. S., Wang, H., Schlaitz, A. L., Hao, T., Milstein, S., Fan, C., Tipsword, M., Drew, K., Galli, M., Rhrissorrakrai, K., Drechsel, D., Koller, D., Roth, F. P., Iakoucheva, L. M., Dunker, A. K., Bonneau, R., Gunsalus, K. C., Hill, D. E., Piano, F., Tavernier, J., van den Heuvel, S., Hyman, A. A., and Vidal, M. (2008) A protein domain-based interactome network for C. elegans early embryogenesis *Cell* **134**, 534–45.
11. Braun, P., Tasan, M., Dreze, M., Barrios-Rodiles, M., Lemmens, I., Yu, H., Sahalie, J. M., Murray, R. R., Roncari, L., de Smet, A. S., Venkatesan, K., Rual, J. F., Vandenhaute, J., Cusick, M. E., Pawson, T., Hill, D. E., Tavernier, J., Wrana, J. L., Roth, F. P., and Vidal, M. (2009) An experimentally derived confidence score for binary protein-protein interactions *Nat Methods* **6**, 91–7.
12. Simonis, N., Rual, J. F., Carvunis, A. R., Tasan, M., Lemmens, I., Hirozane-Kishikawa, T., Hao, T., Sahalie, J. M., Venkatesan, K., Gebreab, F., Cevik, S., Klitgord, N., Fan, C., Braun, P., Li, N., Ayivi-Guedehoussou, N., Dann, E., Bertin, N., Szeto, D., Dricot, A., Yildirim, M. A., Lin, C., de Smet, A. S., Kao, H. L., Simon, C., Smolyar, A., Ahn, J. S., Tewari, M., Boxem, M., Milstein, S., Yu, H., Dreze, M., Vandenhaute, J., Gunsalus, K. C., Cusick, M. E., Hill, D. E., Tavernier, J., Roth, F. P., and Vidal, M. (2009) Empirically controlled mapping of the Caenorhabditis elegans protein-protein interactome network *Nat Methods* **6**, 47–54.
13. Venkatesan, K., Rual, J. F., Vazquez, A., Stelzl, U., Lemmens, I., Hirozane-Kishikawa, T., Hao, T., Zenkner, M., Xin, X., Goh, K. I., Yildirim, M. A., Simonis, N., Heinzmann, K., Gebreab, F., Sahalie, J. M., Cevik, S., Simon, C., de Smet, A. S., Dann, E., Smolyar, A., Vinayagam, A., Yu, H., Szeto, D., Borick, H., Dricot, A., Klitgord, N., Murray, R. R., Lin, C., Lalowski, M., Timm, J., Rau, K., Boone, C., Braun, P., Cusick, M. E., Roth, F. P., Hill, D. E., Tavernier, J., Wanker, E. E., Barabasi, A. L., and Vidal, M. (2009) An empirical framework for binary interactome mapping *Nat Methods* **6**, 83–90.
14. Lievens, S., Van der, H. J., Vertenten, E., Plum, J., Vandekerckhove, J., and Tavernier, J. (2004) Design of a fluorescence-activated cell sorting-based Mammalian protein-protein interaction trap *Methods Mol Biol* **263**, 293–310.
15. Lievens, S., Vanderroost, N., Van der Heyden, J., Gesellchen, V., Vidal, M., and Tavernier, J. (2009) Array MAPPIT: high-throughput interactome analysis in mammalian cells *J Proteome Res* **8**, 877–86.
16. Ziauddin, J., and Sabatini, D. M. (2001) Microarrays of cells expressing defined cDNAs *Nature* **411**, 107–10.
17. Lamesch, P., Li, N., Milstein, S., Fan, C., Hao, T., Szabo, G., Hu, Z., Venkatesan, K., Bethel, G., Martin, P., Rogers, J., Lawlor, S., McLaren, S., Dricot, A., Borick, H., Cusick, M. E., Vandenhaute, J., Dunham, I., Hill, D. E., and Vidal, M. (2007) hORFeome v3.1: a resource of human open reading frames representing over 10,000 human genes *Genomics* **89**, 307–15.

Chapter 18

MAPPIT as a High-Throughput Screening Assay for Modulators of Protein–Protein Interactions in HIV and HCV

Bertrand Van Schoubroeck, Koen Van Acker, Géry Dams, Dirk Jochmans, Reginald Clayton, Jan Martin Berke, Sam Lievens, José Van der Heyden, and Jan Tavernier

Abstract

The discovery of novel antivirals for HIV and HCV has been a focus of intensive research for many years. Where the inhibition of critical viral enzymes by small molecules has proven effective for many viruses, there is considerable merit in pursuing protein–protein interactions (PPIs) as targets for therapeutic intervention. The mammalian protein–protein interaction trap (MAPPIT) is a two-hybrid system used for the study of PPIs. The bait and prey proteins are linked to deficient cytokine receptor chimeras, where the bait and prey interaction and subsequent ligand stimulation restores JAK-STAT signaling, resulting in reporter gene expression controlled by a STAT3-responsive promoter. We report the use of MAPPIT as a high-throughput screening assay for the discovery of inhibitors or stimulators of the Vif–APOBEC3G interaction and the reverse transcriptase heterodimerization (RTp66-RTp51) for HIV and the NS4A–NS3 interaction for HCV.

Key words: Protein–protein interaction, Vif–APOBEC3G, NS4A–NS3, RTp66–RTp51, MAPPIT, High-throughput screening assay

1. Introduction

Scientists have been successful in the search for antiviral compounds to combat the diseases caused by human immunodeficiency virus (HIV) and hepatitis C virus (HCV), however, the disadvantages of current therapies highlight the need for new antiviral compounds. For HIV, therapeutic intervention through inhibition of essential viral enzymes has proven highly effective, but the emergence of

Bernhard Suter and Erich E. Wanker (eds.), *Two Hybrid Technologies: Methods and Protocols*, Methods in Molecular Biology, vol. 812, DOI 10.1007/978-1-61779-455-1_18, © Springer Science+Business Media, LLC 2012

resistant viruses against all antiretrovirals is seen (1). For HCV, the current standard of care (interferon/ribavirin) may result in adverse effects and has limited efficacy in certain HCV genotypes (2). Although enzyme inhibition is a proven approach for viral inhibition, essential protein–protein interactions (PPIs) offer a wealth of additional possibilities as therapeutic targets and are therefore of considerable interest.

The mammalian protein–protein interaction trap (MAPPIT) technology was first described by Eyckerman et al. (3). It is a two-hybrid system used to study PPIs. The chimeric receptor which is pivotal in MAPPIT consists of the extracellular part of the erythropoietin (EPO) receptor and the intracellular part of the leptin receptor in which the tyrosine residues were mutated to phenylalanine to prevent phosphorylation of the receptor. The bait is C terminally fused to the cytosolic portion of this chimeric receptor. The prey consists of the protein of interest fused to a portion of gp130, the signaling subunit of the interleukin-6 (IL-6) receptor containing four tyrosine motifs. Upon binding of EPO, the cytosolic portions of the receptor dimerize, allowing cross-phosphorylation of the associated Janus kinase 2 (JAK2) kinases. When bait and prey interact, the tyrosines of the gp130 subunit of the prey fusion protein are brought in close proximity to the activated JAK2 enzymes and will be phosphorylated, providing docking sites for Signal Transducer and Activator of Transcription 3 (STAT-3) proteins. These subsequently become phosphorylated and migrate as dimers to the nucleus where they activate the responsive rPAP1 promoter to express a reporter protein (Fig. 1).

Several PPIs have been described as being essential for viral replication and inhibitors of these interactions would be valuable additions to the antiviral armamentarium. MAPPIT technology is well suited to study such PPIs and to screen for inhibitors of the interaction.

The antiviral protein APOBEC3G (apolipoprotein B mRNA-editing enzyme, catalytic polypeptide-like 3G) was first described by Sheehy et al. (4). APOBEC3G deaminates cytidine to uracil during reverse transcription of the viral antisense RNA strand resulting in glycine to alanine hypermutation in the viral sense DNA strand (5), eventually giving rise to replication-deficient virus particles. This antiviral action of APOBEC3G is inhibited by the viral infectivity factor (Vif) which interacts with APOBEC3G in an E3-ligase complex targeting APOBEC3G for degradation by the proteasome (6). Hence, inhibition of the Vif–APOBEC3G interaction would result in the incorporation of APOBEC3G in the virion resulting in replication-incompetent virus particles, and is therefore considered an attractive target for antiviral drug intervention.

The non-nucleoside reverse transcriptase inhibitor (NNRTI) efavirenz (EFV) increases processing of the HIV-1 GAGPOL polyprotein (7). RTp66–RTp51 dimerization is important during GAGPOL

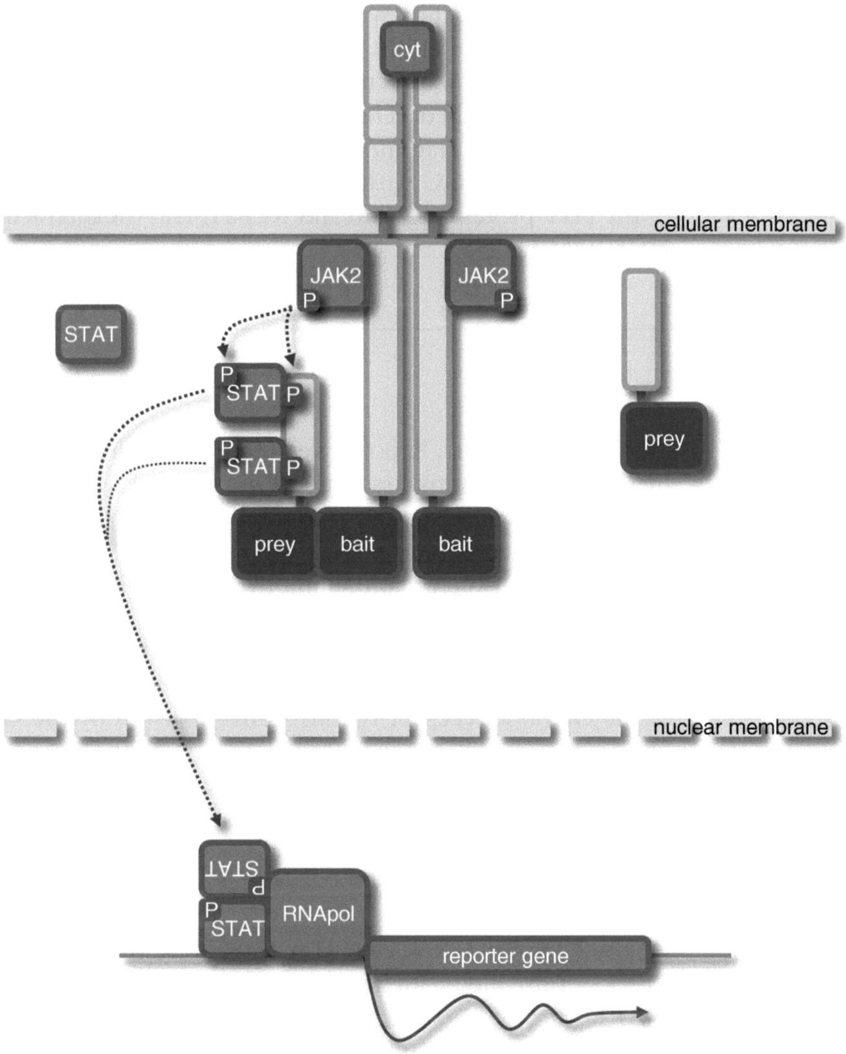

Fig. 1. MAPPIT assay principle. The MAPPIT technology was first described by Eyckerman et al. (3). HEK293 (human embryonic kidney 293) cells are used and the main processes occur close to the plasma membrane.

polyprotein processing (7). The usability of MAPPIT technology in the study of EFV-stimulated RTp66–RTp51 dimerization was first demonstrated by Pattyn et al. (8), highlighting the application of this valuable technique for the discovery of novel stimulators of RTp66–RTp51 dimerization.

The HCV NS3 protein is a multifunctional enzyme with serine protease activity responsible for the proteolytic cleavage of the NS3/NS4A, NS4A/NS4B, NS4B/NS5A, and NS5A/NS5B links in the HCV polyprotein (9). NS4A is a cofactor required for optimal NS3 protease activity (10). Inhibition of NS4A–NS3 PPI alone or in combination with interferon-α (IFN-α) has been shown to have an antiviral effect (11, 12).

Here, a configuration of the MAPPIT assay for high-throughput purposes is described. The resulting platform was used to search small molecules that modulate the interactions of the human protein APOBEC3G with the HIV protein Vif, the dimerization of the HIV proteins RTp66 and RTp51, and the interaction of the HCV proteins NS3 and NS4A.

2. Materials

2.1. Plasmid Constructs

pXP2d2-rPAP1-Luc: Reporter vector expressing luciferase controlled by the rat Pancreatitis Associated Protein-1 (rPAP1) promoter in a vector background engineered to have a very low constitutive promoter activity.

pCEL: Bait vector containing the extracellular part of the EPO receptor fused to the intracellular part of the leptin receptor which is made signaling-deficient by mutation of the three tyrosines to phenylalanines. A bait is fused to the C terminus and expression of the chimeric receptor is constitutively driven by the CMV promoter (3).

pMG2: Prey vector containing sequences of the STAT3 recruitment sites from the gp130 receptor fused to the prey of interest. The expression of the fusion protein is constitutively driven by the SRα promoter (3).

2.2. Stable Cell Lines

Hek293 Flp-In T-Rex cells (Invitrogen) were transfected with a STAT3-controlled luciferase reporter construct (pXP2d2-rPAP1-Luc). A neomycin resistance marker plasmid was co-transfected at a five times submolar ratio. G418 (500 μg/mL) resistant colonies were selected for the largest window of luciferase expression after Leukemia inhibitory factor (LIF) stimulation (1 ng/mL). The selected cell clone (clone 17) was subsequently transfected with the pCEL–p66, pCEL–Apobec3G, or pCEL–NS4A bait construct by integration in the Flp-In expression cassette guided by the Flp recombinase expressed by the co-transfected pOG44 plasmid (Invitrogen). The isogenic pools of transfectants were selected in hygromycin (100 μg/mL) and subsequently transfected with respectively the pMG2–p51, pMG2–VifSLQ/AAA, or pMG2–NS3 prey constructs together with the puromycin resistance marker. After selection in 500 ng/mL puromycin, individual colonies were isolated and tested for luciferase induction with EPO. Western blot analysis for prey expression was performed and the cell clones with the highest ratio for luciferase induction over prey expression were selected.

2.3. Cell Culture

1. Dulbecco's phosphate-buffered saline (D-PBS) (1×), liquid (Gibco).
2. Dulbecco/Vogt modified Eagle's minimal essential medium (DMEM) (Lonza).
3. Fetal calf serum (FCS) fetalclone II (Hyclone).
4. L-Glutamine (Gibco).
5. Gentamycine (Gibco).
6. Puromycine (Invivogen).
7. Hygromycine (Invivogen).
8. Geneticine (Gibco).
9. Isotone II dilution (Beckman Coulter).
10. Nunclon™ TripleFlask, 500 cm², straight neck, filter cap, (Nunc).
11. 0.05% Trypsin EDTA (Invitrogen).

Selection medium. DMEM, 10% FCS, 1% L-glutamine, gentamycine 250 μg/mL, geneticine 500 μg/mL, puromycine 1 μg/mL, hygromycine 100 μg/mL.
Cell culture medium. DMEM, 10% FCS, 1% L-glutamine, gentamycine 250 μg/mL.

2.4. Luciferase Signal Testing

1. D-PBS: (1×), liquid (Gibco).
2. DMEM (Lonza).
3. FCS fetalclone II (Hyclone).
4. L-Glutamine (Gibco).
5. Gentamycine (Gibco).
6. Isotone II dilution (Beckman Coulter).
7. Steady lite plus (Perkin Elmer).
8. Brite lite (Perkin Elmer).

2.5. Compounds

Efavirenz (EFV), Beviramat, Thiophene 2, and Saquinavir (SQV) were synthesized in-house. Jak 2 inhibitor II and STAT3 Inhibitor V were supplied by Calbiochem, Ach-806 was obtained from Acme Biosciences, HIV T-20 from Roche, HIV C34 from Abgent, and BMS806 from Asinex. The 50,000 compounds that were screened are J&J property.

3. Methods

3.1. Cell Culture

Cells are cultured twice a week for maintenance:

1. Discard the medium.
2. Wash once with D-PBS.

3. Add 5 mL of trypsin.
4. Shake the flask to spread the trypsin solution over the flask.
5. Discard the trypsin.
6. Incubate for 2 min at 37°C.
7. Add selection medium and split 1/10 in triple flask.
8. Add selection medium until the medium can be equally spread over the three layers of the triple flask.

Culturing cells for an experiment (see Note 1):

1. Discard the medium.
2. Wash once with D-PBS.
3. Add 5 mL of trypsin.
4. Shake the flask to spread the trypsin solution over the flask.
5. Discard the trypsin.
6. Incubate for 2 min at 37°C.
7. Resuspend in culture medium (see Note 2).
8. Count the number of cells.
9. Centrifuge the desired amount of cells.
10. Resuspend in the appropriate volume of culture medium to have the right cell density.
11. Pipette or drop the cells in a white 384-well plate.

3.2. Screening Protocol

Different parameters are evaluated during the optimization of the assay, including cell density, EPO concentration, DMSO concentration, incubation time, and substrates with the intention of obtaining a robust assay combining optimal luciferase signal with the lowest possible cell density. Due to the fact that there are pre-existing complexes between bait and prey at the time cells and compounds are mixed, a 24-h preincubation of the cells with the compounds is performed.

In Fig. 2 the fold induction for all three assays are summarized. The fold induction values are calculated as the ratio between the positive controls (cells with Epo) and the negative controls (cells without Epo).

The screening protocol is a 48-h (Fig. 3) assay, where the total volume per well of a 384-well plate is 45 μL. Two compound plating protocols are used:

Protocol 1:

1. White 384-well Lumitrac plates are prepared in which 10 μL of compound is plated at a fourfold concentration in cell culture medium at a DMSO concentration of 2%; control wells are filled with 10 μL of cell culture medium with 2% DMSO (see Note 3).

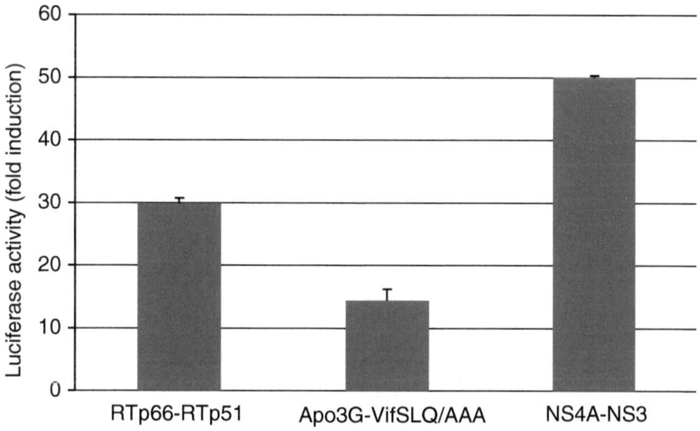

Fig. 2. Induction using final experimental conditions of 750 cells per well, 0.5% DMSO, 2 U/mL EPO final concentration and readout after 48 h. The fold induction values were calculated as the ratio between the positive controls (cells with Epo) and the negative controls (cells without Epo) of two independent experiments each with two plates consisting of 32 positive and 32 negative control wells.

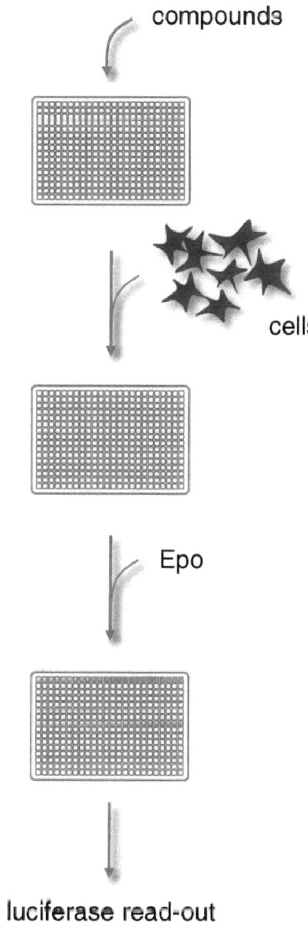

Fig. 3. Assay protocol. Compound plates were prepared (10 μL in 2% DMSO or 200 nL in 100% DMSO), 750 cells/well were plated in every well of the 384-well plate, followed by a preincubation of 24 h at 37°C. EPO was added in a final concentration of 2 U/mL, followed by incubation at 37°C. After 24 h of incubation 45 μL of Steady lite plus (1/1 v/v) was added and the plates were read with the viewlux apparatus at 1 s exposure and binning 6×6.

2. 30 μL of a cell suspension (25,000 cells/mL, 750 cells/well) is added to the plate also to the negative control wells, resulting in a 0.5% DMSO concentration (see Note 4).
3. 24-h preincubation at 37°C.
4. Add 5 μL of an 18 U/mL EPO solution to all wells except to the negative control wells, resulting in a final EPO concentration of 2 U/mL (see Note 5).
5. 24-h incubation at 37°C.
6. Add 45 μL steady lite+, and read the plate with a Viewlux apparatus (Perkin Elmer) at an exposure time of 1 s and binning 6×6.

Protocol 2:

1. White 384-well Lumitrac plates are prepared in which 200 nL of compound is plated at a 200-fold concentration in 100% DMSO.
2. 40 μL of a cell suspension (18,000 cells/mL, 750 cells/well) is added to the plate, including to the negative control wells, resulting in a 0.5% DMSO concentration.
3. 24-h preincubation at 37°C.
4. Add 5 μL of an 18 U/mL EPO solution to all wells except to the negative control wells, resulting in a final EPO concentration of 2 U/mL.
5. 24-h incubation at 37°C.
6. Add 45 μL steady lite+, and read the plate with a Viewlux apparatus (Perkin Elmer) at an exposure time of 1 s and binning 6×6.

The final compound concentration remains identical for both protocols.

3.3. Compound Testing

A significant challenge in the validation of the MAPPIT assay is represented by a lack of reference compounds for both the Vif–APOBEC3G and the NS4A–NS3 interactions. Pattyn et al. (8) demonstrated that EFV is a clear stimulator of RTp66–RTp51 dimerization, and this compound is used for further optimization experiments. Other NNRTIs, kinase inhibitors, and a STAT3 inhibitor are evaluated for the further validation of the assay.

3.3.1. Specificity Testing

During specificity testing, Protocol 1 is applied with a set of compounds with a known mechanism of action enabling prediction of their activity (Table 1). For each compound, nine dilutions are tested in quadruplicate. The compounds are tested in all three MAPPIT assays, and only EFV shows activity in the RTp66–RTp51 cell line as expected (8). In Table 1, the EC_{50} values are shown. The EFV EC_{50} value for the RTp66–RTp51 cell line is the concentration

Table 1
Specificity testing

			NS4A_NS3	APO3G_VifSLQ/AAA	RTp66_RTp51
Name	**Target**	**Reference**	**EC_{50} (μM)**	**EC_{50} (μM)**	**EC_{50} (μM)**
SQV	HIV protease	(13)	31.48	22.52	31.48
EFV (DMP-266)	HIV NNRTI	(14)	9.84	9.84	0.152[a]
HIV-T20	HIV fusion inhibitor	(15)	7.87	7.87	7.87
HIV-C34	HIV fusion inhibitor	(15)	0.98	0.98	0.98
BMS-806	HIV attachment inhibitor	(15)	31.48	31.48	31.48
Thiophene 2	HCV polymerase inhibitor	(16)	31.48	31.48	31.48
Bevirimat	HIV maturation inhibitor	(17, 18)	49.18	49.18	49.18

One experiment, with four replicas per tested concentration were performed
[a]Stimulation

where 50% stimulation above the positive control is observed; for all other compounds the EC_{50} value reflects the concentration where 50% inhibition is seen. None of the other compounds show any activity against one of the other assays which is expected because of their respective modes of action. The SQV EC_{50} value for the APOBEC3G–VifSLQ/AAA cell line is probably due to cytotoxicity because the average toxicity in the 3-day antiviral assay is 22 μM (data not shown).

3.3.2. Repeatability Testing

For the repeatability of the assay, i.e., the intra- and interexperiment variation, Protocol 1 is used with nine dilutions for every compound. Two experiments are performed and each compound is tested twice per experiment in quadruplicate. Table 2 shows the repeatability of the EC_{50} values for the reference compounds. Again for EFV, the EC_{50} value for the RTp66–RTp51 indicates the concentration where 50% stimulation above the positive control is seen; for all other compounds it indicates where 50% inhibition is observed. The EC_{50} value of Ach-806 was in the micromolar range for all three assays which is probably due to cytotoxicity. The JAK2 inhibitor II and the STAT3 inhibitors are active in all three assays because they act on proteins within the signal transduction pathway.

There is no significant difference in EC_{50} values within one experiment and between the two experiments demonstrating robustness and repeatability of the assay.

Table 2
Repeatability

Name	Target	Reference	NS4A_NS3 EC_{50}(μM)	NS4A_NS3 STDEV	APO3G_VifSLQ/AAA EC_{50} (μM)	APO3G_VifSLQ/AAA STDEV	RTp66_RTp51 EC_{50}(μM)	RTp66_RTp51 STDEV
EFV	HIV NNRTI	(14)	42.6	1.8	28.8	2	0.09[a]	0.005
Ach – 806	HCV NS4A	(19)	15.2	3.2	8.2	0.5	9.1	2.8
JAK2 inhibitor II	Janus kinase inhibitor	(20)	37.2	1.8	37.9	1.8	51.6	7.7
STAT3 Inhibitor V	STAT3 inhibitor	(21)	1.9	0.1	1.7	0.3	2.5	0.1
SQV	HIV protease	(13)	41.7	2.3	20.1	3.7	26.8	1.6

Two experiments, each with eight replicas were performed
STDEV standard deviation
[a]Stimulation

Table 3
Inhibitor and stimulator hits

Assay	Inhibitors Hits	Inhibitors Confirmed hits	Stimulators Hits	Stimulators Confirmed hits
RTp66–RTp51	15	0	53	4
APOBEC3G/VifSLQ/AAA	126	18	13	0
NS4A–NS3	29	0	23	0

EC_{50} values were calculated and hits identified by an EC_{50} value that was fourfold smaller than the EC_{50} value of the two other assays

3.3.3. Screening of 50,000 Compounds

For screening purposes, both Protocols 1 and 2 are used depending on the layout of the plates of the particular library. Approximately 50,000 compounds are screened for the three MAPPIT assays in parallel, in which two MAPPIT assays serve as counter screen for the other one (see Note 6).

Hit criteria are defined for inhibitors as well as for stimulators. For inhibitors, the EC_{50} value has to be at least fourfold lower than the EC_{50} of the same compound in the two other assays, or the EC_{50} has to be lower than the lowest test concentration in only one assay. For the stimulators, the maximum response value has to be at least 50% above the positive controls and has to be fourfold higher than the maximum response values of the two other assays.

The screen identified 29 hits for the inhibition of NS4A–NS3 interaction, 126 inhibitors for the APOBEC3G–Vif interaction, and 53 stimulators of RT heterodimerization. Of these, 18 inhibitor hits for APOBEC3G–Vif, and four stimulator hits for RT heterodimerization were confirmed (Table 3).

Table 4
Z′ values

Assays	Pilot Average Z′	Pilot STDEV	Screen Average Z′	Screen STDEV
NS4A_NS3	0.71	0.08	0.73	0.08
APO3G_VifSLQ/AAA	0.69	0.07	0.68	0.12
RTp66_RTp51	0.67	0.14	0.70	0.12

Z′ values of 136 plates per assay for the pilot screen (pilot) where each plate employed 32 positive control wells and 32 negative control wells. For the larger screen of a further ~40,000 compounds, 538 plates were used, where each plate also contained 32 positive control wells and 32 negative control wells
STDEV standard deviation

Because hit confirmation is very low, the *Z′* values (Table 4) of the plates in the different assays are calculated. The *Z′* is calculated according to the formula: $Z' = |\mu_{c+} - \mu_{c-}| - (3\sigma_{c+} + 3\sigma_{c-})/|\mu_{c+} - \mu_{c-}|$, in which μ_{c+} – average value of the positive controls, μ_{c-} = average value of the negative controls, σ_{c+} – Standard deviation of the positive controls, and σ_{c-} = Standard deviation of the negative controls. *Z′* values are found to be acceptable, indicating robustness of the assay.

Further profiling of the antiviral properties of these hits will learn if they could proceed to a hit-to-lead program.

4. Notes

1. Do not let the cells grow to confluency before plating in order to obtain an optimal signal-to-background ratio.
2. In case of aggregation of the cells a cell strainer (BD Falcon catalog #: 352340, 40 μm) can be used to select for single cells, in order to reach the proper cell density in the wells and thus reducing variation.
3. To reduce or to prevent edge effects in the compound plates use the plates as quickly as possible, or freeze the plates at −20°C for longer storage.
4. Thaw the plates for a minimum of 1 h before plating the cells or put them at 4°C the day before, in order to reduce stress or not to kill the cells.
5. Do a titration experiment for every new batch of EPO to check its specific biological activity. All experiments are performed under saturating EPO conditions to reduce variation.

6. During screening, a plate consisting of reference compounds for which the mechanism of action and the activity is known, is used as a technical control.

References

1. Shafer, R. W., and Schapiro, J. M. (2008) HIV-1 drug resistance mutations: an updated framework for the second decade of HAART *AIDS Rev* **10,** 67–84.
2. Shimakami, T., Lanford, R. E., and Lemon, S. M. (2009) Hepatitis C: recent successes and continuing challenges in the development of improved treatment modalities *Curr Opin Pharmacol* **9**, 537–44.
3. Eyckerman, S., Verhee, A., Van der Heyden, J., Lemmens, I., Van Ostade, X., Vandekerckhove, J., and Tavernier, J. (2001) Design and application of a cytokine-receptor-based interaction trap *Nature Cell Biology* **3**, 1114–19.
4. Sheehy, A. M., Gaddis, N. C., Choi, J. D., and Malim, M. H. (2002) Isolation of a human gene that inhibits HIV-1 infection and is suppressed by the viral Vif protein *Nature* **418**, 646–50.
5. Harris, R. S., Bishop, K. N., Sheehy, A. M., Craig, H. M., Petersen-Mahrt, S. K., Watt, I. N., Neuberger, M. S., and Malim, M. H. (2003) DNA deamination mediates innate immunity to retroviral infection *Cell* **113**, 803–9.
6. Marin, M., Rose, K. M., Kozak, S. L., and Kabat, D. (2003) HIV-1 Vif protein binds the editing enzyme APOBEC3G and induces its degradation *Nat Med* **9**, 1398–403.
7. Figueiredo, A., Moore, K. L., Mak, J., Sluis-Cremer, N., de Bethune, M. P., and Tachedjian, G. (2006) Potent nonnucleoside reverse transcriptase inhibitors target HIV-1 Gag-Pol *PLoS Pathog* **2**, e119.
8. Pattyn, E., Lavens, D., Van der Heyden, J., Verhee, A., Lievens, S., Lemmens, I., Hallenberger, S., Jochmans, D., and Tavernier, J. (2008) MAPPIT (MAmmalian Protein-Protein Interaction Trap) as a tool to study HIV reverse transcriptase dimerization in intact human cells *Journal of Virological Methods* **153**, 7–15.
9. Bartenschlager, R., Ahlborn-Laake, L., Mous, J., and Jacobsen, H. (1993) Nonstructural protein 3 of the hepatitis C virus encodes a serine-type proteinase required for cleavage at the NS3/4 and NS4/5 junctions *J Virol* **67**, 3835–44.
10. Tomei, L., Failla, C., Vitale, R. L., Bianchi, E., and De Francesco, R. (1996) A central hydrophobic domain of the hepatitis C virus NS4A protein is necessary and sufficient for the activation of the NS3 protease *J Gen Virol* **77 (Pt 5)**, 1065–70.
11. Lin, K., Kwong, A. D., and Lin, C. (2004) Combination of a hepatitis C virus NS3-NS4A protease inhibitor and alpha interferon synergistically inhibits viral RNA replication and facilitates viral RNA clearance in replicon cells *Antimicrob Agents Chemother* **48**, 4784–92.
12. Lin, K., Perni, R. B., Kwong, A. D., and Lin, C. (2006) VX-950, a novel hepatitis C virus (HCV) NS3-4A protease inhibitor, exhibits potent antiviral activities in HCv replicon cells *Antimicrob Agents Chemother* **50**, 1813–22.
13. Craig, J. C., Duncan, I. B., Hockley, D., Grief, C., Roberts, N. A., and Mills, J. S. (1991) Antiviral properties of Ro 31–8959, an inhibitor of human immunodeficiency virus (HIV) proteinase *Antiviral Res* **16**, 295–305.
14. Young, S. D., Britcher, S. F., Tran, L. O., Payne, L. S., Lumma, W. C., Lyle, T. A., Huff, J. R., Anderson, P. S., Olsen, D. B., Carroll, S. S., and et al. (1995) L-743, 726 (DMP-266): a novel, highly potent nonnucleoside inhibitor of the human immunodeficiency virus type 1 reverse transcriptase *Antimicrob Agents Chemother* **39**, 2602–5.
15. Qian, K., Morris-Natschke, S. L., and Lee, K. H. (2009) HIV entry inhibitors and their potential in HIV therapy *Med Res Rev* **29**, 369–93.
16. Chan, L., Pereira, O., Reddy, T. J., Das, S. K., Poisson, C., Courchesne, M., Proulx, M., Siddiqui, A., Yannopoulos, C. G., Nguyen-Ba, N., Roy, C., Nasturica, D., Moinet, C., Bethell, R., Hamel, M., L'Heureux, L., David, M., Nicolas, O., Courtemanche-Asselin, P., Brunette, S., Bilimoria, D., and Bedard, J. (2004) Discovery of thiophene-2-carboxylic acids as potent inhibitors of HCV NS5B polymerase and HCV subgenomic RNA replication. Part 2: tertiary amides *Bioorg Med Chem Lett* **14**, 797–800.
17. Li, F., Goila-Gaur, R., Salzwedel, K., Kilgore, N. R., Reddick, M., Matallana, C., Castillo, A., Zoumplis, D., Martin, D. E., Orenstein, J. M., Allaway, G. P., Freed, E. O., and Wild, C. T. (2003) PA-457: a potent HIV inhibitor that disrupts core condensation by targeting a late

step in Gag processing *Proc Natl Acad Sci USA* **100**, 13555–60.

18. Zhou, J., Yuan, X., Dismuke, D., Forshey, B. M., Lundquist, C., Lee, K. H., Aiken, C., and Chen, C. H. (2004) Small-molecule inhibition of human immunodeficiency virus type 1 replication by specific targeting of the final step of virion maturation *J Virol* **78**, 922–9.
19. Yang, W., Zhao, Y., Fabrycki, J., Hou, X., Nie, X., Sanchez, A., Phadke, A., Deshpande, M., Agarwal, A., and Huang, M. (2008) Selection of replicon variants resistant to ACH-806, a novel hepatitis C virus inhibitor with no cross-resistance to NS3 protease and NS5B polymerase inhibitors *Antimicrob Agents Chemother* **52**, 2043–52.
20. Sandberg, E. M., Ma, X., He, K., Frank, S. J., Ostrov, D. A., and Sayeski, P. P. (2005) Identification of 1,2,3,4,5,6-hexabromocyclohexane as a small molecule inhibitor of jak2 tyrosine kinase autophosphorylation [correction of autophophorylation] *J Med Chem* **48**, 2526–33.
21. Schust, J., Sperl, B., Hollis, A., Mayer, T. U., and Berg, T. (2006) Stattic: a small-molecule inhibitor of STAT3 activation and dimerization *Chem Biol* **13**, 1235–42.

Chapter 19

Integrated Measurement of Split TEV and *Cis*-Regulatory Assays Using EXT Encoded Reporter Libraries

Anna Botvinik and Moritz J. Rossner

Abstract

Intracellular signaling initiated by extracellular ligands that activate cell surface receptors is a complicated process that involves multiple interconnected biochemical steps. Protein–protein interactions are often regulated by activated kinases via phosphorylation of specific residues. Such transient regulated interactions are central to many signaling cascades. Downstream signaling converges at the level of transcription factors to finally regulate adaptive transcriptional responses. There are powerful methods available to study transcriptional changes even at a global level, however, measuring upstream regulatory mechanisms is still challenging. We designed an experimental approach termed EXTassay that enables the parallel analysis of signaling events upstream of gene expression. We make use of different types of reporter gene assays that are invariably linked to unique *ex*pressed oligonucleotide *t*ags (EXTs) serving as quantitative decoders of respective assays. EXT-reporters can be introduced into living cells and analyzed in pools by microarray hybridization or sequencing.

Key words: Cellular signaling, Regulated protein–protein interactions, Phosphorylation, Split TEV assays, *Cis*-regulatory assays, EXT, Microarray

1. Introduction

Cellular signaling cascades are considered as complex networks of interacting proteins that function at different levels within living cells (1). A prominent canonical example is the Neuregulin–ErbB signaling pathway that has been extensively studied (2). Mainly biochemical methods but more recently also novel techniques such as protein arrays and mass-spectrometry coupled pulldown experiments have been applied to study Neuregulin–ErbB and other signaling cascades (3, 4). These approaches enabled a deeper understanding of receptor activation mechanisms and signaling

Bernhard Suter and Erich E. Wanker (eds.), *Two Hybrid Technologies: Methods and Protocols*, Methods in Molecular Biology, vol. 812, DOI 10.1007/978-1-61779-455-1_19, © Springer Science+Business Media, LLC 2012

by adapter protein recruitment of phosphorylated receptors. These methods can, however, provide only a partial view onto the molecularly very different signaling events that are associated with an extracellular stimulus.

In an attempt to overcome some limitations of existing techniques, we have developed a novel integrated and highly scalable reporter system, which we termed EXTassay. The system relies on different cellular reporter gene assays that can be combined with RNA-based *ex*pressed *t*ag (EXT) reporters in a flexible way. The rational design of highly complex EXT libraries by combinatorial synthesis of defined building blocks (5) ensures that each EXT reporter has identical GC contents, highly similar melting temperatures, and is virtually free of secondary structures (6). As a consequence, EXT reporters have been shown to function with high performance and predictability using DNA microarrays or next-generation sequencing as readout techniques (6). As a biological proof-of-principle showing the applicability of EXTassays, split TEV (7, 8) and *cis*-regulatory reporter gene assays were combined to monitor ERBB receptor activation and downstream signaling simultaneously. Here, we describe a detailed protocol on the synthesis and subcloning strategy of EXT libraries, a simple transfection strategy for multiplexed assays and guidelines for microarray analyses.

2. Materials

2.1. Cloning of EXT Libraries

1. EXT oligodeoxynucleotide synthesis is performed according to the phosphoramidite method on a solid support using phosphoramidite DNA synthesis columns and, e.g., a 394 DNA Synthesizer (Applied Biosystems).
2. PAGE gels and electrophoresis setup (e.g., from Invitrogen).
3. Primers:

EXTlib-F	AGCTAGTTGCTAAGTCTGCCGAGTAGAATTAACCCTCA CTAAAGGGTAGGTGACACTAT
EXTlib-R	TCGTACATGCATTGACTCGCGTCTACTAATACGACTCA CTATAGG
EXT_B3	GGGGCAACTTTGTATAATAAAGTTGAGCTAGTTGCTAA GTCTGCCGAGTAG
EXT_B2	GGGGCCACTTTGTACAAGAAAGCTGTCGTACATGCATT GACTCGCGTCTAC
pENTR_s	CGCGTTAACGCTAGCATGGATCTC
Dec2	TCGTACATGCATTGACTCGCGTCTAC

4. High fidelity or proofreading PCR reagents (e.g., HotStarTaq from Qiagen).
5. Multisite Gateway Pro 3.0 Kit, Gateway BP clonase II and LR clonase II Plus enzyme mix.
6. ElectroMax DH10B or One Shot Mach1 competent cells (all reagents from Invitrogen).
7. LB medium and LB-agar plates (50 μg/ml kanamycin, 100 μg/ml ampicillin, or 100 μg/ml carbenicillin).
8. Donor and destination vectors for entry and expression constructs can be purchased from Invitrogen (pDONR-P1P4, pDONR-P4rP3r, pDONR-P3P2) or generated by cloning using a Gateway recombination cassette (Invitrogen) into any expression vectors of choice (e.g., pGL3 from Promega).
9. Plasmid-DNA purification kit at the Mini and Midiprep scale (Qiagen).

2.2. Transfection and RNA/DNA Isolation

1. PC12 rat adrenal pheochromocytoma cells or any other mammalian cell line or primary cells amenable to transient transfection.
2. Dulbecco's Modified Eagle's Medium (DMEM) (Gibco) supplemented with 10% fetal bovine serum (FBS) (Gibco) and 5% horse serum (HS) (Gibco). Alternatively, use appropriate growth medium for the cell line of choice.
3. Transfection reagents, e.g., Lipofectamine 2000 (Invitrogen) and OptiMEM reduced serum medium (Invitrogen).
4. RNA preparation reagents including the TRIZOL reagent (Invitrogen) and the RNeasy kit (Qiagen).
5. 3 M Sodium acetate (pH 5.2).
6. 7.5 M Ammonium acetate mol. biol. grade.
7. Glycogen.
8. Ethyl alcohol p.a.
9. Reverse transcription reagents: random nanomer primer, Superscript III reverse transcriptase (Invitrogen), deoxyribonucleotides (Roche).
10. PCR reagents, e.g., HotStarTaq (Qiagen).
11. Primers:

Dec1	AGCTAGTTGCTAAGTCTGCCGAGTAG
Dec2	TCGTACATGCATTGACTCGCGTCTAC

2.3. EXT Labeling and Hybridization to EXTarrays

1. T7-RNA polymerase MEGAscript kit (Ambion).
2. 5-(3-Aminoallyl)-UTP (aaUTP) (Ambion).
3. RNAeasy kit (Qiagen).

4. Cy3 and Cy5 monoreactive dyes (GE Healthcare). One tube of the lyophilized dye is dissolved in 55 μl DMSO and used for five labeling reactions. Remaining dye solution can be stored at −20°C protected from humidity for over 6 months.
5. Coupling buffer (Ambion). 0.5 M $NaHCO_3$ pH 9 could also serve as 10× coupling buffer.
6. 4 M Hydroxylamine (Ambion).
7. Custom microarrays from Agilent (http://www.agilent.com) or Roche/Nimblegen (http://www.nimblegen.com).
8. Hybridization reagents (Agilent): HI-RPM microarray hybridization buffer and blocking agent.
9. Formamide (Sigma-Aldrich).
10. 20× SSPE buffer (AppliChem).
11. Hybridization reagents (Nimblegen): Hybridization kit, sample tracking control kit, wash buffer kit.
12. Acetonitryl (J.T. Baker).
13. GenePix 4200A microarray scanner (Axon Instruments) and Analysis Software GenePix Pro 6.1 (Axon).
14. Feature extraction software (Agilent).
15. NimbleScan (Roche NimbleGen).
16. R statistical computing environment (open source at http://www.bioconductor.org).

3. Methods

3.1. Cloning of EXT Libraries and Expression Constructs

The synthesis follows a mix and divide strategy combining defined building blocks (depicted in Fig. 1). The combinatorial synthesis can be performed following standard procedures, e.g., on a solid support using phosphoramidite DNA synthesis columns and any applicable DNA synthesizer.

1. A 15-mer 3′ invariable region (TAGGTGACACTAT) is synthesized in eight cartridges in parallel followed by the first set of the eight different words. Subsequently, by mixing and dividing the samples (see Note 1) another set of the eight different words are synthesized. The cycle is repeated to introduce three additional word blocks (five in total) followed by a symmetrical 9-mer core sequence that are generated by splitting the samples into two equal portions after each next nucleotide (WSWSSSWSW, where W = A/T and S = C/G). The synthesis is completed by five additional word structures and adding a 5′ invariable region (CCTATAGTGAGTCGT) (summarized in Fig. 1).

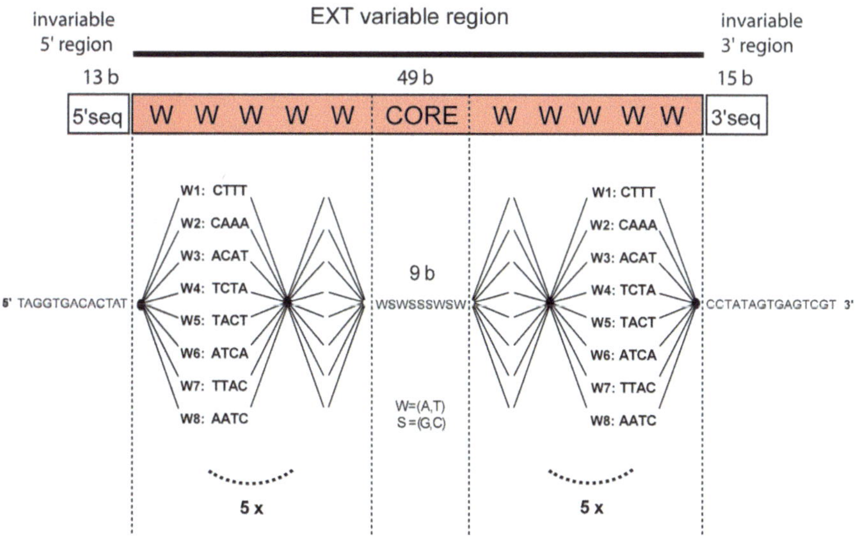

Fig. 1. EXT design. Each EXT contains 5′ and 3′ invariable regions and a central 49-mer variable region. The variable region has a symmetric organization and consists of ten structural modules called "words" (W) that flank a central core region. Each word consists of four nucleotides. Out of all possible four-nucleotide combinations only eight are used that are built of three adenosine (A) or thymidine (T) residues, only one cytosine (C), and no guanine (G) residues at all. Each of the eight possible words (CTTT, CAAA, ACAT, TCTA, TACT, ATCA, TTAC, and AATC) occurs randomly at each of the ten possible positions. The core region comprises nine bases of alternating A, T (W) or G, C (S) residues with three centrals G, C (S) residues. The total theoretical complexity of an EXT library following this design can be calculated as: $8^{10} \times 2^{9} \approx 5.5 \times 10^{11}$.

2. The resulting library of 77-mer oligonucleotides is PAGE gel purified and diluted to concentrations not below 50 μM for storage.
3. The EXT oligonucleotide library is diluted to 1 nM and 1 fmol (1 μl) is PCR amplified for 10–15 cycles with the primers EXTlib_F and EXTlib_R complementary to the 5′ and 3′ invariable regions extended by T3 (optional) and T7 promoter sequences for later in vitro transcription (IVT) purposes (see Note 2).
4. The PCR product is reamplified for 10–15 cycles with a second set of primers EXT_B3 and EXT_B2 to introduce attB3 and attB2 recombination sites.
5. PCR products are agarose gel purified and recombined with the pDONRP3-P2 vector using BP Clonase enzyme mix to generate entry shuttle clone libraries (see Fig. 2). The corresponding "BP" reaction is set up in 20 μl volume with 4 μl pDONR-P3-P2 vector (50 ng/μl), 12 μl PCR product (5 ng/μl), and 4 μl BP clonase II enzyme mix.
6. After overnight incubation at 25°C, the recombination product is purified by phenol chloroform extraction and ethanol precipitation, resuspended in 5 μl molecular biology grade H_2O and electroporated into DH10b cells.

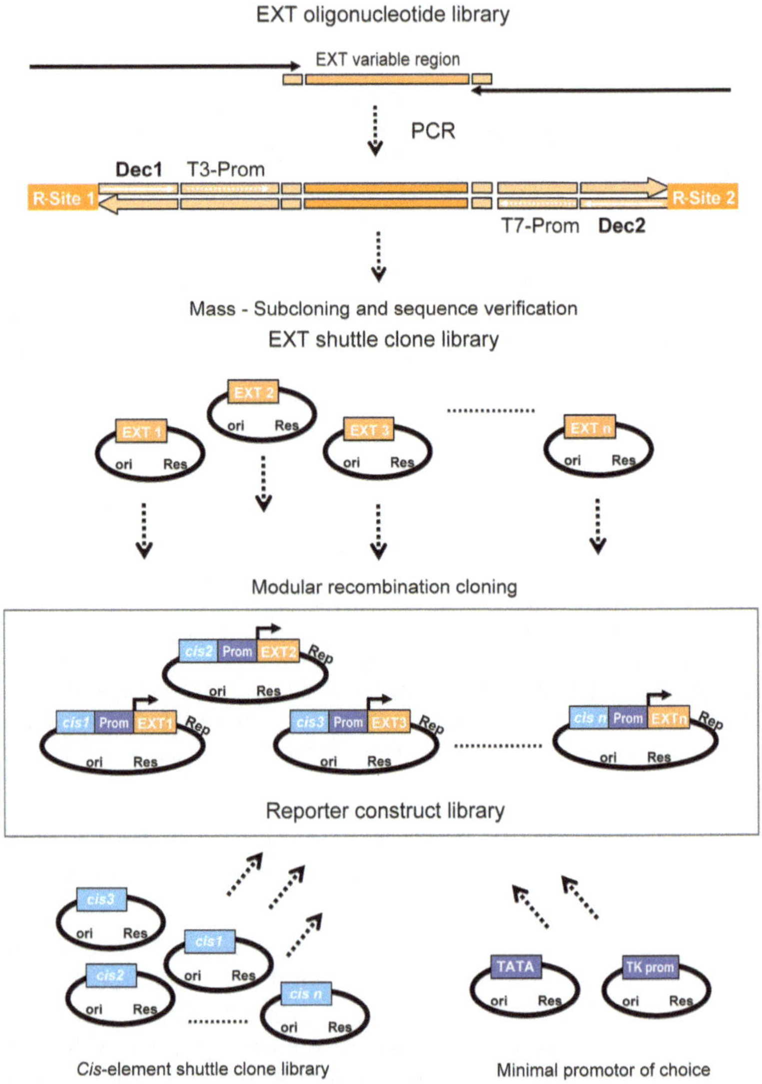

Fig. 2. Subcloning of the EXT library. The EXT oligonucleotide library was PCR amplified and subcloned to generate a pool of shuttle clones. Several functional elements were introduced flanking the EXT sequence: recombination sites (R-Site1/2), "decoder" primer binding sites (Dec1/2), and viral promoters (T3/T7). Three shuttle clones harboring different functional elements were used in one recombination reaction to place a *cis*-regulatory element, a minimal promoter, and an EXT 5′ to a firefly luciferase reporter gene (Rep). Depending on a shuttle clone combination, defined *cis*-regulatory elements were linked to unique EXTs.

7. Electroporation is performed with 1 mm gap cuvettes using, e.g., a BioRad-Gene Pulser II and standard settings. The bacteria are allowed to recover in 2 ml SOC medium for 1 h. 10 and 100 μl of the bacterial suspension is plated to estimate the entry library complexity.
8. Adjust the complexity to exceed your future needs by at least tenfold to avoid unwanted duplicates in further subclonings (see Note 3). Remaining bacteria are transferred to 200 ml LB

medium, 50 μg/ml kanamycin and are allowed to grow at 37°C for 1 h with mild shaking.

9. After subcloning, the EXT library has to be screened for full length unique EXTs to exclude repeating sequences and products of incomplete oligonucleotide synthesis.
10. *Option 1: Analysis of individual entry clones.* A portion of the bacteria are kept at 4°C over-night and appropriate volumes are plated according to the determination of recombinants adjusted to 100–500 clones per 10 cm LB-agar dish.
11. From these plates, e.g., 96 colonies are screened by "colony PCR" by dipping a sterile tooth pick into wells of a 96-well PCR plate containing all reagents including the primers pENTR_s and Dec2. Subsequently, the tooth pick is transferred to the corresponding positions in 96-deep well dishes (with ~1 ml LB/well) and grown at 37°C for at least 16 h. The PCR products are separated on 4% agarose gels and clones are discarded that show reduced insert length (<280 bp). The PCR products of proper size can be directly sequenced using the pENTR_s primer. Bacterial colonies corresponding to perfect clones (expected success rate of nearly 50%) are subjected to plasmid preparations. These can be used in individual subclonings to generate expression clones with predefined EXTs (illustrated in Fig. 2).
12. *Option 2: Mass-Subcloning of entry clones.* The culture is centrifuged and the pellet is used for a midi-sized plasmid DNA preparation and results in an entry-clone library of a given complexity. Imperfect EXTs are excluded during PCR screening and sequencing of final reporter constructs through a similar procedure as described above.
13. The entry clone EXT library (harboring attL3/L2 sites) is subsequently recombined along with *cis*-element (cloned into the pDONR-P1P4 vector) and minimal promoter (cloned into the pDONR-P4P3) entry clones of choice and a destination plasmid harboring a reporter gene such as firefly luciferase (e.g., pDEST_GL3_Luci) to generate a library of *cis*-regulatory EXT reporter constructs (see (6) for selected *cis*-element and minimal promoter sequences) as illustrated in Fig. 2.
14. The "LR" recombination reaction is set up in 20 μl with 40 fmol of the destination vector and 20 fmol of each of the entry clone and incubated at RT overnight followed by a Proteinase K digestion according to the Multisite Gateway manual.
15. The reaction is purified by PEG precipitation and transformed into DH10b (see Note 4). Cells are transferred to 200 ml of the respective LB selection medium and grown for 4 h at 37°C.

10 and 100 μl of the suspension are plated out to estimate the complexity and the rest is kept at 4°C overnight.

16. The next day, the volume corresponding to 2,000–4,000 clones is plated on square (25×25 cm) agar plates with antibiotics (in the case of pDEST_GL3_Luci use 50 μg/ml Carbenicillin) and cultured overnight.
17. The colonies are picked and transferred into 96-well plates containing 100 μl LB per well with antibiotics. Clone picking can be done manually or semiautomated using, e.g., by a QPix Colony Picking Robot (Genetix).
18. The bacteria are grown overnight and frozen at −80°C. The clones are screened by colony PCR using 0.5 μl of the culture as a template. The inserts are amplified with Dec1 and Luc_as primers (when pDEST_GL3_Luci was used) and the PCR products are directly sequenced with the antisense primer.

3.2. Transfection, RNA/DNA Isolation, and cDNA Synthesis

For some multiplexed assays (e.g., *cis*-element activity profiling under defined conditions), the cells are transfected with the entire *cis*-EXT-reporter library using standard protocols. However, for some experimental designs (e.g., integrated measurement of pathway activation including protein–protein interaction assays) multiple transfections are required. To ensure equal experimental conditions and to enable simultaneous analysis of all samples, the cells should be transfected in suspension and combined immediately after transfection. Alternatively, adherent cells could be trypsinized after transfection which, however, increases the workload and may pose additional stress to the cells.

1. For transfection in suspension, e.g., PC12 cells are trypsinized and triturated through a 24-G needle five times to remove any cell clumps and the cell number is determined. Desired numbers of cells are pelleted by centrifugation for 5 min at 800×*g* and resuspended in DMEM 1% horse serum and antibiotics to reach a cell density of 100,000 per 100 μl.
2. The following transfection components are assembled (adjusted to 100,000 cells): Lipofectamine 2000 0.4 μl, plasmid DNA 100 ng, OptiMEM 10 μl. The DNA and the Lipofectamine 2000 reagent each are diluted in one half of the OptiMEM, mixed, and combined together (see Note 5).
3. The Lipofectamine–DNA complexes are allowed to form within 20 min at room temperature and are immediately applied to the cell suspension. The tubes are incubated at 37°C for 4 h without shaking.
4. Treated cells are subsequently centrifuged to remove the transfection mixture, resuspended in fresh growth medium, mixed and plated out according to the experimental design.

5. Following this procedure, cells from several transfections can be combined and cultured under identical experimental conditions. As estimation, 10^5–10^6 cells should be transfected with 2–5 ng of each EXT reporter constructs depending on the overall complexity of the EXTassay. We successfully used 20–100 different plasmids in one transfection although further upscaling may be possible.
6. *Cis*-reporter plasmids at miniprep DNA quality are premixed and used for transfection to minimize the pipetting errors.
7. Plasmids required for the protein–protein interaction assays are derived from midi-prep DNA preparations and 100 ng of DNA is used per 100,000 cells for each multiplexed experiment. Figure 3 illustrates control experiments demonstrating high efficiency of transfection in suspension measured with firefly Luciferase reporter gene assays.
8. RNA and input plasmid DNA can be isolated simultaneously by a combined TRIZOL-RNA column purification (see Note 6). Care should be taken (and properly controlled, see below) to not carry over extensive amounts of supercoiled plasmid into the RNA samples.
9. For the first step of RNA/DNA isolation, you may essentially follow the standard TRIZOL protocol according to the manufacturer (avoid PBS washing steps to reduce risk of cell loss for 96-well plates).

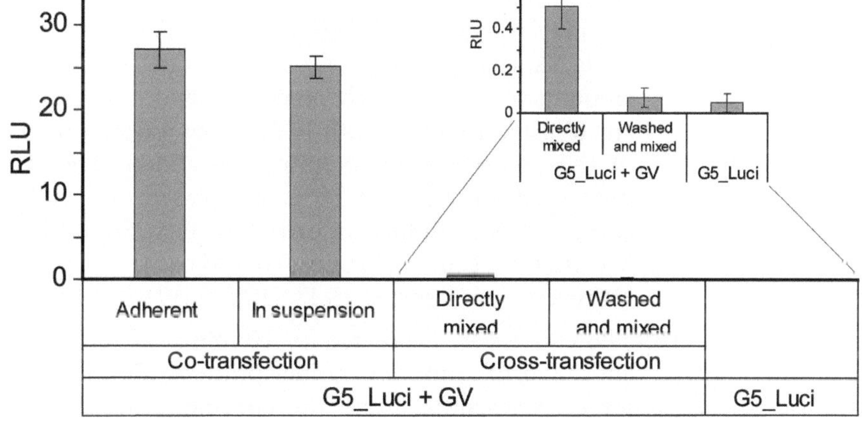

Fig. 3. Transfection of PC12-OFF cells in suspension. PC12-OFF cells were co-transfected with the Gal4-VP16 (GV) synthetic transcription activator and the corresponding GV responsive firefly luciferase reporter construct "G5_Luci." Twenty-four hours after transfection, the cells were lysed and luciferase expression levels were quantified in a chemiluminescence assay. Data are given as relative luminescence units (RLU). The Luciferase assays show that PC12 cells could be transfected on a plate (adherent) or in suspension with similar efficiencies. Independent transfections with GV and G5_Luci and subsequent pooling of cell suspensions after 4 h yielded only slight cross-transfection effects upon direct mixing of both samples. Slight activation of the G5-Luci reporter resulting from the cross-transfection of the GV was completely abolished by introducing a single "washing" step before mixing the samples (see *inlay*).

10. The following amounts of the Trizol reagent and chloroform should be used:

Plate	Area (cm^2)	Cell number	Trizol	Chloroform (μl)
96-Well	0.3	50,000	125 μl	25
24-Well	2	250,000	500 μl	100
6-Well	10	1,000,000	1 ml	200

11. TRIZOL reagent is added to the wells, left for at least 1 min and lysates are transferred into 2 ml test tubes, vortexed to ensure complete lysis and 1/5th volume of chloroform is added.
12. Each sample is vortexed thoroughly and centrifuged for 15 min at 4°C, 12,000×*g*. Aqueous phase of the TRIZOL lysate is transferred to a new tube. One-third of the sample is used for DNA and two-third for RNA isolation.
13. The DNA is precipitated with sodium acetate (f.c. 0.3 M) and 1 volume isopropanol.
14. The RNA sample is subjected to an RNeasy column purification including an on-column DNase treatment, according to the manufacturer's protocol.
15. The RNA is precipitated with ammonium acetate salt (f.c. 2.5 M), as carrier glycogen and 2 volumes ethyl alcohol,
16. The RNA is washed with 100 μl of 70% EtOH, shortly dried and resuspended in 4–6 μl of H_2O and quantified with appropriate instruments (e.g., nanodrop or bioanalyzer).
17. For cDNA synthesis, 0.5–1 μg total RNA is reverse transcribed using Superscript III reverse transcriptase (see Note 7). Equal amounts of RNA for each sample are used. The sample volumes are adjusted to 4.5 μl with RNase-free water, 1 μl (120 pmol) of the random nanomer primer is added. The samples are denatured for 2 min at 70°C and placed on ice. Then other components are added in order: 5× First Strand buffer, 2 μl; 0.1 M DTT, 1 μl; dNTP mix (10 mM each), 0.5 μl; Superscript III reverse transcriptase (RT) (200 U/μl) 1 μl and 10 μl.
18. The samples are incubated for 10 min at 25°C to allow for the annealing of random nanomer primers, then for 1 h at 50°C for cDNA synthesis and the enzyme is heat inactivated for 5 min at 85°C.
19. The cDNA should be diluted 1:10 with water and 1–2 μl should be used for PCR amplification.
20. EXTs are amplified from both cDNA and plasmid DNA samples with Dec1 and Dec2 primers and standard PCR reagents (to avoid unnecessary overcycling quantitative PCR, e.g., using SYBR Green could be performed).

21. In parallel, a PCR with Dec1 and Dec2 primers is run on the RT-control samples to determine the success of the cDNA synthesis and potential carry over of plasmid DNA during RNA purification.

3.3. EXT Labeling and Hybridization to EXTarrays

Unpurified PCR products carrying T7-promoter sequences can be used without further purification as a template for IVT incorporating 5-(3-aminoallyl)-modified UTP as illustrated in Fig. 4. The IVT reaction volume can be downscaled relative to the manufacturer's protocol (MEGAscript Kit, Ambion) still yielding sufficient material.

1. For the IVT reaction assemble the following reagents: PCR product 4 μl; 10× Buffer 1 μl; ATP (7.5 mM) 1 μl; CTP (7.5 mM) 1 μl; GTP (7.5 mM) 1 μl; UTP (7.5 mM) 0.5 μl; 5-(3-aminoallyl-UTP) (50 mM, Ambion) 0.75 μl; T7 RNA Polymerase 1 μl.
2. Incubate the reaction overnight at 37°C using a heated lid at 42°C to prevent evaporation.
3. Degrade the template DNA subsequently by adding 1 μl DNase to each sample and incubation at 37°C for 15 min.
4. The typical yield of an EXT-IVT is 5–10 μg of aminoallyl-modified RNA (aaRNA) at 111 bases in length.
5. The aaRNA is purified over an RNeasy column (Qiagen) using the following modified protocol (see Note 8). The volume of all samples is adjusted to 100 μl with RNase-free water. Per each sample, 100 μl buffer RLT is added and the samples are mixed. Then, 400 μl of isopropanol is added, the samples are quickly mixed by vortexing, immediately transferred to the RNeasy columns, and centrifuged for 15 s at 10,000 ×*g*. The membranes are washed twice with 500 μl RPE buffer and dried by centrifugation for 2 min. The aaRNA is eluted in two steps with 50 μl H_2O each time to maximize the yield.
6. To generate target samples for hybridization to custom EXT decoding microarrays (EXTarrays), aaRNA is coupled to cyanine dye esters Cy3 and Cy5 (we routinely label cDNA derived samples with Cy5 and plasmid DNA derived with Cy3).
7. For each labeling reaction, 5–10 μg of the aaRNA is lyophilized and resuspended in 4.5 μl coupling buffer. 5.5 μl of the Cy3/Cy5 dye is added to each sample. After 30 min incubation at room temperature in the dark, the reaction is neutralized by the addition 2.3 μl 4 M hydroxylamine.
8. The tubes are incubated in the dark for additional 15 min and the labeled RNA is purified using the modified RNeasy protocol as described above.

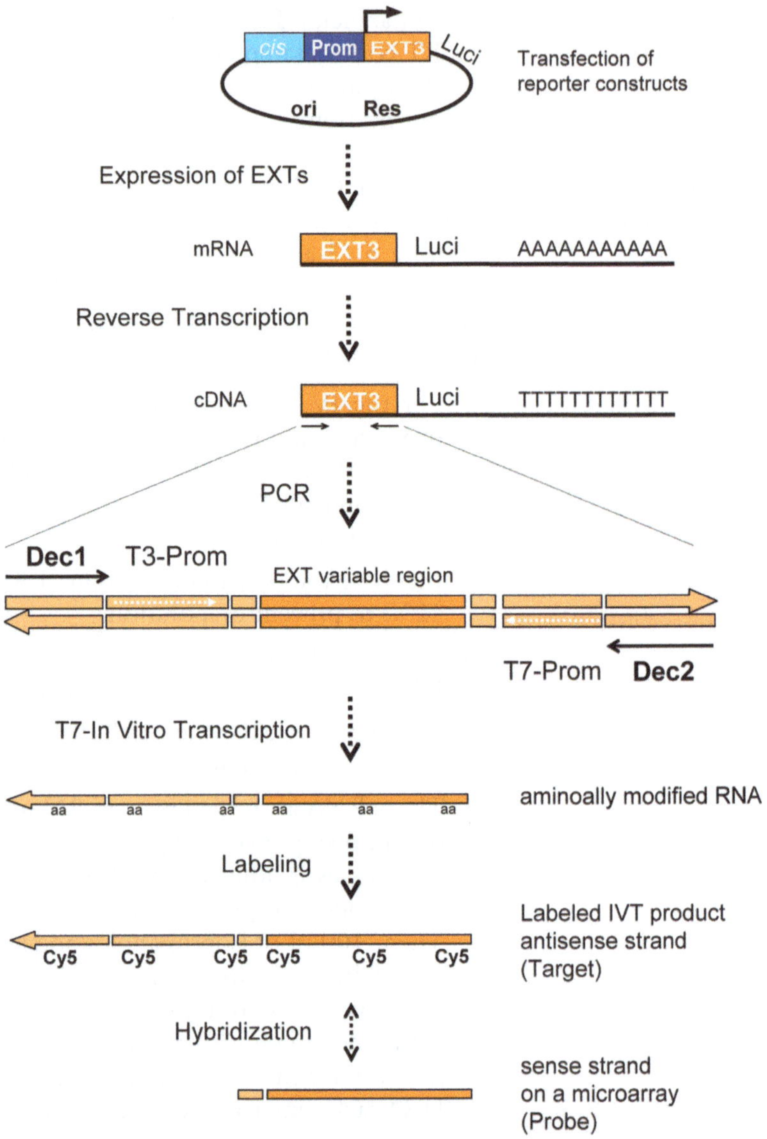

Fig. 4. Workflow of the EXT isolation, amplification, and labeling. After transfection, regulated expression of the *cis*-EXT-reporter constructs generates mRNAs carrying *cis*-element specific EXT and the firefly Luciferase reporter gene. To quantify the expression levels of the individual reporters total RNA is isolated. After reverse transcription, expressed EXT reporters are amplified by PCR with Dec1/Dec2 primers. To generate single-stranded targets for microarray hybridization, T7 in vitro transcription (IVT) is performed in the presence of 5-(3-aminoallyl)-UTP. The resulting aminoallyl modified RNA (aaRNA) is labeled by coupling of cyanine dye esters (Cy3 or Cy5) to the aminoallyl moieties. The labeled product (=target, antisense strands) is hybridized to a microarray harboring complementary probes (sense orientation).

9. The dye incorporation efficiency (E) can be calculated as follows: $E\,(\text{pmol}/\mu\text{g}) = \text{Dye}\;(\text{pmol}/\mu\text{g}) / \text{RNA Conc.}\,(\mu\text{g}/\mu\text{l})$. Successful labeling should result in a dye incorporation of 50–100 pmol/μl, corresponding samples can be used for the microarray hybridizations.
10. The amount of the target should be adjusted to the amount of the dye and not the amount of RNA in order to compensate for the slight differences in labeling efficiencies. The amount of target should correspond to 10–30 fmol of the Cy-dye per each unique EXT.
11. The hybridization cocktail for an 8-plex Agilent custom microarray (see Note 9) contains: 2× hybridization buffer 25 μl; 10× blocking agent 5 μl; formamide (f.c. 15%) 7.5 μl; target 12.5 μl (or water up to 50 μl final volume). The samples are mixed by pipetting, centrifuged for 1 min at 10,000 × *g*, and 40 μl are loaded per subarray. After hybridization at 65–68°C for 16–20 h the arrays are washed according to the manufacturer's recommendations and dried by dipping for 30 s into 250 ml acetonitrile.
12. NimbleGen 4-plex arrays (see Note 9) are hybridized according to the standard protocol provided by the manufacturer. Each target should comprise of 10–30 fmol of the Cy-dye per EXT. The arrays are hybridized at 48–50°C. Washing procedures are carried out using the NimbleGen wash buffer kit following the manufacturer's recommendations.
13. The slides are dried by dipping for 30 s in acetonitrile and scanned, e.g., with the GenePix 4200A microarray scanner or any other scanner with at least 5 μm resolution. Data were extracted according to the manufacturer's specifications.
14. Raw data analysis can be performed, e.g., with "R" statistical computing environment (http://www.r-project.org) or any other suitable analysis platform. Microarray data are normalized using, e.g., the RMA algorithm.
15. The average probe intensities are calculated across all EXT replicate features depending on the design of the array. Average Cy5 signals (RNA/cDNA derived) represent the relative EXT expression levels and are normalized to the Cy3 signals originating from the plasmid DNA input. Plasmid input normalized EXT expression data are either given as relative differences (fold-changes) between samples or as arbitrary normalized expression values.

4. Notes

1. Before starting with the synthesis of the combinatorial EXT oligonucleotide library, ensure that the repeated opening and tight closure of the coupling cartridges is possible. After each synthesis step, the resins from all cartridges are mixed on a clean glass surface and split into eight equal portions, e.g., using razor blades. These portions are reloaded to the cartridges for the next synthesis step.
2. When amplifying the EXT olgonucleotide library by PCR, do not use more than 1 fmol of template to avoid competition of the library oligonucleotides with the amplification primers.
3. The number of colonies to be screened should be determined by the final size of the library. Calculate with a maximum of 25% correct final expression clones.
4. Commercial electrocompetent DH10b from Invitrogen (Electromax) perform by far better compared to our self-made competent cells according to standard procedures. Bacteria from the XL-1 blue strain may not compatible with the Multisite Gateway system but chemically competent MACH1 as well as DH10b cells (Invitrogen) can be used.
5. In suspension transfection conditions are only given for PC12 cells. Principally, the described multiplexed assays should be applicable to any genetically amenable cell type; however, optimal conditions need to be adjusted with appropriate control experiments. In any case, circumvent shaking of cells during transfection to avoid increased cell death.
6. Avoid unnecessary washing step while harvesting cells to reduce losses particularly when using 96- or 384-well plates.
7. To control the cDNA synthesis with respect to plasmid DNA contamination, always analyze for each sample a control lacking reverse transcriptase (containing all reagents except reverse transcriptase).
8. The amplified antisense RNA has to be purified over an RNeasy column using a modified protocol for small-sized species. Do not change the conditions as you may reduce the yield or even loose the RNA.
9. Custom microarray should contain complementary EXT replicates (optimally $n > 4$ features per EXT) and a set of hybridization specificity controls including EXT sequence permutations carrying mismatches at different sites along the sequence and varying numbers of mismatches to determine optimal hybridization stringency (see Botvinnik et al., 2010).

References

1. Papin JA, Hunter T, Palsson BO, Subramaniam S (2005) Reconstruction of cellular signalling networks and analysis of their properties. Nat Rev Mol Cell Biol **6**, 99–111.
2. Citri A, Yarden Y (2006) EGF-ERBB signalling: towards the systems level. Nat Rev Mol Cell Biol **7**, 505–516.
3. Jones RB, Gordus A, Krall JA, MacBeath G (2006) A quantitative protein interaction network for the ErbB receptors using protein microarrays. Nature **439**, 168–174.
4. Schulze WX, Deng L, Mann M (2005) Phosphotyrosine interactome of the ErbB-receptor kinase family. Mol Syst Biol **1**, 2005 0008.
5. Brenner S, Williams SR, Vermaas EH, Storck T, Moon K, McCollum C, et al. (2000) In vitro cloning of complex mixtures of DNA on microbeads: physical separation of differentially expressed cDNAs. Proc Natl Acad Sci USA **97**, 1665–1670.
6. Botvinnik A, Wichert SP, Fischer TM, Rossner MJ (2010) Integrated analysis of receptor activation and downstream signaling with EXTassays. Nat Methods **7**, 74–80.
7. Wehr MC, Laage R, Bolz U, Fischer TM, Grunewald S, Scheek S, et al. (2006) Monitoring regulated protein-protein interactions using split TEV. Nat Methods **3**, 985–993.
8. Wehr MC, Reinecke L, Botvinnik A, Rossner MJ (2008) Analysis of transient phosphorylation-dependent protein-protein interactions in living mammalian cells using split-TEV. BMC Biotechnol **8**, 55.

INDEX

Bernhard Suter and Erich E. Wanker (eds.), *Two Hybrid Technologies: Methods and Protocols*, Methods in Molecular Biology, vol. 812, DOI 10.1007/978-1-61779-455-1, © Springer Science+Business Media, LLC 2012

V

X

Y

MIX
Papier aus verantwortungsvollen Quellen
Paper from responsible sources
FSC® C105338

If you have any concerns about our products,
you can contact us on
ProductSafety@springernature.com

In case Publisher is established outside the EU,
the EU authorized representative is:
Springer Nature Customer Service Center GmbH
Europaplatz 3, 69115 Heidelberg, Germany

Printed by Libri Plureos GmbH
in Hamburg, Germany